Organic Chemistry Laboratory

Oscar R. Rodig
University of Virginia

Charles E. Bell, Jr.
Old Dominion University

Allen K. Clark
Old Dominion University

Standard and Microscale Experiments

SAUNDERS COLLEGE PUBLISHING
Philadelphia Ft. Worth Chicago San Francisco
Montreal Toronto London Sydney Tokyo

Text Typeface: Times Roman
Compositor: EPS Group, Inc.
Acquisitions Editor: John Vondeling
Developmental Editor: Kate Pachuta
Managing Editor: Carol Field
Project Editor: Janet B. Nuciforo
Copy Editor: Charlotte Nelson
Manager of Art and Design: Carol Bleistine
Art Director: Christine Schueler
Art Assistant: Doris Bruey
Text Designer: Edward A. Butler
Cover Designer: Lawrence R. Didona
Text Artwork: Rolin Graphics Inc.
Director of EDP: Tim Frelick
Production Manager: Bob Butler

Cover Credit: Photograph by Robert K. Ander

Printed in the United States of America

ORGANIC CHEMISTRY LABORATORY Standard and Microscale Experiments

0-03-012644-4

Library of Congress Catalog Card Number: 89-043220

0123 071 987654321

Preface

In recent years, there has been a growing tendency to carry out elementary organic laboratory experiments on a so-called "micro" scale; that is, to use milligram quantities of materials rather than grams. The advantages of such small scale experiments are obvious: the use of smaller amounts of materials, normally a major cost item in such courses; generally shorter reaction times; greatly diminished waste disposal problems; lower potential safety hazards; and a practical accession to experiments utilizing relatively expensive starting materials.

Recently, several laboratory texts have appeared which are structured around the microscale concept. We have felt strongly for some time that a well-rounded training course in elementary organic laboratory principles should include certain "macro" sized experiments since techniques used in these are often quite different from those using micro quantities. Thus, students should certainly learn techniques such as distillation, extraction, and filtration, and a sampling of synthetic procedures on a larger scale since they surely will meet such situations in the typical organic laboratory they may encounter in their future.

In this book, we describe three types of experiments—those that teach valuable techniques but are not practical on a very small scale, those that can be carried out on either scale and where the choice is left to the instructor or the student; and finally those that are practical only on a micro scale, generally because of the prohibitive costs of materials or excessively long reaction times if carried out in larger quantities. By far, the largest number of experiments are of the dual type.

In general, the microscale experiments were developed using what we consider practical amounts of materials. At times, the scale may be slightly larger than those described by others, but our experiences have shown that with very small amounts, the slight additional saving in chemical costs did not justify the increased frustrations the students experienced when working with such small quantities. In addition, the microscale procedures are based on techniques used for many years in research laboratories, and can be carried out using readily available standard laboratory equipment, such as the centrifuge and test tube, Pasteur pipet, Hirsch funnel, septum, syringe, and ordinary small-scale glassware, thus eliminating the need for specialized apparatus or micro kits.

We have also attempted to include some more unusual experiments, such as Aromatic Nucleophilic Substitution (Chapter 26), Heterocyclic Syntheses (Chapter 37), a comparison of Acid Catalyzed and Enzymatic Hydrolyses (Chapter 42), and the Analysis of Fats and Oils (Chapter 43). A discussion of ^{13}C NMR is included in Chapter 10, and on-line literature searching in Chapter 11. To increase a student's exposure to instrumentation, we describe experiments utilizing gas chromatography in Chapters 6, 13, 16, and 43, and the optional use of high performance liquid chromatography (HPLC) in Chapter 13. The use of a pH meter is included in Chapter 22, a visible spectrometer or colorimeter in Chapter 34, and a simple polarimeter in Chapters 39 and 42. In the macroscale experiments, we have attempted to use products from one experiment as the starting material for another where practical, thus further reducing chemical costs and waste disposal problems.

Safety is all-important in any laboratory course. It is stressed in a general section in Chapter 1 and throughout the text in the form of **Safety Notes** associated with each experiment. Whenever possible, chemicals classed as hazardous by OSHA have been eliminated. In the few instances where they do appear, they and the associated hazards are clearly identified so that adequate safety precautions can be taken. Further advice on the safe dispensing of reagents and other general information on the experiments can be found in the accompanying Instructors Manual.

A difficulty frequently encountered in organic laboratory texts is that early experiments do not correlate well with material covered in the lecture course. To avoid this problem, the first eight chapters emphasize laboratory techniques, followed by two on instrumentation and one on literature searching. Only then are experiments described where an understanding of the chemistry makes the laboratory experience most meaningful.

A number of the experiments contain unknowns; thus the answers are not forthcoming until the experiments are performed. We find this approach heightens student interest and emphasizes the importance of working carefully to obtain the correct result.

In any laboratory course, the student's work efficiency is greatly enhanced if he or she ponders the material beforehand. To this end, we have included prelaboratory questions for all chapters except 1 and 46. In addition, we have added questions of variable difficulty at the ends of the chapters, which can be best answered after the experiments have been completed.

All of the experiments can be carried out in one (or two) three-hour laboratory period, and have been repeatedly tested in our classes at the University of Virginia and Old Dominion University. In this regard, we wish to express our gratitude to Robin E. Fretwell, Andrea L. Cox, Rashmi A. Majali, Charles G. diPierro, Broderick C. Bello, and Joseph C. Pedulla, who spent considerable time and effort in developing the microscale procedures. Finally, we would like to thank Professors Exum D. Watts, of Middle Tennessee State University; Charles E. Sundin, of the University of Wisconsin, Platteville; Kenneth W. Raymond, of Eastern Washington University; Lisa McElwee-White, of Stanford University; and Arlyn Myers, of the University of California, Berkeley for their helpful comments and criticisms; Project Editor Janet Nuciforo for her patience and helpful suggestions during the preparation of the manuscript; and John Vondeling, Kate Pachuta, and others on the staff of Saunders College Publishing, for their help, cooperation, and encouragement.

Oscar R. Rodig
Charles E. Bell, Jr.
Allen K. Clark

Contents

1

Introduction

Laboratory work is an integral and essential part of any chemistry course. Chemistry is an experimental science—the compounds and reactions that are met in lecture and classroom work have been discovered by *experimental observations*. Organic compounds exist as gases, liquids, or solids with characteristic odors and physical properties. They are synthesized, distilled, crystallized and chromatographed, and then transformed by reactions into other compounds. The purpose of laboratory work is to provide an opportunity to observe the reality of compounds and reactions and to learn something of the operations and techniques that are used in experimental organic chemistry and in other areas in which organic compounds are encountered.

LABORATORY SAFETY

Along with the opportunity to learn at first hand about the properties and reactions of compounds and the manipulation of laboratory equipment, there must be proper concern for safety. Most organic compounds are flammable, and they are toxic or irritating to a greater or lesser degree. Many organic reactions are potentially violent. Laboratory work in organic chemistry is not a dangerous occupation, however, nor is the laboratory a perilous place to be, provided that some simple precautions and safety rules are followed. There are some potential hazards that must be recognized and avoided; accidents can and do occur when these hazards are ignored. Throughout this book, **Safety Notes** are included for each experiment. They emphasize the specific hazards that may be encountered and must be kept in mind. In the following section, some general precautions are discussed, and a few rules are given that must be observed during any laboratory work.

RULES FOR PERSONAL SAFETY

Avoiding injuries is largely a matter of good sense. Carelessness can lead to accidents and injuries to yourself and others. The following rules cover some important general precautions and should be observed at all times.

1. **Eye Protection.** Approved eye protection must be worn at all times in the laboratory, regardless of what is being done. In many locations, chemical safety goggles are required by law. Safety glasses (either prescription or plain) with side shields to protect from splashes also offer good eye protection.

   Contact lenses should *not* be worn in the laboratory. If a splash occurs, the lens can act as a trap for corrosive materials and greatly increase the likelihood of permanent injury. Moreover, low levels of vapors that are present in the laboratory can cause severe irritation by the lens.
2. Never work in a laboratory without another person being present or within calling distance. Minor accidents can become disasters if help is not available.
3. Do not carry out any reaction that is not specifically authorized by the instructor.
4. Never taste a compound; never pipet a chemical by mouth; do not eat, drink, or smoke in the laboratory.
5. Avoid contact of the skin with any chemical. If a substance is spilled on your hands, wash them thoroughly with soap and water. Do not rinse them with a solvent, since this may cause more rapid absorption.
6. Long hair should be tied back. Shoes must be worn to prevent injury from spilled chemicals or bits of glass. Avoid loose-fitting sleeves and clothing that leave expanses of skin unprotected.
7. Never heat a flask or any apparatus that is sealed or stoppered (i.e., a closed system)—make certain that there is an opening to the atmosphere.
8. When inserting glass tubing into a rubber stopper or rubber tubing, lubricate it first with a drop of glycerin and protect your hands with a towel. The same is true for inserting thermometers into stoppers. Thin-walled transfer pipets should not be used as connectors for rubber tubing or stoppers; they are fragile and very easily crushed.
9. Some experiments require the use of a well-ventilated hood. These experiments should not be attempted in the open laboratory.
10. **Peroxides.** A major safety concern with certain organic substances is the buildup of potentially explosive peroxides. These can form by the light-catalyzed autoxidation of ethers, aldehydes, and alkenes and aromatic compounds having allylic or benzylic hydrogen atoms, and even alcohols. The danger is particularly acute when these materials are used as reaction solvents or in extractions and then are concentrated in the workup. This is one reason why liquids should never be distilled to dryness. Such commonly used substances as the diethyl and diisopropyl ethers, tetrahydrofuran, cumene, tetralin, and 2-butanol (Fig. 1.1) should be checked for peroxides if they have been stored for an extended period of time, especially in a partially empty container. Particular care

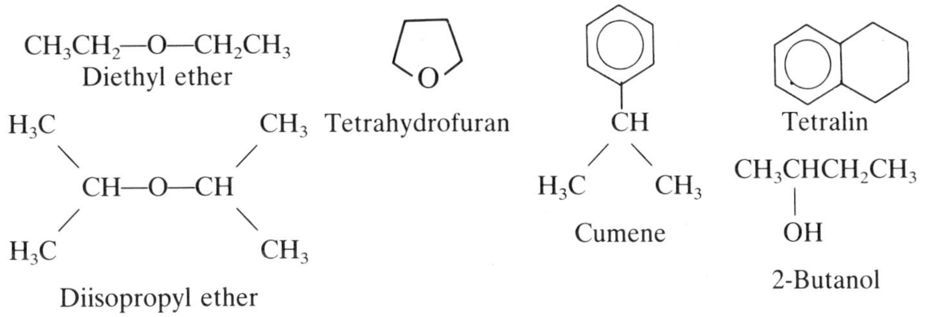

$CH_3CH_2\text{—}O\text{—}CH_2CH_3$
Diethyl ether

O
Tetrahydrofuran

$$\begin{array}{ccc} H_3C & & CH_3 \\ & CH\text{—}O\text{—}CH & \\ H_3C & & CH_3 \end{array}$$

Diisopropyl ether

$$\begin{array}{c} CH \\ H_3C \quad CH_3 \end{array}$$

Cumene

Tetralin

$$\begin{array}{c} CH_3CHCH_2CH_3 \\ | \\ OH \end{array}$$

2-Butanol

Figure 1.1 Some substances that can form peroxides by autoxidation.

should be taken with diethyl ether because of its wide use as an extraction solvent.

Peroxides can be detected with starch iodide paper or by adding 1 mL of the suspected material to 1 mL of glacial acetic containing 0.1 g of sodium or potassium iodide. A yellow-to-brown color indicates the presence of peroxides. A blank should be run to confirm the validity of the test.

Peroxides can be removed from ethers by shaking with a 30% solution of aqueous ferrous sulfate or by percolating through a column of alumina. The latter method also removes traces of water. Additional information on peroxides can be found in the references cited at the end of the chapter.

FIRST AID

1. **Emergency Equipment.** Learn the location of safety showers, eyewash fountains, and fire extinguishers, and know how this equipment is used.
2. **Chemical Spills on the Skin.** Immediately flush the skin with running water for several minutes; if the eyes or face are involved, use an eyewash fountain or the nearest faucet. For minor acid burns, use a paste or ointment of sodium bicarbonate after the flushing. For bromine, apply a compress soaked in 0.6 *M* sodium thiosulfate solution. For any severe chemical burn, consult a physician as soon as possible after the initial thorough water flushing.
3. **Fire.** If clothing is ignited, immediately extinguish it with a safety shower, by rolling on the floor, or, if necessary, with a coat or anything else available. Do not allow a person with burning clothing to run—such action only fans the flames. Even standing should be avoided because of the danger of inhaling hot fumes.*

---

*Fire blankets are no longer recommended for use with fires. They can act as a chimney, thus encouraging the fire. Such blankets can be useful, however, for covering victims in shock.

4. In any severe case of chemical spill, a burn, or a cut, the affected person should be escorted to a physician or a hospital emergency room.

CHEMICAL TOXICITY AND CARCINOGENS

It has long been recognized that certain chemicals—for example, phosgene or the nerve gas isopropyl methylfluorophosphonate—are highly toxic substances that are lethal in extremely small amounts. In recent years there has been a considerable increase in awareness and concern about the toxicity of all chemicals encountered in laboratory and manufacturing environments. Major efforts are now being made to identify toxic chemicals and avoid exposure to them.

The National Institute of Occupational Safety and Health (NIOSH) has prepared a registry of a large number of compounds for which some data on toxic effects are available. Another government agency, the Occupational Safety and Health Administration (OSHA), issues regulations governing permissible limits of exposure to chemicals, with particular attention being placed on compounds that are commonly encountered as air contaminants. Many of the compounds for which limits have been set, such as diethyl ether and ethanol, have relatively low toxicity, but limits are nevertheless placed on prolonged exposure. On the other hand, compounds that may have high acute toxicity can be transferred and used in a laboratory experiment with simple precautions to avoid contact.

A major concern in recent years has been the carcinogenicity of organic compounds, that is, their ability to induce cancer. For certain compounds that were used industrially for many years, there is a clear link between exposure to the compound and the incidence of certain types of cancer in the workers who handled them, just as there is between cigarette smoking and lung cancer. More recently, evidence has been found for the occurrence of tumors in experimental animals exposed to very large doses of a wide variety of other organic compounds. A much-publicized example is the artificial sweetener saccharin, which was used for many years in low-calorie beverages.

Government agencies, such as the Carcinogen Assessment Group (CAG) of the Environmental Protection Agency (EPA) and NIOSH have compiled lists of suspected carcinogens, and laboratory chemical catalogs often identify these as "cancer suspect agents." The National Institutes of Health has published guidelines for the laboratory use of certain chemical carcinogens, and OSHA, in turn, has issued regulations on exposure to a number of carcinogenic substances, most of which are in commercial use.

A brief list of carcinogenic compounds that are frequently encountered in laboratory work is given in Table 1.1. It should be pointed out that this list contains only a very small fraction of the compounds for which data on carcinogenic properties are known. Remember that carcinogenic activity is usually based on tests in animals at high doses for prolonged duration, and

Table 1.1 Partial List of Chemical Carcinogens

acetamide	CH_3CONH_2	dimethyl sulfate	$(CH_3)_2SO_4$
acrylonitrile	$CH_2{=}CHCN$	dioxane	$O(CH_2CH_2)_2O$
aminobiphenyl	$NH_2C_6H_4C_6H_5$	ethyl carbamate	$NH_2CO_2C_2H_5$
aziridine	CH_2CH_2NH	hydrazine	NH_2NH_2
benzene	C_6H_6	methyl iodide	CH_3I
benzidine	$NH_2C_6H_4C_6H_4NH_2$	1-naphthylamine	1-$C_{10}H_7NH_2$
benzpyrene	$C_{20}H_{12}$	2-naphthylamine	2-$C_{10}H_7NH_2$
bis(chloromethyl) ether	$(ClCH_2)_2O$	4-nitrobiphenyl	$NO_2C_6H_4C_6H_5$
t-butyl chloride	$(CH_3)_3CCl$	nitrosomethylurea	$NH_2CON(CH_3)NO$
carbon tetrachloride	CCl_4	phenylhydrazine	$C_6H_5NHNH_2$
chloroform	$CHCl_3$	thiourea	NH_2CSNH_2
chromic anhydride	CrO_3	*o*-toluidine	$CH_3C_6H_4NH_2$
diazomethane	CH_2N_2	trichloroethylene	$CHCl{=}CCl_2$
dibromoethane	$BrCH_2CH_2Br$	vinyl chloride	$CH_2{=}CHCl$

the risk from occasional brief exposure is unknown. The chief concern with carcinogens, as with other toxic substances, is with permissible levels for continuous exposure, not occasional use. The important consideration for laboratory work is to avoid any unnecessary exposure and to use these compounds only when essential, with due care and protection. In particular, benzene, chloroform, and carbon tetrachloride should not be used as solvents for extraction or column chromatography.

In the experiments in this book, all compounds that have been implicated as carcinogens have been eliminated wherever possible; in the remaining few cases they have been identified as such. It must be emphasized that, with present knowledge, these compounds should *not* be considered as deserving any more concern than many others. Careless handling of simple acids or solvents normally presents a much greater safety hazard.

DISPOSAL OF CHEMICALS

In recent years there has been an ever-increasing awareness of the pollution of our water supplies. Many areas have strict laws and/or regulations concerning the disposal of chemicals in sinks. It is best to check with local authorities before disposing of *any* chemicals in sinks. Water-immiscible solvents or other organic liquids should be discarded in a designated waste solvent container and should not be poured into a sink. Residues and vapors can remain in the sink trap; moreover, any chemicals discharged in a drain eventually add to pollution of surface or ground water.

Chemicals that react vigorously with water, such as acid chlorides, metal hydrides, or alkali metals, should be decomposed in a hood in a suitable way, such as by controlled reaction with alcohol. Strong acids or strong bases should be neutralized before they are discarded.

Solids or glass should be disposed of in a nonmetallic chemical waste jar; do not throw them into a wastepaper basket.

FIRE HAZARDS

Fire hazards are present in any organic laboratory because of the frequent use of volatile, flammable solvents. By far the greatest risk of fire is associated with gas burners for heating, and this form of heating should be avoided whenever possible. On the other hand, electric heating devices may contain exposed elements and are not totally immune from being fire hazards.

Organic vapors are heavier than air and flow downward; they diffuse rapidly and can be ignited several feet away from the source. The following precautions should always be observed:

1. Organic liquids should only be heated to boiling or distilled on a steam bath or under a condenser. When refluxing a liquid, particularly over a flame, be certain that the condenser is tightly fitted to the flask.
2. Before lighting a flame, if one must be used, check to see that volatile liquids are not being poured or evaporated in your vicinity. Conversely, before pouring or evaporating a liquid, be certain that none of your neighbors is using a flame. Always turn off a burner as soon as you are finished using it—never leave it on unnecessarily.
3. Smoking creates an avoidable fire hazard and is not permitted in the laboratory.

LABORATORY EQUIPMENT AND TECHNIQUES

For successful laboratory work it is essential that before beginning an experiment you understand what you are going to do and why and how you are going to do it. Study the assigned experiment in advance and plan your operations. If required by your instructor, answer in writing the prelab questions at the end of the experiment. These questions provide background for the work and will acquaint you with points that can and should be understood before you actually do the experiment.

Experimentation in organic chemistry calls into play a number of operations and techniques and a rather large assortment of apparatus. Detailed instructions in the techniques and equipment used for various separation and purification methods are given in Chapters 2 through 8 in conjunction with actual experimental procedures. A few points of general practice that apply to nearly all experiments are covered in this section.

EQUIPMENT

Glassware

In this book you will learn two major laboratory techniques for carrying out chemical reactions. Those on a *macroscale* normally involve gram quantities of materials, and typical glassware used for this is shown in Figure 1.2. The glassware is equipped with standard-taper ground joints that permit quick and secure assembly of apparatus. It is expensive and must be handled with care. *Microscale* experiments are carried out on milligram amounts of material and generally demand more careful manipulation of reagents and products. Often these experiments can also be carried out with the glassware shown in Figure 1.2, but on a reduced scale.

Some examples of glassware designed specifically for microscale use are shown in Figure 1.3. This glassware contains joints with external threads and open-top caps that allow them to be screwed together. Since the pieces cannot pull apart, the assemblies require considerably less clamping. In addition, the inner parts of the joints are also standard-taper ground glass that allow their attachment to other ground joint apparatus. Although convenient, such equipment is not required for the experiments in this book.

Other equipment that will be needed include Erlenmeyer flasks, beakers, and test tubes in various sizes, Buchner & Hirsch funnels, graduated

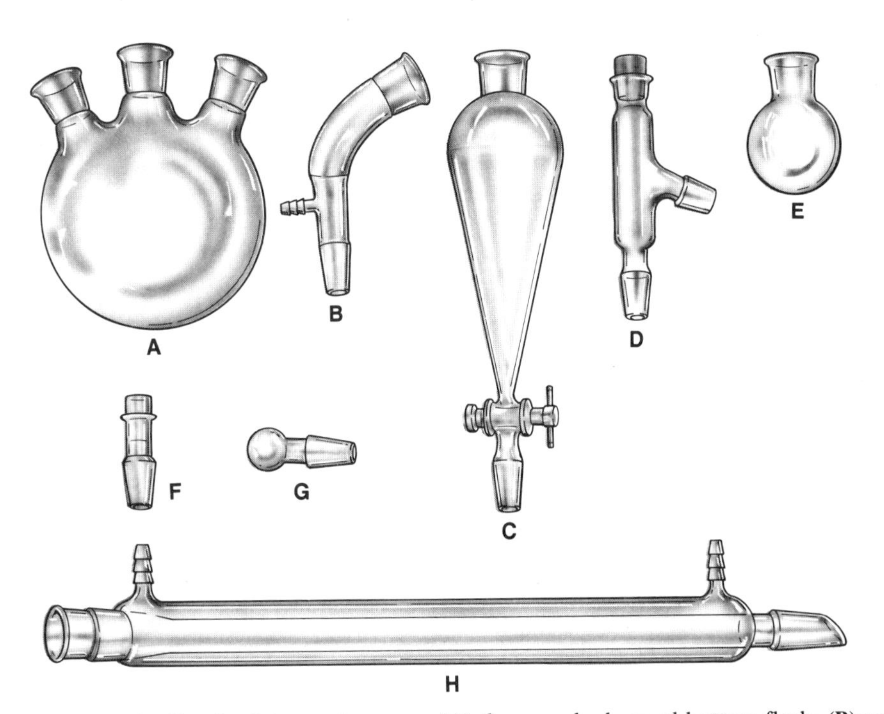

Figure 1.2 Standard-taper glassware. **(A)** three-necked round-bottom flask; **(B)** vacuum take-off adapter; **(C)** dropping funnel; **(D)** distillation head with rubber connector; **(E)** round-bottom flask; **(F)** tubing connector; **(G)** standard-taper stopper; **(H)** condenser.

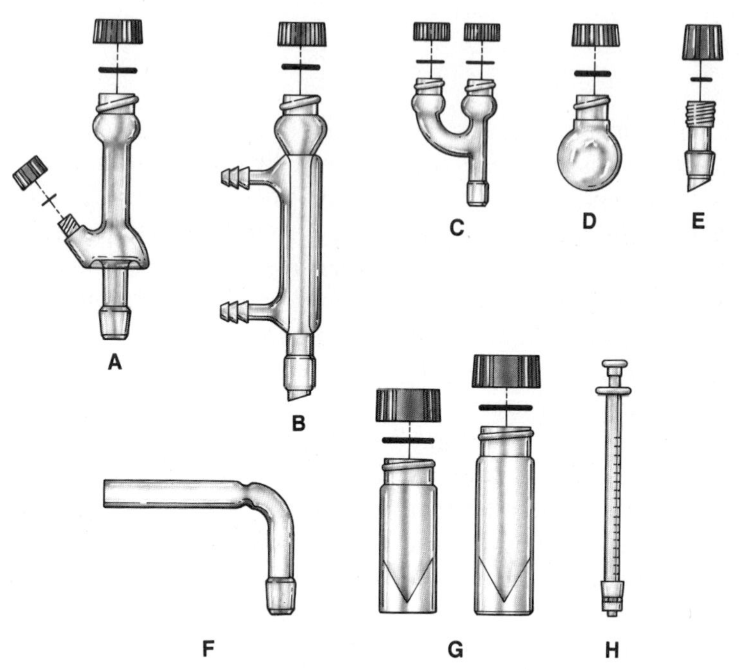

Figure 1.3 Microscale glassware. **(A)** Hickman still head; **(B)** condenser; **(C)** Claisen head adapter; **(D)** small round-bottom flask; **(E)** thermometer adapter; **(F)** drying tube; **(G)** 3- and 5-mL conical reaction vials; **(H)** 1-mL sample syringe.

cylinders, pipets, centrifuge tubes (preferably graduated), thermometers, spatulas, septums, syringes, and assorted clamps and other hardware.

For effective laboratory work it is most important that you develop good working habits, learn the proper equipment for a given purpose, and know how to use it. Maintain a well-organized locker or equipment drawer; keep your equipment clean and as conveniently located as possible. Make a practice of washing or rinsing glassware as soon as it has been used so that it will be clean and dry the next time you need it. Take enough time to clean up and store equipment properly before you leave the laboratory.

Reaction Setups

Many reactions involve the combining of two reactants at a controlled rate. Simple reactions on a small scale require only a pipet and centrifuge or test tube, which can be swirled by hand. In other cases, on both macroscale and microscale, the reactants are combined and heated in a simple setup, such as that shown in Figure 1.4. The temperature is controlled by a heating unit or by the boiling point of a reactant or solvent. Mixing, if needed, is provided by the turbulence of boiling. If the reaction is exothermic, cooling occurs by the heat being transferred to the condenser by the refluxing solvent, or, if necessary, the flask can be immersed in an ice bath.

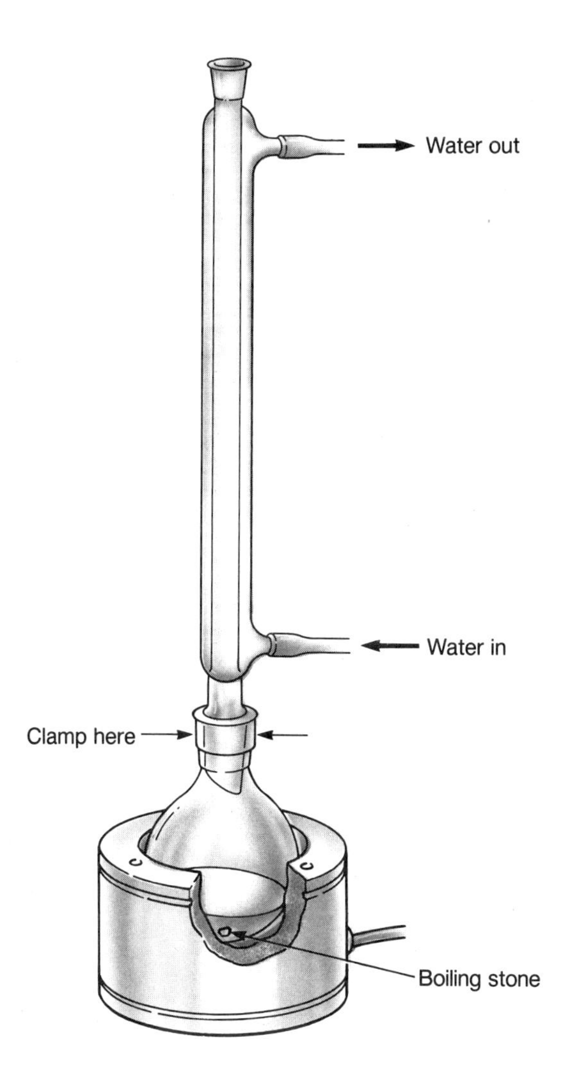

Figure 1.4 Simple reflux apparatus.

For reactions on a large scale or those requiring special conditions, a more elaborate setup such as that shown in Figure 1.5 may be needed. The separate necks for the dropping funnel, stirrer, and condenser permit flexible assembly and easy control of addition and mixing.

Stirring is important if a reactant must be added and dispersed at a steady rate or if the reaction must occur between separate phases. For relatively small flasks and for reaction mixtures that are not viscous, a bar magnet with an inert coating is placed in the flask and is spun by a motor-driven magnet below the flask (Fig. 1.6). Alternatively (as seen in Fig. 1.5), a rod with a small propeller or paddle, turning in a closely fitting sleeve, is driven by an electric or air-powered motor.

The atmosphere in a reaction may be critical. If the reactants or products are sensitive to water, a drying tube filled with desiccant attached to the condenser may be sufficient. For scrupulous removal of moisture or

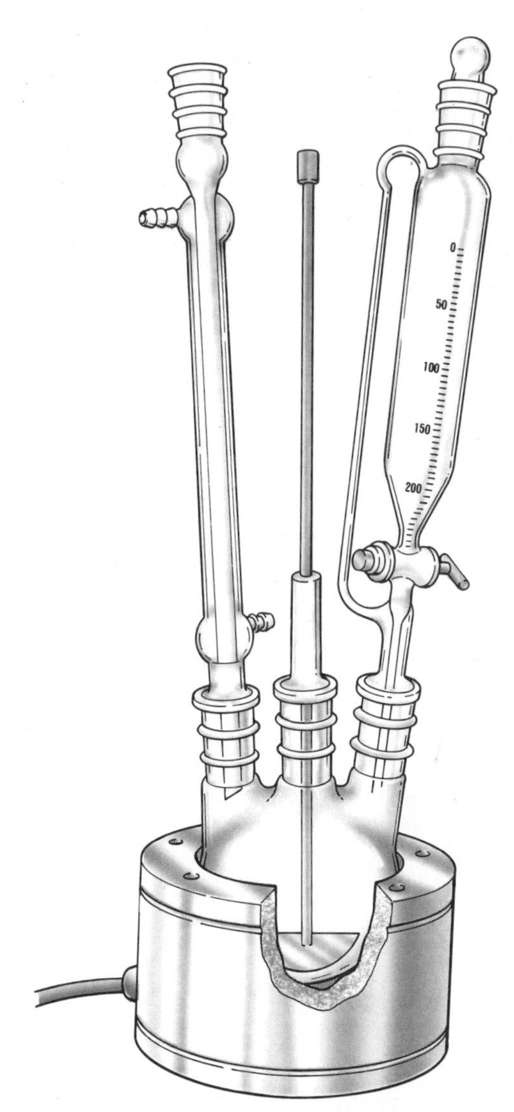

Figure 1.5 Reaction assembly with heating mantle, stirrer, and pressure-equalizing dropping funnel.

when oxygen must be excluded, a dry inert gas such as nitrogen is passed over the reaction mixture and out through a trap. With this setup, the dropping funnel must be equipped with a pressure-equalizing arm (see Fig. 1.5).

Heating Sources

The sources of heat in most undergraduate laboratories are electric mantles or hot plates and steam baths. Bunsen burners are also used, but they are the most frequent cause of laboratory fires and must be used with great care, as noted in the section on Fire Hazards. When a flask must be heated

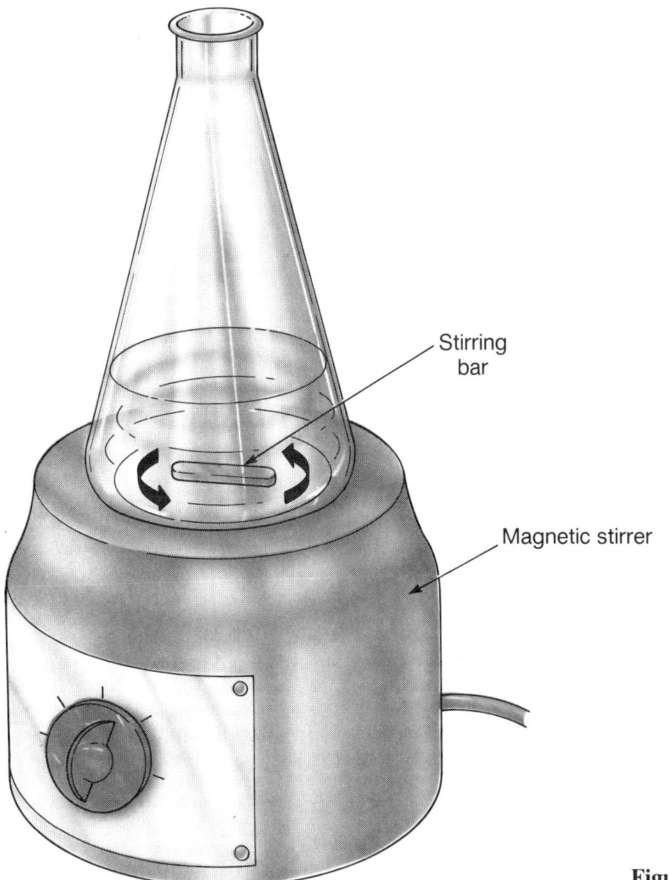

Figure 1.6 Magnetic stirrer.

over a burner, a piece of wire gauze placed under the flask will distribute the heat more evenly; one must be ever-vigilant about organic vapors.

When possible, the steam bath (Fig. 1.7) should be used for heating or evaporating an organic liquid. Connect the bath with the steam inlet in the *upper* side arm and the condensate outlet in the *lower* side arm. Place the flask to be heated on the largest ring that will support it. Do not remove the rings or set the flask on the bottom. Turn on the steam until it just begins to escape; no benefit is gained from billowing clouds.

Electric heating elements, controlled by a variable transformer, are available in a variety of forms. In a mantle (see Fig. 1.4) the element is embedded in a hemispherical shell that fits around a spherical flask. Hot plates are convenient for heating a flat-bottomed vessel such as a beaker or an Erlenmeyer flask. Hot plates and mantles may contain exposed contacts or elements and should not be used to vaporize flammable liquids without a condenser.

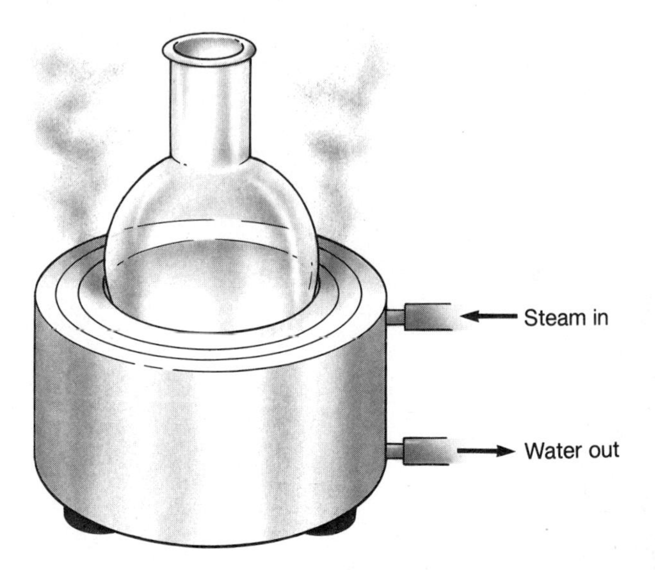

Figure 1.7 Steam bath.

For experiments on a microscale, a sand bath is a convenient heat source, and is readily made by filling a 100-mL mantle with sand. Ceramic lined mantles work better than those of fiberglass, since the latter tend to burn out when used in this manner.

A recently introduced heating method uses an aluminum block that is placed on a hot plate and contains various sized holes for flasks, test tubes, and a thermometer (Fig. 1.8A). Auxiliary aluminum blocks can also be placed around flasks and test tubes to improve further the heat transfer. Such an assembly is shown in Figure 1.8B using a conical reaction vial attached to a Hickman still head, an apparatus used to carry out microscale distillations.

TECHNIQUES

Use of the Aspirator

An aspirator provides a convenient source of reduced pressure for vacuum operations. Water should be turned on to full capacity when an aspirator is used. Splashing may be a problem, but it can usually be avoided by attaching some rubber tubing, tying a piece of rag around the outlet, or wedging a piece of wire gauze in the trough below the stream. Heavy-wall rubber tubing should be used when connecting apparatus to the aspirator, since regular tubing will collapse and pinch off the system. If the water pressure drops, or the water is turned off while the aspirator is in use, water tends to be drawn back through the arm into the evacuated system. Most aspirators contain a check valve to prevent this, but a trap should always be connected between the aspirator and the evacuated vessel (Fig. 1.9) as

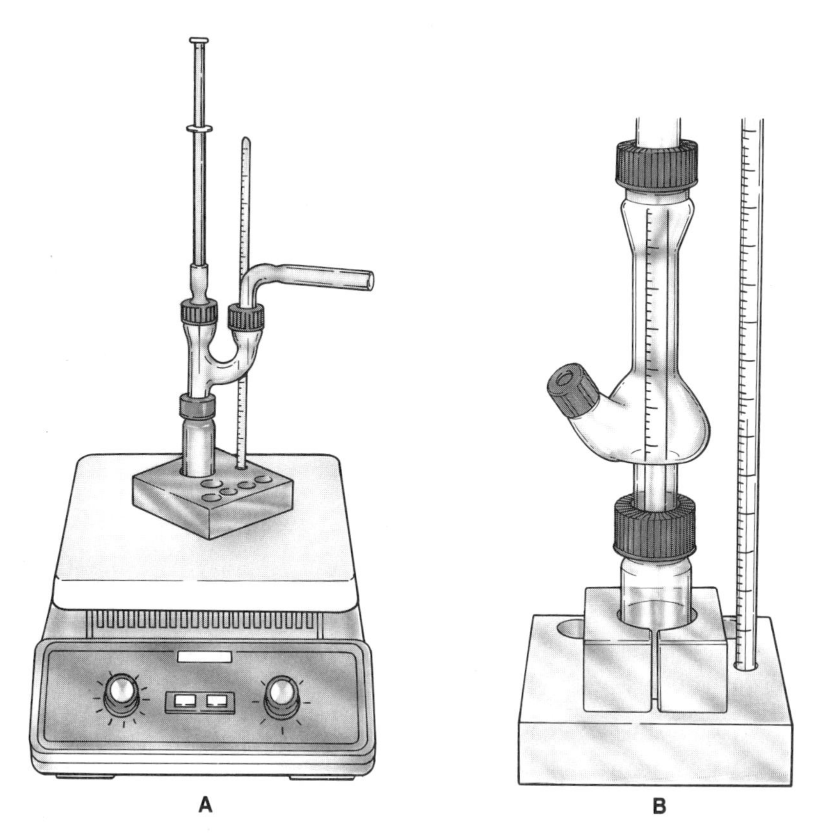

Figure 1.8 Aluminum heat transfer devices.

a safety precaution. The vacuum should always be released *before* turning off the water.

Handling and Measuring Chemicals

In most preparative experiments, solid starting materials and reagents can be weighed on a beam or top-loading balance to ± 0.1 g for the macroscale experiments and a top-loading electronic balance (Fig. 1.10) with a 100-g capacity and capable of weighing to ± 0.001 g for the microscale experiments. A beaker is the most convenient vessel for handling more than a few grams. For weighing small quantities of solid reagents or products, weighing cups or glassine paper (not filter paper) should be used. The sample can then be placed in a vial or added to a solution simply by picking up the paper by opposite edges or corners and using it as an open funnel. Finely divided solids can be transferred quite completely by gently scraping with a spatula. Metal spatulas should be wiped clean after use and polished occasionally.

It is usually unnecessary to weigh certain reagents, such as activated

Figure 1.9 Aspirator and safety trap.

carbon, salt, or drying agents; the appropriate amount often can be estimated by bulk. Small amounts of sodium hydroxide or potassium hydroxide normally need not be weighed. The pellets are fairly uniform and weigh roughly 0.1 g each. If more accurate measurement of a small quantity is required, as in some microscale experiments, an aliquot of a solution of known concentration should be used. Never attempt to weigh hydroxide pellets on paper or a watch glass; a few inevitably roll away and liquify into a highly corrosive puddle.

Liquid reagents and starting materials are conveniently measured by volume in a graduated cylinder or sometimes with a graduated pipet or syringe, but these are accurate to only a few percent. For more accurate macroscale measurements, liquids should be weighed, particularly if they are rather viscous. For transfer of small volumes of liquids and very approximate measurements, thin-wall, soft-glass transfer (Pasteur) pipets are particularly useful. A supply of these should be maintained; they can be rinsed and reused repeatedly. Rubber bulbs of 1- and 2-mL capacity will

Figure 1.10 A top-loading electronic balance.

fill these pipets to about half and full capacity, respectively; the drawn-out section of these pipets contains about 0.2 mL (see inside back cover). There will be many occasions to obtain approximate volumes of liquid in this way. A detailed procedure for transferring small volumes of organic solutions is described in Chapter 4.

Submission of Samples and Reports: Percentage Yield

In many of the described experiments, a sample of compound is to be turned in or a specific result may be called for in written form. Samples should be submitted in an appropriate-sized plastic-cap vial and neatly labeled with the student's name, the name of the product, its weight, percentage yield, melting point or boiling point, and any other data specified by the instructor.

The *yield* in a reaction is the amount of product of acceptable purity actually obtained (the term should not be used for the product itself). The percentage yield is the ratio of this amount to that theoretically obtainable × 100. In a reaction involving more than one starting material, the theoretical or 100% yield is determined by the reactant that is used in the smallest stoichiometric quantity. In calculating a percentage yield, therefore, one must first determine the molar amounts of each starting compound and then the theoretical amount of product that could be obtained from the limiting reactant.

As an example, consider the esterification of succinic acid in which 5.0 g of the acid is heated with 100 mL of ethanol and 6.2 g of ester is isolated.

$$\underset{\text{succinic acid}}{HOCOCH_2CH_2CO_2H} + 2\ \underset{\text{ethanol}}{C_2H_5OH} \xrightarrow{H^+} \underset{\text{diethyl succinate}}{C_2H_5OCOCH_2CH_2CO_2C_2H_5} + 2\ H_2O$$

The molar amounts are as shown.

SUCCINIC ACID	ETHANOL	DIETHYL SUCCINATE
5.0 g; mol wt 118 = 0.042 mole	100 mL; d, 0.79 g/mL = 79 g; mol wt 46 = 1.72 moles	6.2 g; mol wt 174 = 0.036 mole

The ethanol is present in large excess (0.084 mole required) and the yield of ester product is thus based on the acid. The theoretical amount of diethyl succinate is the same as that of acid, 0.042 mole, and the percentage yield is

$$\frac{0.036}{0.042} \times 100 = 86\%$$

In the foregoing example, if 0.40 g of unreacted succinic acid were recovered from the reaction in usable form, the yield might alternatively be based on "unrecovered starting material," that is, 5.0 − 0.4 = 4.6 g; the percentage yield on this basis would be 92%.

Records

Equally as important as the actual product or the final report is the **laboratory notebook**. An important objective of this course is to develop good practices and habits in keeping permanent notes of experimental work and in working in an orderly, systematic way.

The laboratory notebook must be bound with a hard cover and used only for experimental data in this course. It must, of course, be at hand at all times while you are in the laboratory. The notebook is the fundamental record of actual laboratory operations and observations. It should provide an account of *what* was done, *how* it was done, and *what happened.* The apparatus used, the sequence of steps, all measurements, significant time intervals, changes in appearance, and other relevant data should be recorded. It should be possible for someone else to repeat the experiment as *you* did it and obtain the same results. The notebook should *not* be a verbatim transcription of a procedure from the laboratory manual or any other source that you then purport to follow, nor should it be an after-the-fact scrapbook of recollections and miscellaneous jottings. It is impossible to reconstruct an accurate record from isolated numbers and memory; on the other hand, your experimental data cannot be recorded before they exist. It is necessary, therefore, to record the salient operations, measurements, and observations *as they are done or made*, insofar as possible. This procedure usually will not result in a flawless copybook record, but the notebook should be coherent and legible.

Each experience should be dated (each day if protracted) and placed on a separate page or pages. It is a good idea to use only right-hand pages for the actual write-up; calculations and the like can be recorded on the facing page.

A common objection raised by students is "Why must I write down in a notebook all of the setups in a procedure that I am following in a lab manual?" The reason for doing this is to provide an orderly account into which you can incorporate your own data and observations. Moreover, it is a habit that must be acquired. In later experiments, you will be adapting a general procedure to your own situation; your own specific case is unique, and the record of what you do is vital.

An illustrative example of an actual notebook page that may serve as a guide to the type of record that should be kept in a typical experiment is shown in Figure 1.11.

Diels Alder Reaction of Cyclopentadiene and Maleic Anhydride Sept. 6, 1990

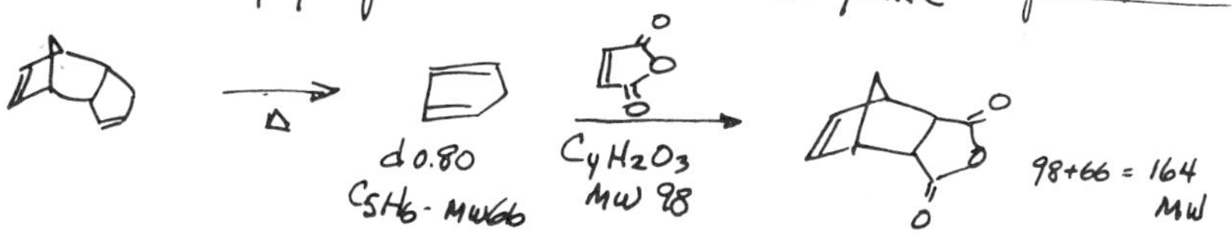

d 0.80 C_5H_6 - MW 66

$C_4H_2O_3$ MW 98

98+66 = 164 MW

Prep calls for:

20 ml "dicyclopentadiene"

6 ml cyclopentadiene = 6 × 0.80 = 4.8 g → $\frac{4.8}{66}$ = 0.073 moles

6 g maleic anhydride = 6/98 = 0.061 moles

20 ml of dicyclopentadiene was placed in 100 ml r.b flask. Fractionating column and condenser attached .. ice-cooled 50 ml rb used as receiver, with vacuum take-off adapter & $CaCl_2$ tube: Had to distill very slowly to avoid liquid frothing over. Thermometer fluctuated around 40–42°. Distilled about 1/3 – 1/2 of total -- stopped 2:15. Distillate pretty clear, but added about 8–10 lumps of $CaCl_2$.

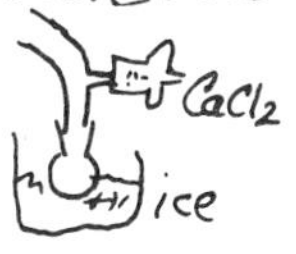

$CaCl_2$ ice

43.6
37.6 tare
6.0 g Mal. anhy.

6.0 g Maleic anhydride in 50 ml Erlenmeyer. Dissolved in ~~20~~ 22 ml ethyl acetate (Warmed to dissolve, 10 ml hexane was added; started to xslize again at room temp.

Added 6.0 ml of the distilled cyclopentadiene (still cold). Soln became perceptibly warm and then tremendous crystallization! Warmed on steam bath to dissolve and cooled slowly -- beautiful long xstls. Collected on Büchner -- washed with a little cold EtOAc–hexane (1:1). Dried in air: 3.64 g of endo-norbornene-5,6-dicarboxylic anhydride, mp 162–164° (lit 164–165°). TLC of ML (CH_2Cl_2):

4.18
.54
3.64 g

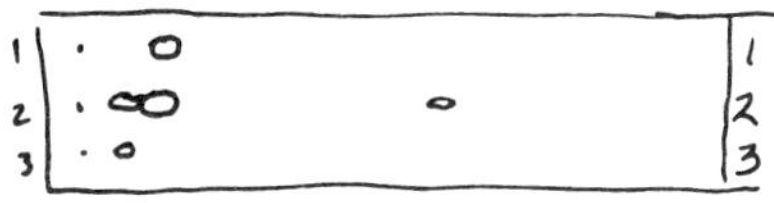

1 xstls mp 162
2 ML
3 Mal. anhydride

Still more product in ML but not time to isolate

$\frac{3.64}{164}$ = 0.022 mole

$\frac{.022}{.061}$ × 100 =

36% yield

Figure 1.11 A typical notebook write-up.

REFERENCES

Laboratory Safety

Armour, M. A.; Brown, L. M.; Weir, G. L. *Hazardous Chemical Information and Disposal Guide*, 3rd ed.; The Lab Store: Milwaukee, 1987.

Documentation of the Threshold Limit Values for Substances in Workroom Air; American Conference of Governmental Industrial Hygienists: Cincinnati, OH, 1978.

Green, M. E.; Turk, A. *Safety in Work with Chemicals*; Macmillan: New York, 1978.

Hazards in the Chemical Laboratory, 4th ed.; Bretherick, L., Ed.; Royal Society of Chemistry: London (distributed by the American Chemical Society, Washington, DC), 1986.

Introduction to Safety in the Chemical Laboratory; Freeman, N.; Whitehead, J., Eds.; Academic Press: San Diego, 1983.

NIOSH Registry of Toxic Effects of Chemical Substances; Lewis, R. J., Sr., Ed.; National Institute for Occupational Safety and Health: Cincinnati, OH, 1978.

Petersen, D. *The OSHA Compliance Manual*, Rev. ed.; King Publications: Northbrook, IL, 1979.

Safe Storage of Laboratory Chemicals; Pipitone, D. A., Ed.; John Wiley: New York, 1984.

Safety in Academic Chemistry Laboratories; American Chemical Society; Washington, DC, 1985.

Sax, N. I. *Cancer Causing Chemicals*; Van Nostrand Reinhold: New York, 1981.

Sax, N. I. *Dangerous Properties of Industrial Materials*, 6th ed.; Van Nostrand Reinhold: New York, 1984.

Sax, N. I.; Lewis, R. J. *Hazardous Chemicals Desk Reference*; Van Nostrand Reinhold: New York, 1987.

Laboratory Equipment and Techniques

Elvidge, J. A.; Sammes, P. G. *A Course in Modern Techniques of Organic Chemistry*, 2nd ed; Butterworth; London, 1966.

Kanare, H. M. *Writing the Laboratory Notebook*; American Chemical Society: Washington, DC, 1985.

Lodwig, S. N. The Use of Solid Aluminum Heat Transfer Devices in Organic Chemistry Laboratory Instruction and Research; *J. Chem. Ed.*, **1989**, *66*, 77.

Vogel, A. I. *Practical Organic Chemistry*, 5th ed.; Longman/Wiley: New York, 1989.

Wiberg, K. *Laboratory Technique in Organic Chemistry*; McGraw-Hill: New York, 1960.

2

Melting Points

The melting point of a solid is defined as the temperature at which the solid and liquid phases are in equilibrium. The equipment and time necessary to obtain such an equilibrium value usually are not available to organic chemists; therefore, the melting-point range of temperatures between the first sign of melting and the complete melting of the solid is taken. A *narrow range* indicates high purity of the sample, whereas a *broad range* usually indicates an impure sample.

To determine a melting point range, a small sample of the solid in close contact with a thermometer is heated in an oil bath or metal heating block so that the temperature rises at a slow, controlled rate. As the thermal energy imparted to the substance becomes sufficient to overcome the forces holding the crystals together, the substance melts. The rate of heating should be controlled so that the melting range is as narrow as possible. The temperature is recorded when the first melting appears and when the last solid disappears. A sharp melting point is generally accepted to have a range of 1 to 2°C. Impurities will usually cause the melting-point range to become wider and melting to occur at lower temperatures than that of a pure compound. A familiar example of this is the lowering of the melting point of ice by the addition of salt. This phenomenon results from the fact that both the liquid and solid are in equilibrium with the vapor. An impurity dissolved in the substance lowers the vapor pressure of the liquid, causing the solid to melt and restore the equilibrium among the three phases. The amount of lowering will depend on several factors, among which are the molal freezing point lowering constant (K_f), the concentration, and whether the solute is ionic or not. A typical curve for the lowering of the melting point of substance A by added amounts of substance B is shown in Figure 2.1.

The melting point of pure A is 120°C and that of B is 110°C. The upper curves connecting points *A* with *E* and *E* with *B* are the boundary above which mixtures of A and B of any composition are completely melted. The lower horizontal line through point *E* is a boundary representing the temperature (99°C) below which the sample is completely solid at any composition. In the areas between these phase boundaries are mixtures of one

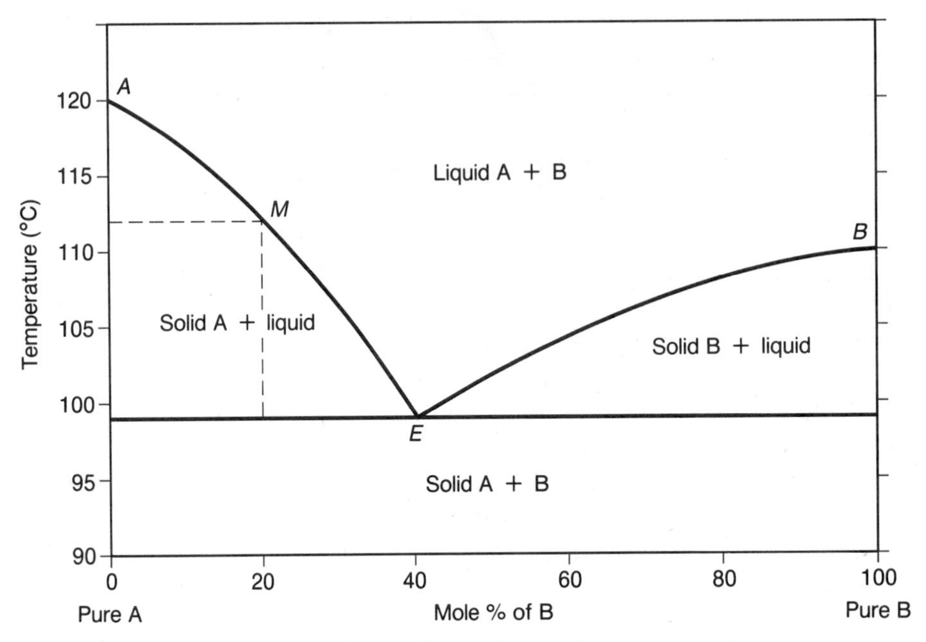

Figure 2.1 Solid-liquid phase diagram for a mixture of two compounds.

or the other pure solids plus liquid. A sample of composition M for example, will show a melting point range between the temperature at *M* (112°C) and that at line *E* (99°C). Point *E*, where the lines intersect, is called the **eutectic point**. At this composition (40% B) the mixture would melt sharply at 99°C.

The phase diagram in Figure 2.1 represents a simple or limiting case in which no solution of B in A or A in B occurs in the solid phase. More complex phase diagrams are possible, in which a solid solution does occur or in which there are two eutectic points with a maximum between them, corresponding to the formation of a complex or compound between the two components. Even in these cases, the melting point of the pure compound is generally lowered by the presence of a second component.

The number of exceptions to the rule is small, and chemists are generally safe in using melting point determinations as criteria for purity of known compounds. However, the use of melting points as criteria of purity for unknown substances must be tempered by the knowledge that exceptions do exist.

MELTING POINT BEHAVIOR

Many solids undergo some degree of decomposition or unusual behavior prior to melting. There may be changes in appearance of the sample, such as loss of luster or darkening, before the sample actually begins to melt.

Some compounds melt with decomposition, as evidenced by bubbling or formation of a dark char. In such cases, the observed melting point (or decomposition point) frequently depends on the rate of heating.

Often compounds begin to soften, shrink, or appear moist just prior to melting. These changes are not the beginning of melting and are sometimes referred to as sintering; the actual melting begins when the first drop of liquid is visible and is completed when the last solid disappears.

If a compound is appreciably volatile, determination of the melting point may be accompanied by sublimation, wherein the solid vaporizes and disappears before it melts. With a capillary tube, sublimation can be prevented by sealing the top of the tube after filling it (see Chap. 18). On an open block, the sample will seem to shrink into an increasingly small circle and may disappear before melting occurs. It may be possible to observe the melting point by using a larger than normal sample and placing it on the block at a temperature not too far below the expected melting point.

MIXTURE MELTING POINTS

In identifying an unknown, a useful practice is to take the melting point of the unknown and, at the same time, to take that of a mixture of the unknown and a known compound suspected to be the same as the unknown. If the known and unknown samples are the same substance, there will be no depression of the melting point; however, if they are different, a depression of the melting point will occur. To make sure that the mixture was not one of the exceptions mentioned above, another ratio of known to unknown should be melted.

To prepare a sample for determination of mixture melting point, place approximately equal amounts of the two compounds on a piece of glassine paper or a watch glass and mix them together. Then crush the crystals to a powder and grind them thoroughly with a spatula. When the pile of crystals is spread out to a thin layer, scrape the powder together and grind again. Fill the melting point capillary in the usual way and fill other capillaries with samples of the two individual compounds that have been ground to the same degree of fineness. Place all three samples together, with the mixture in the middle, in the bath or block and observe the melting points simultaneously. The temperature can be raised rapidly to about 20° below the expected melting point, then more slowly.

APPARATUS

A thermometer and a means of heating the sample in close contact with the thermometer bulb at a steady, controlled rate, are required for the

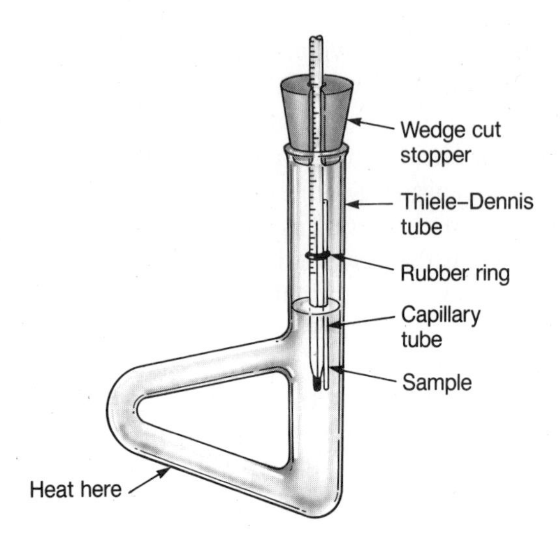

Figure 2.2 Thiele–Dennis tube melting point bath.

determination of the melting point. In the most commonly used method, the sample is enclosed in a glass capillary tube that is sealed at one end. Commercially available 2-mm capillaries are somewhat thicker than necessary but are satisfactory for most purposes. Thinner-walled tubes of smaller diameter, which permit better heat transfer, can be pulled from a soft-glass disposable pipet.

The capillary tube containing the sample is heated either in a liquid bath or a metal block. The bath is usually in the form of a vertical tube with a side loop to provide convection (Thiele–Dennis tube), shown in Figure 2.2. A 150-mL round-bottom flask also can be used. The bath contains a high-boiling mineral oil or silicone fluid that can be heated to 250°C or higher for brief periods without decomposition. The capillary is held on the thermometer with a small ring sliced from a piece of rubber tubing, with the sample as close to the thermometer bulb as possible. Thermometer and capillary are clamped in the bath or held in place by a cork with a wedge cut out to permit the thermometer to be seen and to prevent pressure from building up. The bulb and sample should be just below the upper arm of the loop. The bath is heated on the loop with a microburner; a current of heated fluid circulates upward in the loop and down the tube to provide mixing and an evenly rising temperature.

A typical melting-point block for use with capillary tubes is illustrated in Figure 2.3. The block is heated by electrical resistance, with a variable voltage control to regulate the rate of heating. A slot in the front of the block can hold three capillary tubes adjacent to the thermometer. The samples are illuminated and can be observed through a magnifying lens.

In another method for determination of melting point, the sample is

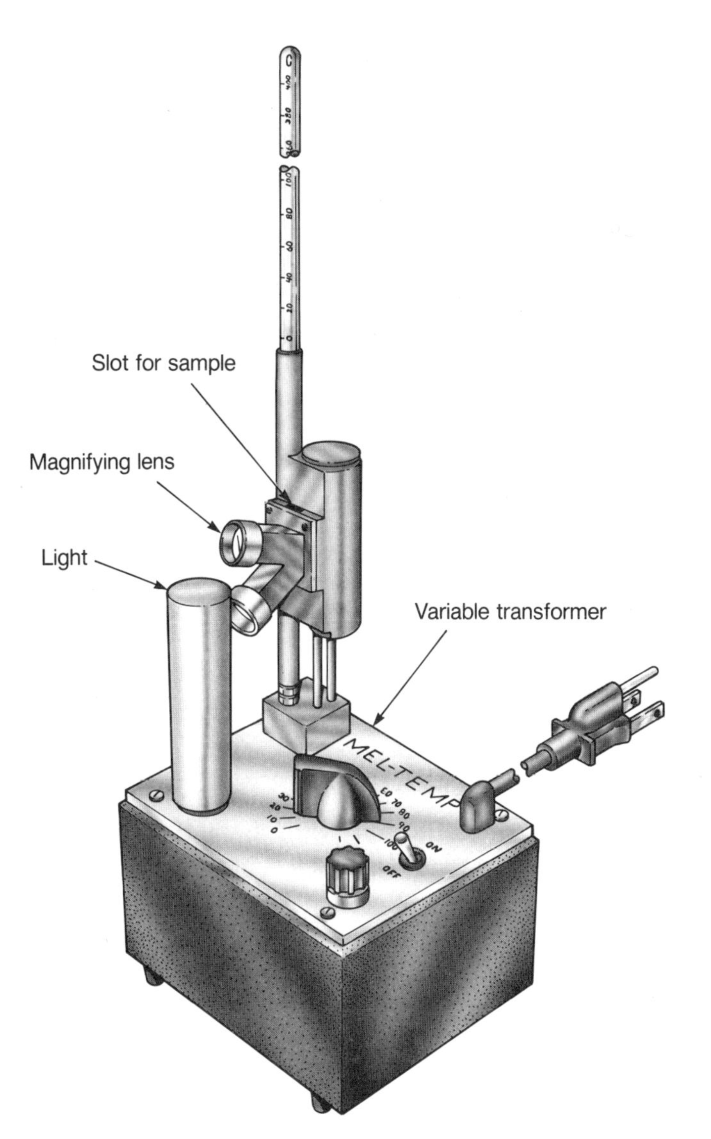

Figure 2.3 Mel-Temp capillary melting point apparatus. (Courtesy of Laboratory Devices, Cambridge, MA)

sandwiched between thin cover glasses, which are then placed directly on an electrically heated block (Fig. 2.4). The sample is illuminated and viewed through a lens above the block. A smaller sample (only a few tiny crystals) can be used with this type of block than with a capillary tube.

Either of the melting-point blocks shown in Figure 2.3 or 2.4 can be used for temperatures up to 300°C. A block is preferable to a liquid bath for any melting point above 250°C.

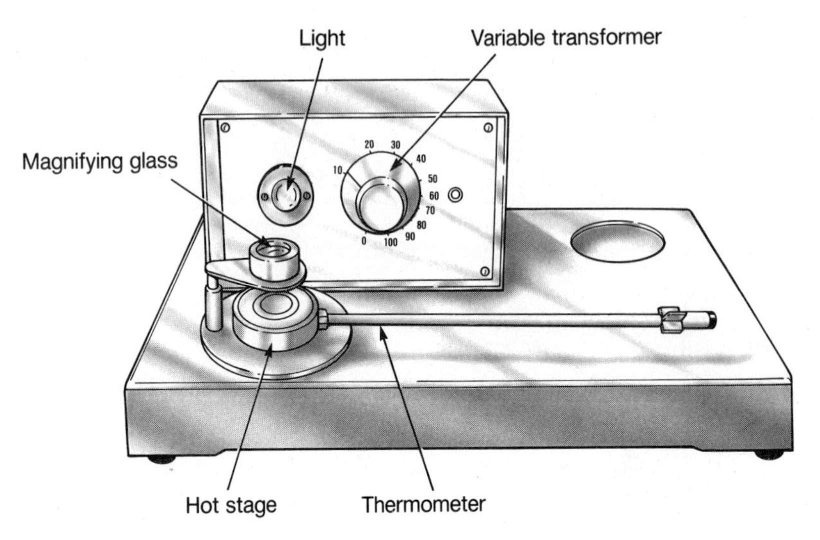

Figure 2.4 Fisher-Johns melting-point block.

Experiments

A. MELTING POINT OF A PURE KNOWN COMPOUND

Fill a melting point capillary tube (2 × 90 mm) with a small portion of sample (a column approximately 5 mm long is ample). This can be accomplished most readily by pressing the open end of the capillary into a small mound of the sample, inverting the capillary so that the closed end is down, and tapping it gently on the bench top or by drawing the flat side of a triangular file across the upper end of the tube. Another method involves dropping it and its contents through a piece of 8 to 10-mm glass tubing 100 cm long onto a hard surface. The tube will bounce off the surface and, in the process, will pack the sample uniformly. Repeat this packing several times, and note the number of times for future reference. All subsequent packing should be done in exactly the same manner. Insert the capillary into the melting point apparatus, and apply heat at a rate that will cause the temperature to rise about 10 degrees per minute. Watch the sample carefully for any sign of melting. Record the temperatures when melting first occurs and when the sample is all liquid; this is the melting point range. Repeat the procedure with a fresh sample and tube; however, this time when the temperature of the sample reaches 10 degrees below the first temperature you previously recorded, slow the heating rate to about 3 degrees per minute. This slower heating rate is important to allow time for a better equilibrium to be established between sample, heating source, and thermometer. Record the temperatures as before. Duplicate results should agree within 1 to 2°C; the range should also be within a 1- or 2-degree span.

If you do not obtain results within these parameters, check with your instructor.

B. MELTING POINTS AS A THERMOMETER CALIBRATION

The thermometer used for melting points should be calibrated by determining the melting points of three standards of Table 2.1 that melt near 50°, 100°, 150°, and 200°C. Determine the melting points as accurately as possible (within 1 degree) and plot the literature value (x-axis) *vs.* experimental value (y-axis). The resulting graph will allow you to correct any subsequent temperature value that you determine in this course. The same thermometer should be used for all melting point determinations.

Table 2.1 Melting Point Standards

COMPOUND	MP (°C)	COMPOUND	MP (°C)
p-Dichlorobenzene	53	Salicylic acid	159
Acetanilide	114	3,5-Dinitrobenzoic acid	205
Benzamide	130	*p*-Nitrobenzoic acid	241

C. MELTING POINT OF AN UNKNOWN COMPOUND

Obtain an unknown from your instructor and determine its melting point range. The unknown will be one of the substances in Table 2.2. Determine the melting points of mixtures of your unknown with known samples of comparable melting point ranges and use this information to identify the unknown. You should strive to have results in agreement within 1 or 2 degrees. One determination of a melting point is usually insufficient evidence for a positive identification. Before you report your results you should run duplicate samples that agree. Capillary samples should not be used more than once because decomposition may occur during the heating process; always use a fresh sample.

Table 2.2 Compounds for Melting Point Unknowns

COMPOUND	MP (°C)	COMPOUND	MP (°C)
Benzhydrol	68	*o*-Anisic acid	100
Coumarin	69	Phenanthrene	101
Biphenyl	70	*m*-Anisic acid	102
Phenylacetic acid	78	*o*-Acetotoluide	112
Naphthalene	80	Acetanilide	114
Vanillin	81	Fluorene	114

QUESTIONS

1. Why should each sample be tightly packed into the capillary tube prior to heating?
2. Why should the heating rate during melting point determinations be as slow as 2 or 3 degrees per minute?
3. What would be the effect of each of the following conditions on a melting point determination?
   a. the presence of an insoluble impurity such as silica
   b. incomplete drying (i.e., incomplete removal of solvent)
   c. oil level in the Thiele–Dennis tube below the upper connection of the side loop

REFERENCE

Physical Methods of Chemistry; Weissberger, A.; Rossiter, B., Eds.; Wiley-Interscience: New York, 1971; Vol. I, Part V.

Name ______________________ Section ____________ Date ________

2 Melting Points

PRELABORATORY QUESTIONS

1. Define the term "melting point."

2. Give an example of the effect of an impurity on the melting point (or freezing point) of a common substance.

3. Gram for gram, table salt lowers the freezing point of water much more than table sugar. Explain.

4. Give two pieces of information that may be obtained from melting point data on an unknown substance.

5. In determining melting points, why should a sample not be melted more than once?

3

Recrystallization

Of all the techniques studied in the organic chemistry laboratory, the purification of solids by recrystallization from solvents is the one most widely employed and at the same time the most universally misused. Students should master this skill early in their laboratory experiences so that they will not be hampered by it when they should be concentrating on other aspects of subsequent experiments.

In the process of recrystallization, an impure sample is dissolved in an appropriate solvent that differs in its solubility characteristics to the sample and the impurities, respectively. Insoluble impurities may be removed by filtration, whereas the more soluble impurities remain in solution when the purified sample is crystallized. The exact process will depend upon the following criteria:

1. The nature of the differences in structure of the sample and the impurities (if known).
2. Differences in solubility of the sample and impurities in various hot and cold solvent systems. For ease of method, the sample should be more soluble in the hot solvent than in the cold solvent; the impurities should be either highly soluble or extremely insoluble.
3. The solvent should be readily volatilized after collection of the purified sample crystals. Moreover, the solvent should be unreactive towards the sample.
4. The solvent may be a pure substance, such as diethyl ether or ethyl alcohol, or it may be a mixture of two or more substances that give the desired solubility characteristics. "Ready-mixed" solvents may be available in the laboratory, or the proper solvents can be mixed as needed by adding one of the solvents to the other (in portions) until the desired characteristics are achieved.

For crystallization to occur, the concentration of a dissolved compound must exceed the equilibrium solubility; that is, the solution must be supersaturated at the given temperature. With few exceptions, the solubility of a solid increases with an increase in temperature, often manyfold over a

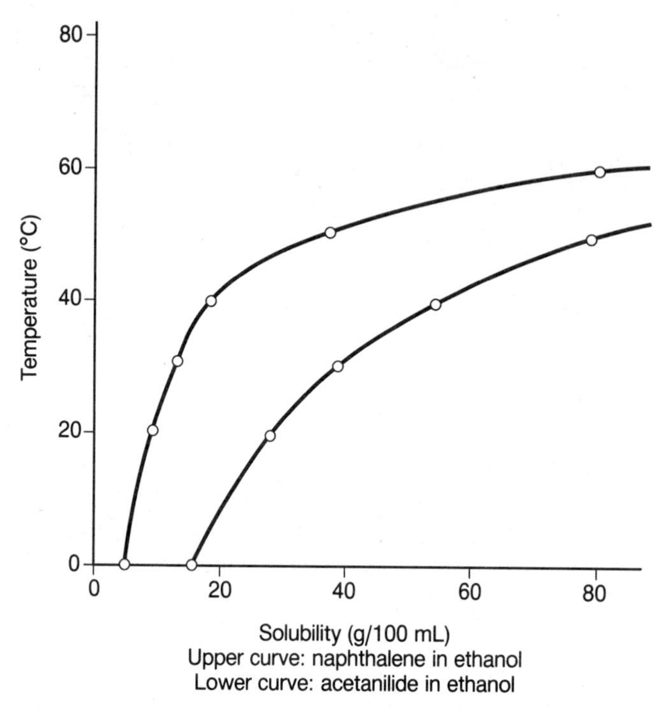

Figure 3.1 Typical temperature-solubility curves.

30° to 40°C temperature increase. Thus, a sample is dissolved in a minimum of hot solvent, and crystallization occurs on cooling. For a proper solvent system, the sample's solubility should increase rapidly as the melting point is approached. Some typical temperature versus solubility curves are shown in Figure 3.1.

The solubility of crystalline organic compounds in a solvent depends on two major factors: (a) the relative polarity of the solvent and solute and (b) the energy of the crystal lattice of the solute. The effect of polarity can be summed up by the statement "like dissolves like." Thus, compounds that contain one or more polar groups, such as —OH, —NH_2, —CO_2H, or —$CONH_2$, are usually more soluble in hydroxylic solvents (e.g., water or alcohols) than in hydrocarbons, such as toluene or hexane. Conversely, the latter more easily dissolve compounds of low polarity. Within a series of compounds of the same type, the relationship is that the higher the melting point (high crystal lattice energy), the lower the solubility in a given solvent. This can be seen in a comparison of the melting points and solubilities of the isomeric nitrobenzoic acids (Table 3.1).

The *para-* isomer, having a highly symmetrical structure, can pack into a more stable lattice (strong attraction) than can the others and consequently has a much higher melting point and corresponding lower solubility.

In selecting a solvent for the recrystallization process, the main requirement is suitable solvating power, that is, one that provides high solute

Table 3.1 Melting Points and Solubilities of Nitrobenzoic Acids

	ISOMER		
	ortho- 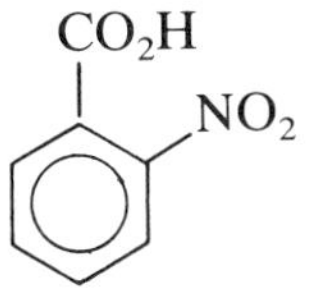	*meta-* 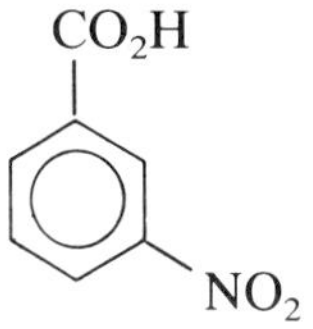	*para-* 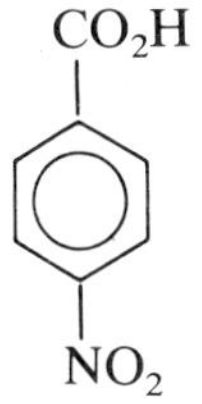
Mp (°C)	147	141	242
Solubility, g/100 mL at 10 to 12°C			
Methanol	43	47	9.6
Ethanol	28	33	2.2
Ether	21	25	0.9

solubility in the hot solvent and a significantly lower solubility at room temperature or below. Frequently, this requirement is met by using a mixture of two solvents in which the solute has quite different solubilities. Solvents that are generally useful for crystallization are listed in Table 3.2 in order of decreasing polarity, together with the solubilities of two representative compounds.

Water is a convenient solvent for recrystallizing some moderately polar compounds with melting points above 100°C. Mixtures of water and alcohols

Table 3.2 Useful Recrystallization Solvents

SOLVENT	BOILING POINT (°C)	SOLUBILITY OF 25°C, g/100 mL SOLUTION	
		Acetanilide (Polar)	*Naphthalene (Nonpolar)*
Water	100	0.53	0.002
Methanol	65	48	9.9
Ethanol	78	30	11.8
Methylene chloride (CH_2Cl_2)	40	(17)*	(55)*
Diethyl ether	35	2.8	57
Hexane	68	0.03	20
Pentane	36	<0.01	

*These values are for chloroform ($CHCl_3$), which has similar solvent properties but is more toxic than CH_2Cl_2.

are widely used. The low boiling points of diethyl ether and methylene chloride (dichloromethane) make them attractive for recrystallizing low-melting solids. Mixtures of low-boiling alkanes (petroleum ether, bp 35 to 50°C, and ligroin, bp 130 to 145°C) have wide utility and are commercially available. These generalizations are illustrated with the two compounds mentioned in Table 3.2. Acetanilide is much too soluble in methanol and ethanol to permit crystallization from these solvents. The solubility in water is much lower, and since the solubility at 100°C is about ten times higher than at 25°C, water is a suitable solvent, although a rather large volume is needed. On the other hand, naphthalene is very soluble in hot ethanol, but the solubility drops to 5% at 0°C. Thus, ethanol is a practical solvent; adding a small amount of water would probably be even better.

The general procedure for recrystallization involves the following steps:

1. Selection of a suitable solvent through experiment or from data on solubility.
2. Dissolution of the material in the hot solvent (near the boiling point).
3. Filtration of the hot solution to remove insoluble impurities or impurities adsorbed on activated carbon. (This step is sometimes omitted—see below.)
4. Crystallization of the solute from the cool solution.
5. Collection of the purified crystals.
6. Washing and drying the product.

FILTRATION

Insoluble matter is removed from the hot solution by gravity filtration (Step 3). This step may be omitted if the solution is clear and obviously free of insoluble particles, fibers, and the like. Hot filtration must be carried out with precaution to avoid premature crystallization in the funnel or on the filter paper because of the cooling of the solution. Thus, a conical funnel with a short stem or no stem should be used. The funnel is placed in a beaker or flask containing a small amount of the boiling solvent (Fig. 3.2). This arrangement allows the funnel to be bathed in the hot solvent vapors during the process. The solution to be filtered should be maintained close to the boiling temperature, and small amounts of additional solvent should be added as necessary. Rapid filtration can be achieved using fluted filter paper, which provides a maximum amount of surface area for the solvent to pass through. Details for proper folding to flute the filter paper are shown in Figure 3.3.

Note that the above procedure works well on samples as small as 100 mg. Smaller amounts should be handled by using a cotton plug or specially adapted pipet. A small cotton wad the size of a pea stuffed into the funnel stem very often results in fast filtration and eliminates the losses on the

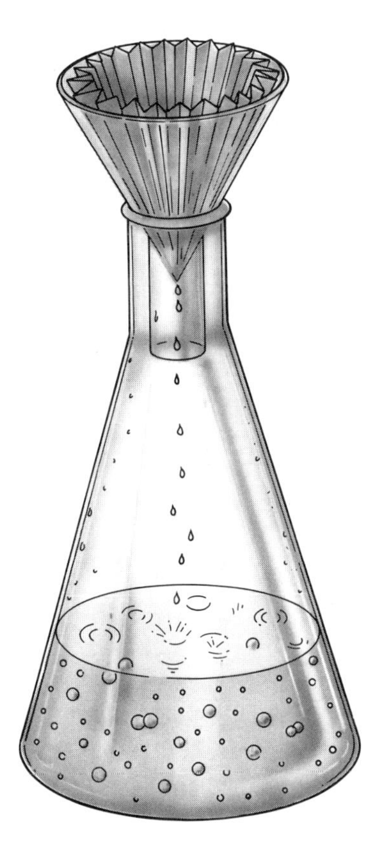

Figure 3.2 Filtration of a hot solution.

filter paper. This method is particularly convenient for microscale procedures to filter a few milliliters of solution or less. A disposable (Pasteur) pipet may serve as the funnel (Fig. 3.4). With the tip of a second pipet, a very small pellet of *loose* cotton is pushed through the top and wedged

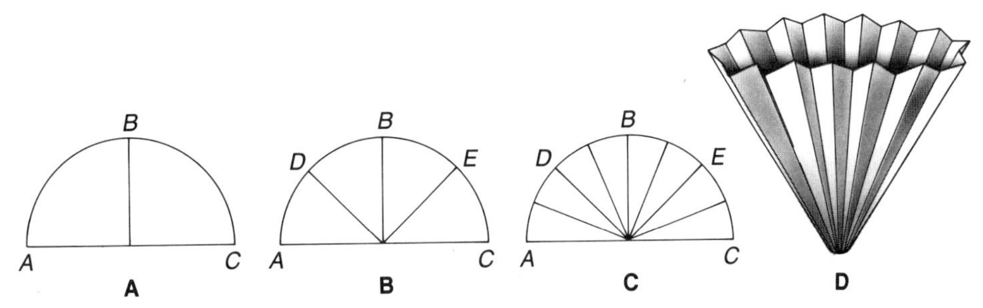

Figure 3.3 Folding Fluted Filter Paper

On a hard surface, fold a piece of 11- or 15-cm circular filter paper in half and then in quarters. Open the second fold like a book (**A**), and fold corners *A* and *C* to meet point *B*. Reopen it to a semicircle (**B**), and fold corner *A* first to point *D* and then to point *E*. Similarly, fold corner *C* to *D* and *E*, reopening it to the semicircle after each fold. The paper should now look like (**C**), with the folds bending the semicircle into a partially open cone. In each of the eight segments, make a fold in the center, alternating with and opposite to the direction of the previous folds. This accordion-like arrangement will open into a fluted funnel (**D**).

gently in the constriction in the pipet. Snip off the narrow part of the pipet to a length of about 2 cm and transfer the solution to be filtered through the top with another pipet.

For the filtration of finely dispersed or gelatinous solids, it may be necessary to use a filter aid such as Celite, using a Büchner funnel and a thick-walled side-arm flask. The filter aid retains the fine particles more effectively than the paper alone. If the filtration becomes slow because of the build-up of a coating on the surface of the Celite, fresh surface can be exposed by gently scraping with a spatula. Suction filtration of a hot solution may cause rapid evaporation and cooling, leading to crystallization and clogging in the funnel. To avoid this problem, the original solution can be diluted with extra solvent, or a steam-jacketed funnel may be used.

Highly colored contaminants may be removed from solution by adding activated carbon (Norit, Darco, or other trade-named samples of finely

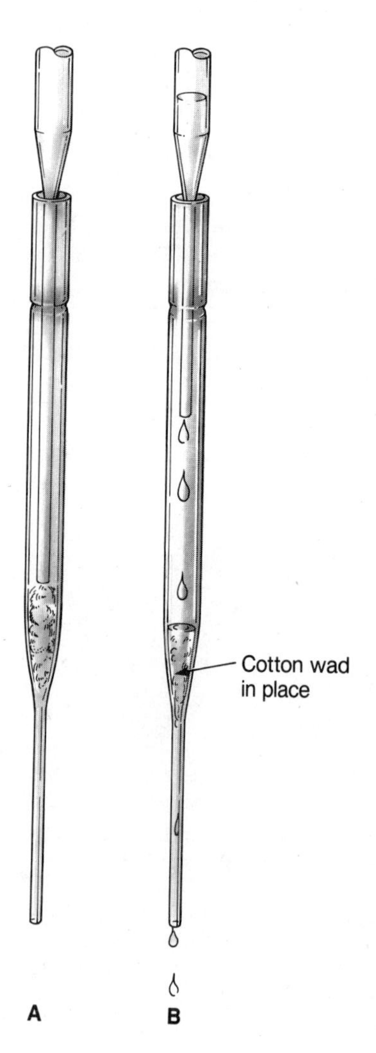

Figure 3.4 Filtration using a pipet as a funnel. (A) Cotton stuffed in pipet to be used as funnel. (B) Filtration in progress.

divided carbon particles, or an easily filtered granulated form, which works especially well with microscale experiments) to the warm (not hot) solution, heating the solution to boiling and filtering while hot. Polar substances with chromophores that are usually responsible for imparting a color to the solution are strongly adsorbed on the surface of the carbon. The use of carbon is by no means a panacea for removing impurities; sometimes it is ineffective and can reduce the yield of product because of adsorption. If the compound to be purified is known to be colorless but the solution is dark, it will often be worthwhile to try the effect of carbon on a small portion. Activated carbon is usually most effective in polar solvents, such as water or alcohol.

CRYSTALLIZATION

In the optimum case, crystallization occurs spontaneously when the solution is cooled to room or ice-bath temperature; however, in some cases, crystallization has to be induced. Sometimes the volume of solvent is too large and the solution must be concentrated. Crystal growth often can be initiated by scratching the inside wall of the flask with a glass rod or spatula or by removing a few drops of the solution and rubbing it with a spatula on a watch glass as the solvent evaporates. Crystals obtained in this way can then be used as a "seed" to induce crystal growth in the main solution.

The size of crystals depends on the rate of formation and the number of particles present. Rapid cooling and shaking or stirring the solution lead to the rapid formation of small crystals. This is normally desirable, since it helps to eliminate solvent molecules entrapped in the crystal lattice. In delicate selective crystallization, it may be crucial to allow crystals to form slowly and without agitation.

COLLECTION AND WASHING OF CRYSTALS

When a crystallization is complete, the solid is scraped loose from the wall of the flask and collected in a porcelain Büchner or Hirsch funnel using vacuum (Figs. 3.5 and 3.6). If a small volume of liquid is to be filtered and retained, a test tube can be placed inside the filter flask or side-arm test tube and the stem of the funnel directed into the tube (see Fig. 3.6). These porcelain funnels are fitted with a circle of filter paper that should lie flat and cover the perforations, which permits liquid to drain rapidly into the evacuated flask. The paper is wet with a few drops of solvent and the flask is connected to suction. This will seat the filter paper firmly on the funnel.

After transferring the mixture to the funnel, any solid clinging to the wall of the flask is scraped loose and rinsed into the funnel with a small portion of the filtrate (mother liquor) or cold solvent. In microscale crys-

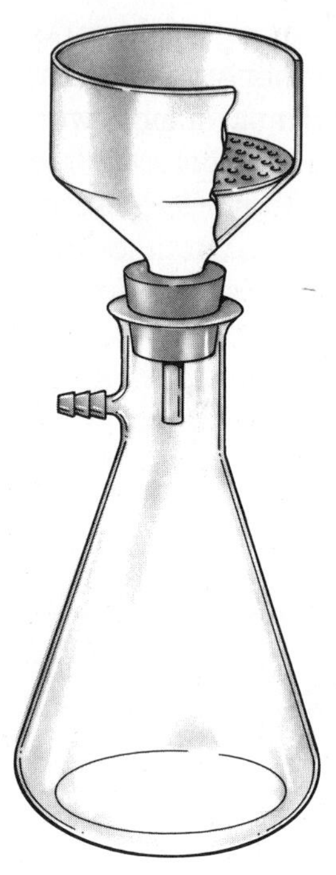

Figure 3.5 Büchner funnel and suction flask.

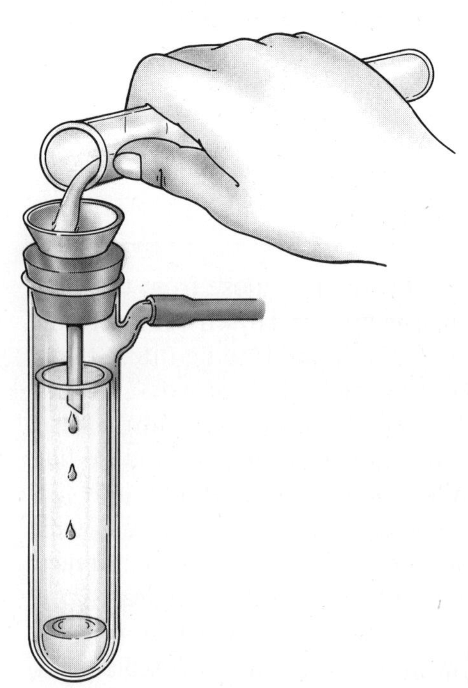

Figure 3.6 Hirsch funnel and side-arm test tube.

tallizations in a test tube, crystal growth may occur entirely on the walls, thereby permitting the mother liquor to be decanted cleanly without the necessity of transferring the solid. Alternately, crystallization can be performed in a centrifuge tube, and the crystals centrifuged to the bottom of the tube for easy removal of the supernatant mother liquor with a pipet. Normally, the crystals are washed free of the mother liquor using small portions of chilled solvent. This insures the removal of impurities that might be carried along in the mother liquor. If a more thorough washing to remove oily mother liquor is needed, the suction should be broken and the crystals stirred well on the funnel before again applying suction and allowing the liquid to drain through. If a solvent with a high boiling point has been used, a final rinse with a more volatile solvent, such as ether or pentane, will facilitate rapid drying of the crystals.

The crystal mass may be compacted with a cork or spatula to hasten the removal of solvent, and it is often possible to remove traces of solvent by drawing air through the funnel using suction. The crystals are transferred to a watch glass, filter paper, or glassine paper for further drying in air or in an oven with the temperature well below the melting point of the sample.

FILTER AID

A filter aid such as Celite is used to remove very small particles that otherwise would pass through normal filters. This procedure involves the laying down of a thin layer (about 0.5 cm) of filter aid over the filter paper in a Büchner or Hirsch funnel as follows: A piece of filter paper is placed in a Büchner or Hirsch funnel attached to a vacuum flask. Solvent is added until the funnel is approximately half filled and sufficient dry filter aid is added. The mixture is stirred until the filter aid is evenly suspended and suction is applied slowly. When the solvent has drained, the suction is broken and the flask is removed, cleaned, and reattached to the funnel. The funnel is now ready for use.

Experiments

A. RECRYSTALLIZATION OF BENZOIC ACID (Microscale)

Add 60 mg of benzoic acid and approximately 2 mL of water to a 10 mL Erlenmeyer flask, add one boiling stone, heat to boiling on a hot plate, and continue to add boiling water in small increments until all of the benzoic acid is dissolved (be patient; organic compounds often do not dissolve instantaneously). Record the total amount of water used. Remove the flask from the heat and allow it to cool to room temperature. If crystals do not appear, induce their formation by scratching the inside of the flask with a glass rod or spatula. Filter the crystals with suction using a Hirsch funnel, side-arm test tube, and aspirator (see Fig. 3.6). Allow the sample to air dry, and weigh and record the amount of benzoic acid recovered. Calculate the percent recovery.

B. RECRYSTALLIZATION OF PURE ACETANILIDE (Microscale)

Weigh 50 mg of pure acetanilide in a small test tube or centrifuge tube. Add 2 mL of methylene chloride, and with stirring, heat to boiling on a steam bath. If the sample does not dissolve, add small portions of solvent and reheat. Keep adding solvent until the sample dissolves completely in the hot solvent. Allow the mixture to cool to room temperature and record the solubility characteristics in your notebook. If crystals do not form spontaneously, add a seed crystal or scratch the inside of the tube to induce crystal growth. If this fails, evaporate half of the solvent (hood) and chill the tube again. Repeat the recrystallization using diethyl ether, ethanol, and petroleum ether in order to determine which of the solvents gives the best percent recovery and the purest acetanilide. The most difficult part of this procedure is determining the proper solvent and the proper ratio of solvent to solute. To reiterate, the solvent should dissolve the solute when hot and give a high recovery of good crystals when cold. Sufficient solvent should be used to keep impurities dissolved but not so much that it is impossible to recover the solute in good yield. Check your decision with the instructor before you proceed with the next part.

C. CRYSTALLIZATION OF IMPURE ACETANILIDE (Macroscale)

From the data you obtained in Experiment B, predict the approximate amount of solvent you will need to recrystallize a 2.0-g sample of acetanilide containing impurities that may be solvent-soluble or solvent-insoluble. A general approach to the process of purification by recrystallization is demonstrated with the following procedure.

Weigh 2.0 g of impure sample into an Erlenmeyer flask of the size that will hold all of the solvent you have calculated that will be required. Add approximately half of the solvent and a boiling stone (used to prevent superheating of the solvent that might result in a violent eruption of the solution). Place the flask on a hot plate and slowly heat the contents to boiling. Stir the contents occasionally, and when the solvent starts to boil, add small portions of the solvent until you cannot detect any further solution. Add a small amount of activated carbon to remove any highly colored impurities. (*Note:* Do not add the carbon to the hot solution; rather, cool the solution slightly or it may boil over when the carbon is added.) Reheat to boiling and filter the solution into a second Erlenmeyer flask through a double layer of fluted filter paper in a funnel heated with boiling solvent (see Fig. 3.2). If the mixture is still gray or highly colored, you may have to repeat the filtration process after reheating the solution. Set the flask aside to cool. When it has cooled to room temperature, chill it in an ice bath to induce crystallization. When crystallization appears to be complete, filter the mixture using suction. Partially dry the crystals on the filter by drawing air through the vacuum system for several minutes, place them on a large watch glass or glassine or filter paper, and allow them to dry in air for approximately an hour. Determine the melting point range; when the correct melting point is reached (an indication that the sample is dry), the crystals can be weighed and stored in a properly labeled sample vial. Calculate the percentage yield and submit the sample and required data to your instructor.

D. RECRYSTALLIZATION OF DIMETHYL TEREPHTHALATE (Microscale)

Dimethyl terephthalate is used as an intermediate in the synthesis of the polyester Dacron (see Chapter 44). As such, it must be in a very high state of purity. It is conveniently recrystallized from ethanol and in this experiment serves as an introduction to recrystallization on a microscale.

Procedure

Weigh out 30 mg of crude dimethyl terephthalate and place it into a 10-mL or 15-mL centrifuge tube, add 10 drops of 95% ethanol and warm on a steam bath until the solvent simmers. If necessary, add additional drops of ethanol until no more solid dissolves. (*Note:* All of the solid may not dissolve as it may contain an insoluble impurity.) Add 5 drops of ethanol in excess of the amount needed for solution (in order to ensure that the product does not crystallize in the next operation). Fit a Pasteur pipet with a cotton plug in the tip and allow the tip to be warmed by the ethanol vapors in the centrifuge tube as the latter is being heated. Place the pipet to the bottom of the tube and draw off the liquid, leaving behind any undissolved solids. Transfer the contents of the pipet to a clean, dry centrifuge tube, let it cool, and allow the dimethyl terephthalate to crystallize. If crystals do not form, warm the tube gently on the steam bath using a boiling stick, until approximately 20% of the solvent evaporates. Recool the tube and allow the crystals to form; then place the tube in an ice bath. When crystallization is complete,* withdraw the supernatant "mother liquor" with another Pasteur pipet fitted with a cotton plug, retaining the crystals in the tube. Add 2 or 3 drops of ice-cold ethanol and stir to wash the crystals in the tube.* Withdraw the supernatant liquid as before and

*If a centrifuge is available, spinning the tube for 1 minute will help compact the solid.

allow the crystals to dry in a stream of air or under vacuum. Scrape the crystals onto a piece of glassine paper and weigh. Calculate the percent recovery.

E. PREPARATION OF ACETANILIDE (Macroscale)

$$C_6H_5NH_2 + (CH_3CO)_2O \longrightarrow C_6H_5NHCOCH_3 + CH_3CO_2H \quad [3.1]$$

aniline C_6H_7N; acetic anhydride $C_4H_6O_3$ (d. 1.08 g/mL); acetanilide C_8H_9NO; acetic acid

In this experiment, aniline is acetylated, and the product, acetanilide, is purified by recrystallization from water.

SAFETY NOTE ANILINE IS A VERY TOXIC COMPOUND; TAKE CARE TO AVOID CONTACT WITH YOUR SKIN. ACETIC ANHYDRIDE IS IRRITATING TO THE EYES AND NASAL MEMBRANES. IF EITHER COMPOUND IS SPILLED, WASH THE AFFECTED AREA THOROUGHLY WITH SOAP AND COLD WATER. IF EITHER IS SPLASHED INTO THE EYES, TREAT IT AS A GENUINE EMERGENCY AND SEEK A PHYSICIAN'S HELP IMMEDIATELY.

Procedure

Measure 4.0 mL of technical grade aniline in a 10-mL graduated cylinder, weigh the cylinder to the nearest 0.1 g, and pour the contents into a 250-mL Erlenmeyer flask. Reweigh the cylinder to determine accurately the amount of aniline used in the reaction, and add 30 mL of water to the flask. While swirling the flask, add exactly 5.0 mL of acetic anhydride in 1-mL portions. Record your observations as the crude acetanilide forms.

The product is recrystallized in the same flask. Add 100 mL of water and one boiling stone, and heat the flask on a hot plate until all of the solid and oil are dissolved. Pour a few milliliters of the hot solution into a small beaker and set it aside to cool. (This will be used to assess the effectiveness of the activated carbon treatment of the main batch.) Before boiling resumes, add about 0.1 g of activated carbon to the main solution. Do *not* add it to a boiling solution, or frothing and possible spillage will result. Swirl the mixture and boil it gently for a few minutes. Meanwhile, prepare the hot filtration apparatus, complete with conical funnel and a double layer of fluted filter paper arranged to filter into a 250-mL flask (see Fig. 3.2). Have available approximately 50 mL of boiling water for washing.

Warm the funnel by pouring a few milliliters of hot water through it, and then filter the solution a little at a time. Keep the solution boiling gently until it is poured completely into the funnel. If crystallization occurs in the funnel, add hot water and set the flask containing the filtrate on the hot plate. When filtration is complete, set the flask aside to cool; finally, chill it in an ice bath. When crystallization is complete, collect the product by suction filtration using a Büchner funnel, and dry it as thoroughly as possible by drawing air through the funnel by suction. Spread the crystals on a sheet of paper on a watch glass, cover them with another watch glass, and allow them to dry until the next class period.

Using a Hirsch funnel, collect the "mini" batch of crystals that were not treated with activated carbon and compare the color and melting point with the main batch.

Weigh the main batch; calculate the percent yield, including the "mini batch"; and determine the melting point. Transfer the product to a labeled vial and submit it for grading.

F. PREPARATION OF ASPIRIN (Microscale)

$$o\text{-}HOC_6H_4COOH + Ac_2O \xrightarrow[\text{acetic anhydride}]{H^+} o\text{-}CH_3C(=O)OC_6H_4COOH + CH_3-C(=O)OH \qquad [3.2]$$

Of all of the pharmacological agents developed, none has had the wide spectrum of uses as does aspirin (acetylsalicylic acid). For many years it was the primary treatment for fever, mild pain, such as headaches, and mild inflammatory disorders, including some forms of arthritis. Now, many of these conditions are treated with acetaminophen or ibuprofen; however, recent evidence indicates that new uses may be found for aspirin in the treatment of cardiovascular diseases. Despite the fact that aspirin has been used as a drug for over 90 years, it still retains some of its "wonder drug" reputation.

Procedure

Place 100 mg of salicylic acid and one drop of 85% phosphoric acid in a 13 × 100-mm test tube, rinse down the sides of the tube with 0.25 mL of acetic anhydride, and immerse the test tube in a water bath at 80 to 85°C for 10 minutes. Carefully add 1 mL of water (the reaction of the water with the excess acetic anhydride may generate additional heat) and allow the tube to cool. When crystals start to form, place the tube into an ice bath for approximately 15 minutes to complete the crystallization process. Filter the crude product using a Hirsch funnel and wash the crystals with 0.2 to 0.3 mL of ice water. Draw air through the funnel until the crystals are dry. Weigh the crude product.

Place the product into a 10-mL Erlenmeyer flask, add approximately 1 mL of a 25% ethanol–water solution, and warm the flask until the solvent boils. A wood applicator stick or a closed microcapillary tube ("boiling bell") should be used to prevent the solution from "bumping." If the product does not completely dissolve, add a little more solvent, but do not exceed an additional milliliter. If a residue persists, filter the solution through a disposable pipet fitted with a loose cotton plug (see Fig. 3.4). Keep the pipet warm by setting it in a flask with a small amount of boiling solvent. Allow the filtrate to cool, and when crystallization is complete, collect the purified product on a Hirsch funnel. Thoroughly dry the product, weigh it, and determine its melting point. Place the product in a small vial and submit it for grading.

QUESTIONS

1. Describe how you would purify the major component in the following mixtures using the recrystallization procedure.
   a. 9 g of *para*-nitrobenzoic acid and 1 g of the *ortho*- isomer.
   b. 8 g of *meta*-nitrobenzoic acid and 2 g of the *para*- isomer.
2. During recrystallization, an orange solution of a compound in hot alcohol was treated with activated carbon and then filtered through fluted paper. On cooling, the filtrate yielded gray crystals, although the compound was reported to be colorless. Explain why the crystals were gray, and describe what steps you would take to obtain a colorless product.
3. Assume that 8 mL of aniline and 9 mL of acetic anhydride were used in the preparation of acetanilide (Equation 3.1). What is the limiting reagent? What is the theoretical yield of acetanilide?
4. The solubility of compound A in ethanol is 0.4 g per 100 mL at 0°C and 5.0 g per 100 mL at 78°C. What is the minimum amount of solvent needed to recrystallize an 8.0-g sample of compound A? How much would be lost in the recrystallization?
5. Compound B is quite soluble in toluene, but only slightly soluble in petroleum ether. How could these solvents be used in combination in order to recrystallize B?

REFERENCE

Tipson, R.S. In *Techniques of Organic Chemistry*, 2nd ed.; Weissberger, A., Ed.; Interscience: New York, 1956; Vol. III, Part I, Chapter 3.

Name ______________________ Section ____________ Date ________

3 Recrystallization

PRELABORATORY QUESTIONS

1. Describe the requirements for a "good" recrystallizing solvent.

2. Describe how you would separate a mixture of sand and sugar.

3. What is the purpose of adding a small amount of activated carbon to a recrystallizing solution?

4. What are boiling stones?

5. Describe how seed crystals are used in recrystallization.

6. Explain the adage "like dissolves like."

4

Extraction

One of the main concerns of experimental organic chemistry is the separation of mixtures and the isolation of compounds in as pure a form as needed for subsequent use. Crystallization is a useful method for purification and isolation, but it is restricted to solids, and as indicated in Chapter 3, it is a relatively inefficient way to separate a mixture of very similar compounds.

Several more general separation methods are described in this and the following three chapters. All depend in some way on the *partitioning* of the compounds to be separated *between two distinct phases.* By choosing the phases so that the different compounds are *unequally distributed* between them, **fractional separation** of the compounds is effected. The various methods are mechanically quite different, but they all depend on this common principle, and it should be thoroughly understood. The principle is most easily illustrated by the general method of extraction, and this method is therefore covered in some detail in this chapter.

THEORY OF EXTRACTION

Extraction is the general term for the recovery of a substance from a mixture by bringing it into contact with a solvent that preferentially dissolves the desired material. The initial mixture may be a solid or liquid, and various techniques and apparatus are required for different situations. In synthetic organic chemistry, the reaction product is frequently obtained as a solution or a suspension in water along with inorganic and other organic by-products and reagents. By shaking the aqueous mixture with a water-immiscible organic solvent, the product is transferred to the solvent layer and may be recovered from it by evaporation of the solvent.

The extraction of a compound from one liquid phase into another is an equilibrium process governed by the solubilities of the substance in the two solvents. The ratio of the solubilities is called the *distribution coefficient*, $K_d = C_1/C_2$, and is an equilibrium constant with a characteristic value for any compound and pair of solvents at a given temperature.

Let us consider the following situation: A 100-mL aqueous solution contains 1 g each of compounds A and B, the solubilities of which in water and ether are given below.

COMPOUND	SOLUBILITY IN WATER C_{water}	SOLUBILITY IN ETHER C_{ether}	$K_{ether/water}$
A	10 g/100 mL	1 g/100 mL	$\frac{1}{10} = 0.1$
B	2 g/100 mL	10 g/100 mL	$\frac{10}{2} = 5.0$

By the foregoing definition, the distribution coefficients of A and B (C_{ether}/C_{water}) are 0.1 and 5, respectively. If the aqueous solution is shaken with 100 mL of ether, the amount of each compound transferred to the ether phase, X, can be calculated as follows:

Compound A

$$\frac{C_{ether}}{C_{water}} = \frac{X_A/100}{(1 - X_A)/100} = 0.1$$

$$X_A = 0.091 \text{ g in ether}$$
$$1 - X_A = 0.909 \text{ g in water}$$

Compound B

$$\frac{C_{ether}}{C_{water}} = \frac{X_B/100}{(1 - X_B)/100} = 5$$

$$X_B = 0.833 \text{ g in ether}$$
$$1 - X_B = 0.167 \text{ g in water}$$

If the same aqueous solution is extracted with the same amount of ether, but in four 25-mL portions, we have for the first extraction:

$$\frac{X_A/25}{(1 - X_A)/100} = 0.1 \qquad \frac{X_B/25}{(1 - X_B)/100} = 5$$

(1) $X_A = 0.0244$ g in ether, leaving 0.9756 g in water; $X_B = 0.556$ g in ether and 0.444 g in water

For the second 25-mL extraction:

$$\frac{X_A/25}{(0.9756 - X_A)/100} = 0.1 \qquad \frac{X_B/25}{(0.444 - X_B)/100} = 5$$

(2) $X_A = 0.0238$ g in ether, leaving 0.9518 g in water; $X_B = 0.247$ g in ether and 0.197 g in water

After the third and fourth extractions:

(3) $X_A = 0.0232$ g in ether; $X_B = 0.109$ g in ether
(4) $X_A = 0.0226$ g in ether; $X_B = 0.049$ g in ether

Totals:

$$X_A = 0.094 \text{ g A in ether} \qquad X_B = 0.961 \text{ g B in ether}$$
$$1 - X_A = 0.906 \text{ g A in water} \qquad 1 - X_B = 0.039 \text{ g B in water}$$

It can be seen from these values that even with a relatively small distribution coefficient ($K_d = 5$) virtually complete extraction of compound B can be effected and that several extractions with small volumes of extractant are more efficient than a single extraction with the same total volume in one portion.

In the extractions described above, a significant amount of compound A was also transferred to the ether layers. If the ether were removed from the solution at this point, the residue would be 1.055 g of material that is only 92% pure B. However, if the ether is shaken (back-extracted) with 50 mL of water before evaporation, the amounts (Y) of A and B removed from the ether solution can be calculated as follows:

$$\frac{C_{\text{ether}}}{C_{\text{water}}} = \frac{(0.094 - Y_A)/100}{Y_A/50} = 0.1 \qquad \frac{(0.961 - Y_B)/100}{Y_B/50} = 5$$

$Y_A = 0.078$ g in water, leaving 0.016 g in ether

$Y_B = 0.087$ g in water and 0.874 g in ether

Evaporation of the ether now leaves a residue of 0.890 g that is more than 98% pure B. Some B was lost in this process, but that which remains may be pure enough for its intended use.

APPLICATIONS OF EXTRACTION

LIQUID–LIQUID EXTRACTION

As described in the preceding section, extraction of one liquid with another is a standard operation that is used for a number of purposes. Simple extractions on a macroscale are carried out in a separatory funnel, which is a conical vessel with a stopcock at the bottom (Fig. 4.1). The two layers are mixed by shaking to permit transfer from one layer to another; after they separate, the lower layer is drained out through the stopcock.

In the most common everyday applications of liquid–liquid extraction, such as the recovery of an organic compound from an aqueous solution containing inorganic acids, bases, or salts, the distribution coefficient is very favorable. In this case, one or two extractions with an organic solvent, followed by the back-extraction or "washing" of the solvent layer with water, is often sufficient. If the procedure does not specify the number of extractions, it is normally understood by organic chemists to be three, followed by one back-extraction.

If the distribution coefficient is unfavorable, that is, if the compound has a high solubility in water, extraction with many small portions of solvent becomes impractical. For such extractions, an apparatus such as that shown in Figure 4.2 is used. The extracting solvent is continuously distilled into the condenser, from which the condensed solvent returns in a stream that

Figure 4.1 Separatory funnel.

passes down through the solution. The extracted material collects and becomes concentrated in the flask containing the boiling extraction solvent.

A common and important use of extraction is in the separation of acidic, basic, and neutral organic compounds. An organic acid, RCO_2H, or a base, such as RNH_2, is usually much more soluble in organic solvents than in water. However, the salts of these compounds, for example, $RCO_2^-Na^+$ or $RNH_3^+Cl^-$, have much higher solubilities in water, since they are ionic substances. To separate an acid and a neutral compound, for example, the mixture is dissolved in ether and extracted with an aqueous solution of a base such as sodium hydroxide or sodium bicarbonate. The acid is converted to the salt (Eq. 4.1) and is extracted into the water layer. After separation of the water layer, the acid is recovered by reacidification of the aqueous solution (Eq. 4.2).

$$RCO_2H + NaHCO_3 \rightarrow RCO_2^-Na^+ + CO_2 + H_2O \quad \textbf{[4.1]}$$

$$RCO_2^-Na^+ + HCl \rightarrow RCO_2H + Na^+Cl^- \quad \textbf{[4.2]}$$

By the same principle, a base can be separated from neutral compounds by extraction with aqueous acid (Eq. 4.3). After separation of the aqueous solution of the salt, the addition of NaOH yields the organic base (Eq. 4.4).

$$RNH_2 + HCl \rightarrow RNH_3^+Cl^- \quad \textbf{[4.3]}$$

$$RNH_3Cl + NaOH \rightarrow RNH_2 + H_2O + NaCl \quad \textbf{[4.4]}$$

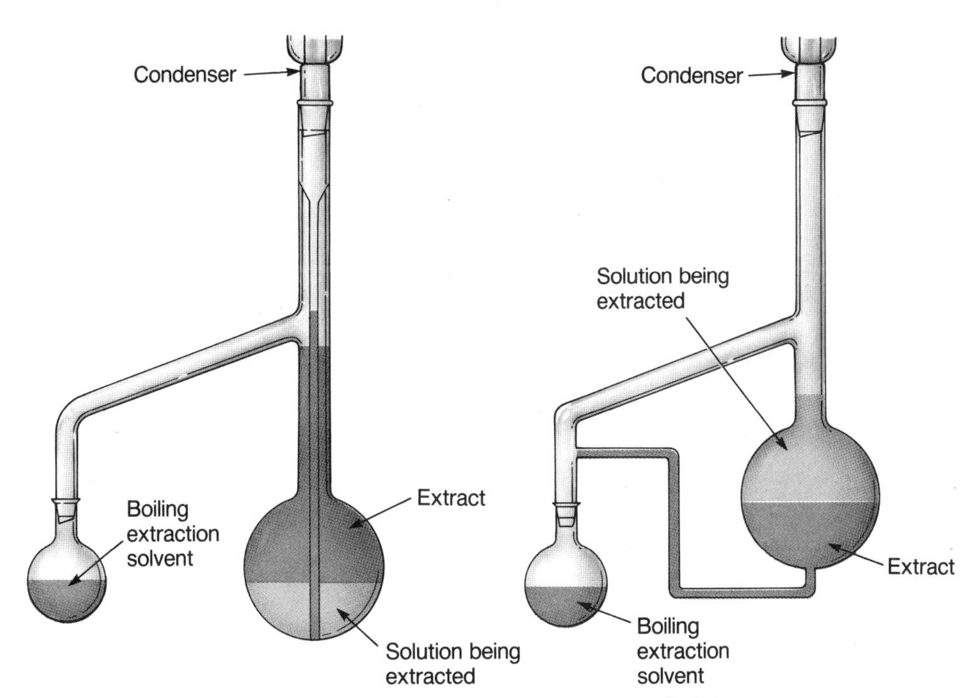

Figure 4.2 Continuous liquid–liquid extraction apparatus. *Left*, lighter-than-water solvent. *Right*, heavier-than-water solvent.

The separation scheme is outlined in the following diagram.

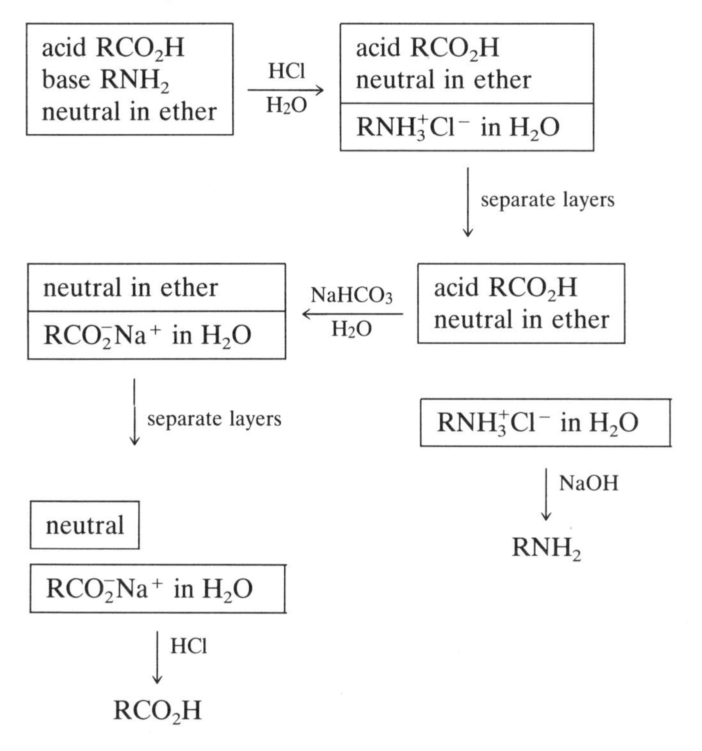

SOLID–LIQUID EXTRACTION

Another important use of extraction is in the isolation of a compound from a solid. In the study of naturally occurring compounds in plants, for example, the first step is the extraction of ground leaves, bark, or wood with solvents to remove all soluble compounds before further separation. An extraction of this type is illustrated by the isolation of leaf pigments in Chapter 7. A very familiar example is the brewing of coffee or tea by extraction of the flavors, caffeine, and other substances with hot water. If these compounds from the coffee bean or tea leaf are desired, they can be isolated from the aqueous brew by further extraction using organic solvents.

STEPS AND TECHNIQUE OF EXTRACTION

USE OF THE SEPARATORY FUNNEL

A separatory funnel is expensive and fragile, and when full, it is top-heavy. The funnel should be supported on a ring of the proper size at a convenient height; it should not be propped up on its stem. Before each use, the stopcock should be checked to be sure that it is seated and rotating freely. If the stopcock is glass, a clip or leash should be used to prevent the stopcock from falling out if it is accidentally loosened. A *very light* film of stopcock lubricant should be applied around the stopcock in bands on each side of the hole. Excess grease is to be avoided, since it will be washed away by organic solvents and contaminate the solution. (Teflon stopcocks require no lubricant.)

The separatory funnel should be filled to no more than approximately three fourths of the total depth, so that thorough mixing is possible. After filling the funnel (check first that stopcock is closed!), it is stoppered with a properly fitting plastic, glass, or rubber stopper. (Before the funnel is used for the first time, it is good practice to shake it with a few milliliters of liquid to make sure that stopcock and stopper are tight.)

Holding the funnel with the stopcock end tilted up, the stopper is kept in place securely with the heel of one hand and the stopcock end is supported in the other hand (Fig. 4.3). As soon as the funnel is inverted, the stopcock is opened to release any pressure. The stem must be pointed away from you and your neighbors, since a small amount of liquid may spurt out. The stopcock is then closed and the funnel is shaken gently in a horizontal position for about 30 seconds. The process is stopped periodically and the stopcock is opened slowly to vent any pressure that may have built up; this step is particularly important in extractions with bicarbonate, when CO_2 pressure may develop during the extraction. The funnel is replaced in the ring and the stopper is loosened immediately. When the phases have separated completely, the lower layer is drawn off through the stopcock.

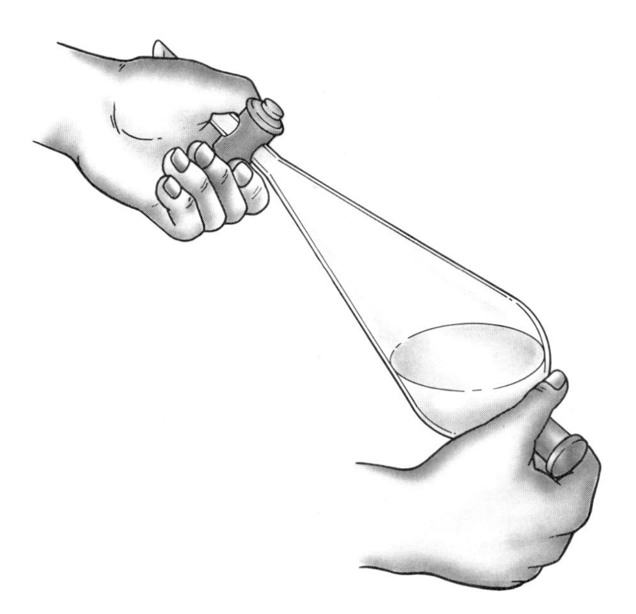

Figure 4.3 Use of a separatory funnel.

In the extraction of an aqueous solution, the solvent may be either less dense (e.g., ether) or more dense (e.g., methylene chloride) than water. In case of doubt as to which layer is the aqueous phase, several drops of the lower layer can be drained into a test tube and checked for miscibility with water.

When the solvent is less dense than water and forms the upper phase, as in an extraction with ether, the aqueous layer must be drained into a receiver (usually the flask in which it was originally contained) and the ether layer transferred to a second flask. The aqueous phase is then returned to the separatory funnel for further extraction with fresh portions of solvent. With a solvent denser than water, the aqueous solution is simply retained in the funnel and shaken with further portions of fresh solvent.

In either case, the organic layers are usually then combined, returned to the funnel, and shaken with a small volume of water to remove traces of the original aqueous phase that are suspended in the organic layer (back-extraction). When the organic layer is less dense (e.g., an ether or pentane layer), it is good practice to remove it from the funnel by pouring it out the top rather than by draining it through the stopcock, to avoid contamination with droplets of the aqueous layer that remain at the bottom of the funnel or in the stem.

A difficulty that is sometimes encountered in extraction is the formation of an **emulsion** of one layer in the other. An example of an emulsion is mayonnaise. Emulsions prevent a sharp boundary between layers and can be frustrating to deal with. They are caused by the presence of colloidal impurities or surfactant materials in the solution being extracted and are very common in the extraction of an aqueous solution of plant or animal tissue. The formation of emulsions is favored by vigorous mixing of the layers, and if emulsification is anticipated, it can often be avoided by mixing the layers with gentle swirling for a longer time than is usually used for shaking. Sometimes emulsions can be cleared or broken up by the addition of an electrolyte, such as saturated salt solution. Another measure that is usually effective is to filter the emulsified portion of the mixture through filter paper or through Celite (see p. 37).

It is important to note that an aqueous solution to be extracted often contains a water-miscible solvent such as alcohol. If the amount of alcohol is significant, an excessive volume of extracting solvent (such as ether) is required to form two liquid phases, and the ether layer will contain a substantial fraction of the third solvent as well as water. If this situation arises, it may be possible to remove some of the alcohol before extraction by distillation or evaporation. Otherwise, a larger-than-normal amount of ether must be used in the extraction, and the ether phase is then thoroughly back-extracted with water.

In many cases the separatory funnel is used simply as a means of recovering a small amount of an insoluble organic product from a large amount of water with minimum mechanical loss. A small volume of ether or other solvent is added to the mixture to permit a sharp separation of layers. Even though the compound may have a negligible solubility in water, a second portion of solvent should be used to rinse the aqueous layer and the separatory funnel.

DRYING AGENTS

After any of these extraction processes, the organic solution is saturated with water; therefore it is desirable to dry the solution before evaporating the solvent. Water is an impurity and should always be removed before a crystallization procedure is carried out or before a liquid is distilled. A number of salts that form hydrates can be used for this purpose. The efficiency of a drying agent depends on the completeness of drying (**intensity**), the degree of hydration (**capacity**), and the rate at which the salt absorbs water. A few of the more commonly used agents are listed here.

1. Magnesium sulfate has high capacity and intermediate intensity. It is inexpensive and rapid in action and is the most generally useful all-purpose drying agent.
2. Sodium sulfate has high capacity but low intensity; it can be used only at room temperature or below.
3. Calcium chloride has high intensity but generally is useful only for hydrocarbons or halides, because of complex formation with most compounds containing oxygen or nitrogen.
4. Calcium sulfate (Drierite) has very high intensity and is rapid, but it has low capacity.

5. Potassium carbonate is often used for drying basic compounds; it has intermediate intensity and capacity.
6. Molecular sieves are complex silicates with a porous structure that selectively entraps water molecules. They are useful for obtaining rigorously anhydrous liquids.

It is difficult to specify how much drying agent to use in any one case but a rough rule of thumb is that for solutions the amount of drying agent usually should be about one twentieth of the liquid volume; a smaller amount is used for a pure liquid.* After allowing the solution to stand over the drying agent for 10 to 20 minutes, with occasional swirling, the drying agent is removed by filtering through a cotton plug and rinsed with a solvent.

EVAPORATION OF SOLVENT

An important operation in extractions and other procedures such as column chromatography (Chap. 7) is the removal of a solvent to permit recovery of a relatively nonvolatile residue. Ether, methylene chloride, and hydrocarbon solvents are flammable and toxic. If more than a few milliliters of solvent is to be removed, provision must be made to avoid the discharge of vapor into the laboratory. Where permitted by safety codes, a convenient way to accomplish this is to sweep the vapor into an aspirator as it distills. This can be done by inserting a piece of glass tubing into the rubber tubing leading to the aspirator and clamping it vertically above the flask on the steam bath, with the glass tubing extending a short distance into the neck (Fig. 4.4). The pressure in the system is not reduced, but most of the vapor will enter the aspirator, where it is diluted with water and discharged into the drain. This technique is inadequate for large volumes of solvent, which must be removed by distillation through a condenser.

For rapid evaporation with minimum heating, the solution is placed under reduced pressure in a round-bottom flask or test tube. This is very easily arranged by fitting a large one-hole rubber stopper on the mouth of the flask (Fig. 4.5A). Atmospheric pressure holds the stopper in place, but it can be quickly released; an alternate procedure, particularly useful for microscale experiments, utilizes the ground glass equipment shown in Figure 4.5B. When this method is used, the flask or test tube must be no more than 10 to 15% full, and it must be agitated constantly, with the pressure controlled by placing a thumb on the tubing, or with the stopcock, to prevent bumping and frothing. This technique requires a little practice, but it will be found to be a great time saver and well worth learning.

---

*Anhydrous sodium sulfate tends to clump when it picks up water. Thus, a useful technique is to add it in small increments with intermittent swirling and standing until added portions remain granular.

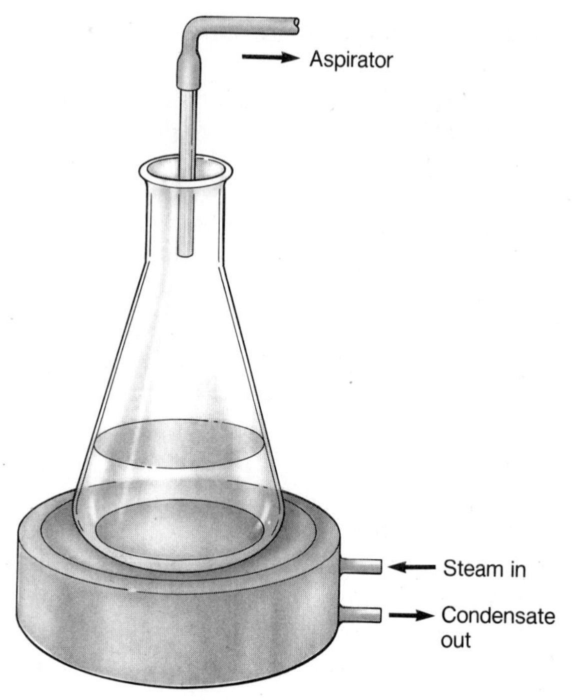

Figure 4.4 Evaporation of solvent on a steam bath.

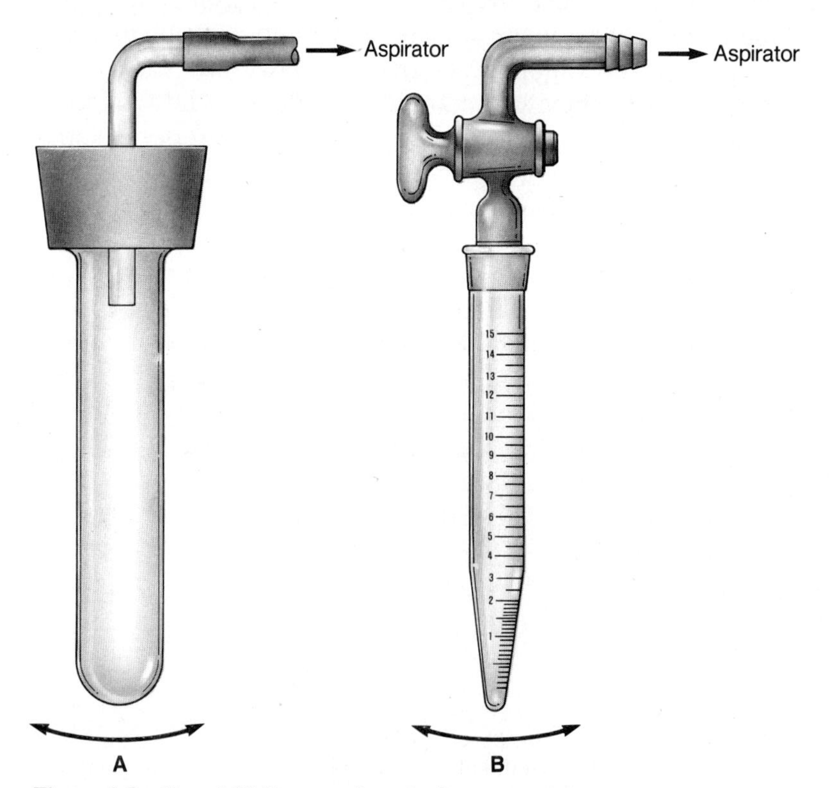

Figure 4.5 (A and B) Evaporation of solvent at reduced pressure.

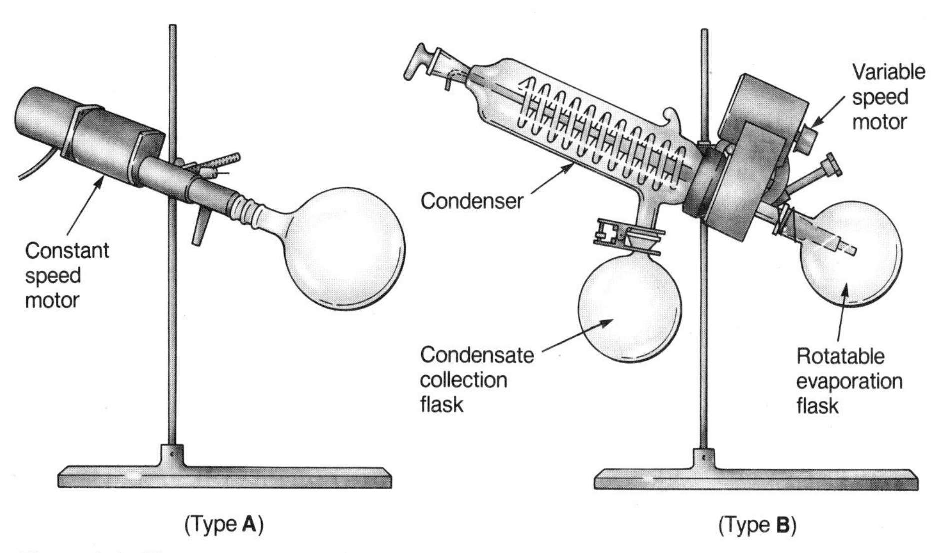

Figure 4.6 Vacuum rotary evaporators.

For rapid evaporation, particularly of larger volumes of solvent, rotary evaporators (of the types shown in Fig. 4.6) are standard equipment in most advanced laboratories. These evaporators are operated at reduced pressure, some with a very efficient condenser (type B). The flask is held at an angle and rotated in the heating bath; this action spreads the liquid in a film on the inside wall and provides a large surface for evaporation.

The correct size and type of glassware for solvent removal deserves comment. There must be room for boiling and agitation, whether under reduced pressure or not, and the flask should never be more than one-fourth to one-third full. Much time can be wasted by gingerly evaporating a solution from a flask that is too full. Evaporation of solvent should never be done in a beaker. Although it has a wide mouth, a beaker is actually a very inconvenient vessel from which to remove a crystalline residue or a liquid, since solvent tends to creep up the walls, and rinsing is less efficient than with a flask or test tube; moreover, a beaker cannot be stoppered, swirled, or evacuated. Vessels containing liquids to be dried or stored should always be securely stoppered with corks to prevent evaporation and contamination. A beaker is really useful only as a container for weighing solids or as a bath for cooling or heating. For practically all other purposes, Erlenmeyer flasks, round-bottom flasks, or test tubes are preferable.

SAFETY NOTE ONLY A *SPHERICAL* OR *CYLINDRICAL* VESSEL (E.G., A TEST TUBE) SHOULD BE PLACED UNDER REDUCED PRESSURE UNLESS IT IS MADE OF HEAVY-WALL GLASS, SUCH AS A FILTER FLASK. A FLAT-BOTTOM SURFACE IS NOT BUILT TO WITHSTAND THE FORCE EXERTED BY NEARLY 1 ATMOSPHERE OF PRESSURE IF THE FLASK IS EVACUATED. IN A ROUND-BOTTOM FLASK OR TEST TUBE, THE PRESSURE IS DISTRIBUTED OVER A CONTINUOUSLY CURVED

SURFACE. MOST MODERN GLASSWARE IS BEING MADE WITH FAIRLY THICK WALLS, BUT IT IS *EXTREMELY* HAZARDOUS TO EVACUATE A LARGE ERLENMEYER FLASK, BECAUSE OF THE DANGER OF IMPLOSION AND RESULTANT FLYING GLASS.

It is frequently necessary to weigh a residue after removing a solvent or decanting a liquid from a solid; if this is the case, the empty weight, or **tare**, of the vessel should be determined in advance.

TRANSFER OF LIQUIDS

A few points of technique are important in manipulating small volumes of organic liquids and solutions. It is usually necessary to transfer a solution to a smaller vessel after solvent removal is about 80% complete, to facilitate removing the last trace solvent and to minimize losses in recovering a crystalline residue.

A small volume of organic liquid should be transferred with a pipet (p. 14). An attempt to pour it will result in a significant fraction of the residual compound being spread over the inside wall, the lip, and the outer surface of the neck of the flask because of the low surface tension of the organic solvent. An excessive amount of solvent is then required for rinsing. In rinsing a relatively large flask with ether or methylene chloride, a useful technique is to add a small amount of solvent and then warm the bottom of the flask so that the condensing vapor rinses down the walls; then, the rinse is concentrated in a small pool for transfer.

Experiments

A. DETERMINATION OF A DISTRIBUTION COEFFICIENT (Macroscale)

In this experiment a K_d value will be determined and the efficiency of single and multiple extractions compared. A solution containing a known amount of benzoic acid ($C_6H_5CO_2H$) in water is extracted with methylene chloride (CH_2Cl_2). The amount of acid remaining in the aqueous layer is determined by titration with base, and the amount of acid in the organic layer is obtained by difference. From these two values, K_d is calculated. Another extraction is then carried out with the same amount of CH_2Cl_2 divided into two portions, and the amount of acid extracted is compared.

Benzoic Acid Solution Weigh out 0.610 g (5.00 mmoles) of benzoic acid and place the acid in a 250-mL Erlenmeyer flask. Add about 150 mL of water and heat the mixture on a hot plate or burner with swirling until the acid is dissolved. Pour the solution into a 250-mL volumetric flask or graduated cylinder and rinse several times with water to transfer all the acid. Cool the solution and adjust the volume to 250 mL to give a 0.020 *M* solution of benzoic acid.

NaOH Solution Dissolve one pellet of NaOH in 100 mL of water and rinse and fill a buret with this solution. Place a 10.0-mL sample of the 0.02 *M* benzoic acid in a small flask and add one drop of phenolphthalein solution. Titrate to a pink end point. Repeat with a second 10.0-mL sample, average the buret readings, and calculate the molarity of the NaOH.

Distribution Coefficient
Place 50 mL of the benzoic acid solution in a separatory funnel, and using a volumetric pipet, add 10 mL of methylene chloride to the funnel. Stopper the funnel and shake vigorously for 30 seconds. Remove the stopper, allow the layers to separate, and drain off the lower (CH_2Cl_2) layer. Swirl the funnel and after a minute or two, a small additional amount of lower layer can be removed.

Pour the aqueous layer into a 250-mL Erlenmeyer flask, rinse the funnel with a few mL of water, and add the rinse to the flask. Add a drop of phenolphthalein solution and titrate with the NaOH solution to the same color end point used in standardizing the base. Calculate the number of mmoles of acid in the water and, by difference, in the CH_2Cl_2 layer, and from these values and the volumes, calculate K_d as described on page 46.

Multiple Extraction
Place another 50-mL sample of the benzoic acid solution in the separatory funnel and extract with 5.0 mL of CH_2Cl_2, drain off the lower layer, and extract with a second 5.0-mL portion of CH_2Cl_2. After the second extraction, remove and titrate the aqueous layer and calculate the amounts of acid in the two CH_2Cl_2 layers. Compare the amount of acid removed by 10 mL of CH_2Cl_2 in one and two portions.

As a check on the accuracy of your experimental values, calculate the amount of acid that should be removed by two 5-mL extractions, using the value of K_d obtained in the 10-mL extraction, and compare it with the amount found.

B. SEPARATION OF ACIDIC, BASIC, AND NEUTRAL COMPOUNDS (Macroscale)

The separation of acids, bases, and neutral compounds (described earlier) by conversion of the acid and base to water-soluble ionic forms and extraction of these into an aqueous layer are illustrated in this experiment. In this procedure, an ether solution of an unknown that may contain an acid, base, or neutral compound, or some combination of these, is first extracted with acid to remove the base. This aqueous extract is then made basic with hydroxide to liberate any organic base. Extraction of the unknown solution with bicarbonate removes any acid, which is then recovered by reacidification. Finally the neutral compound, which remains unextracted, is recovered by removal of the solvent.

SAFETY NOTE NONE OF THE UNKNOWNS IS PARTICULARLY TOXIC OR IRRITATING, BUT NEVERTHELESS CARE SHOULD BE TAKEN TO AVOID SPILLING ETHER SOLUTIONS ON THE HANDS. IF THIS OCCURS, WASH YOUR HANDS THOROUGHLY WITH SOAP AND WATER. ETHER IS HIGHLY FLAMMABLE, AND THERE SHOULD BE *NO FLAMES* OF ANY KIND IN THE LABORATORY. FOLLOW INSTRUCTIONS FOR YOUR LABORATORY ON EVAPORATION, RECOVERY, AND DISPOSAL OF RECOVERED SOLVENTS.

Procedure

Obtain an unknown, dissolve it in 50 mL of ether, and place in a separatory funnel. Extract the solution with three 15-mL portions of 5% aqueous HCl, draining each portion successively into the same 125-mL Erlenmeyer flask. Add 10% NaOH solution to the combined acid extracts until the solution is basic to indicator paper. If a basic compound is present, it will separate at this point. If it is a solid, collect it by filtration on a small Büchner or Hirsch funnel (see Figs. 3.5 and 3.6), wash it on the filter funnel with a small amount of water, and set the filter paper with the product aside to dry.

If the base is an oil, extract the aqueous solution with fresh ether (3×15 mL), combine the ether layers, and dry them over about 2 g of anhydrous Na_2SO_4 or another suitable drying agent for 15 to 30 minutes. (If Na_2SO_4 is used, some granular, uncaked drying agent should remain after the drying time. If this is not the case, add an additional small amount of Na_2SO_4 and let the solution stand for another 10 to 15 minutes [see footnote on page 53]). Remove the drying agent by gravity filtration through filter paper or a cotton plug and evaporate the ether on a steam bath to obtain the basic component.

To isolate the acid component of the mixture, extract the original ether solution with 10% $NaHCO_3$ solution (3×15 mL). If an acid is present, this step will cause the evolution of CO_2 gas, which can build up excessive pressure when the funnel is stoppered and shaken. Therefore, it is good practice first to swirl the funnel unstoppered after the initial addition of the $NaHCO_3$ until most of the CO_2 evolution has taken place. To the combined aqueous bicarbonate extracts, slowly add 10% HCl solution, shaking after each small portion is added to prevent excessive frothing caused by CO_2 evolution. Collect any solid acid that precipitates by suction filtration, wash with water, and air dry. (If the acid separates as an oil, it is recovered by ether extraction in the same way as described previously for a liquid basic component.)

Pour the remaining ether solution from the top of the separatory funnel into an Erlenmeyer flask, rinse the funnel with a few milliliters of ether, and dry the ether solution with Na_2SO_4 as described earlier. Filter through a cotton plug into a dry Erlenmeyer flask and evaporate the ether on a steam bath (see Fig. 4.4). If a neutral compound is present, it will remain as a crystalline or oily residue.

In each fraction in which a crystalline compound is obtained, spread the substance out on glassine paper and finely divide it to dry. Determine the weight and melting point and identify it from the list of unknowns in Table 4.1. If an identification is questionable, it may be desirable to recrystallize the compound or to determine a mixture melting point with an authentic sample, or both. If a residue is noncrystalline, crystallization should be attempted by using the technique described in Chapter 3.

When working with very small quantities of material, the use of a separatory funnel is impractical. Instead, separations are carried out in a centrifuge tube (Fig. 4.7) and transfers are made with a Pasteur pipet (see inside back cover).

C. SEPARATION OF ACIDIC, BASIC, AND NEUTRAL COMPOUNDS (Microscale)

In a centrifuge tube, dissolve 100 mg of an unknown in 2 mL of ether. Add 1 mL of 5% HCl, cap or stopper the tube, and shake gently for about 1 minute. Allow the layers to separate and carefully draw off the bottom layer with a Pasteur pipet,* depositing it in a second (properly labeled) centrifuge tube. Repeat the procedure with two additional 1-mL portions of 5% HCl, combining these with the first portion. Back-extract the combined HCl layers with 1 mL of fresh ether by shaking gently for about 1 minute,* and then carefully draw off the ether layer with a clean pipet. Add this ether layer to the original ether solution.

The combined HCl layers are treated with 10% aqueous NaOH until basic, at which point a precipitate indicates the presence of a basic component in the original unknown. If the precipitate is a solid, it can be isolated by direct filtration using a Hirsch funnel, washed with a small amount of water, and spread out on glassine paper to dry. Alternatively, the aqueous layer can be carefully drawn off with a pipet into which a small amount of cotton or glass wool has been inserted in the tip (Fig. 4.8) to prevent the removal of solid. The solid is washed by adding a small amount of water, and as much of the water removed as possible with the pipet. The centrifuge tube containing the remaining wet solid is then attached to an aspirator and

*An alternate procedure involves the mixing of the layers by using the pipet to draw up and squirt out the mixture several times. The layers are then allowed to separate and the bottom layer drawn off as described.

Table 4.1 Possible Unknowns

ACIDS	MP (°C)	BASES	MP (°C)	NEUTRAL	MP (°C)
o-Toluic acid	104	*p*-Chloroaniline	72	Fluorenone	83
Benzoic acid	122	Ethyl *p*-aminobenzoate	89	Benzil	95
trans-Cinnamic acid	133	2-Aminobenzophenone	106	Phenanthrene	101
m-Nitrobenzoic acid	140	*p*-Phenylenediamine	140	Fluorene	114
m-Bromobenzoic acid	155	4-Aminoacetanilide	162	2,6-Dichloronaphthalene	136
p-Toluic acid	180				

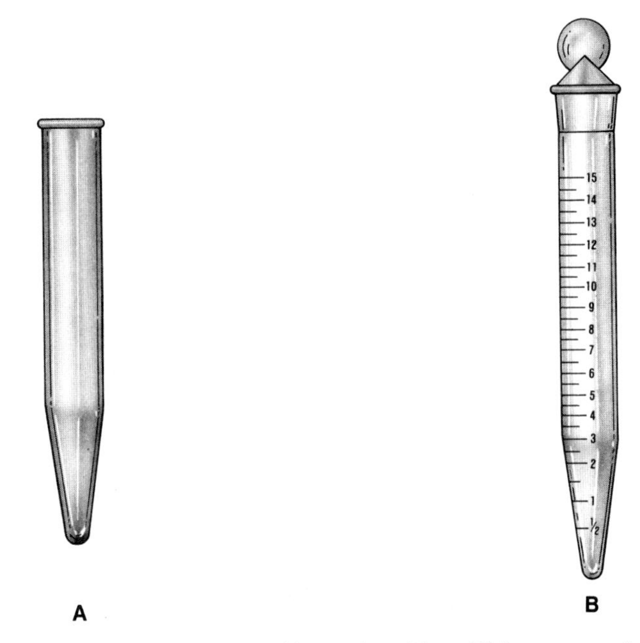

Figure 4.7 Two different types of centrifuge tubes. Type (B) has a ground glass stopper and is also called a distillation receiver.

dried under vacuum (with gentle heating on a steam bath, if necessary). When the solid is dry, it is weighed and the melting point determined.

If the base precipitate is noncrystalline, the mixture is extracted with 3 × 1 mL of ether (in this case the ether layer is carefully removed each time with the pipet). The combined ether extracts are dried by adding a small amount of Na_2SO_4 or other suitable drying agent, separated from the drying agent by transferring to a clean centrifuge tube with a pipet containing cotton or glass wool in the tip, and the ether removed under vacuum with gentle heating on a steam bath and shaking to prevent bumping. If the residue continues to be an oil, crystallization should be attempted by adding a small amount of suitable solvent and following one of the procedures outlined in Chapter 3 under "Crystallization."

An acidic component in the unknown is separated in similar fashion. Extract the original ether solution with 3 × 1 mL of 10% $NaHCO_3$, each time transferring the bicarbonate layer with a pipet to a properly labeled centrifuge tube. If an acid is present, the formation of CO_2 bubbles may be evident at this point. The combined bicarbonate extracts are back-extracted with 1 mL of fresh ether and acidified by adding *cautiously* 10% aqueous HCl (CO_2 evolution). A precipitate indicates the presence of an acidic component. Separate and dry the solid using one of the procedures discussed previously in the basic component isolation step. Weigh, and record the melting point. If the acid separates as an oil, it is recovered and treated as described earlier for a noncrystalline basic component.

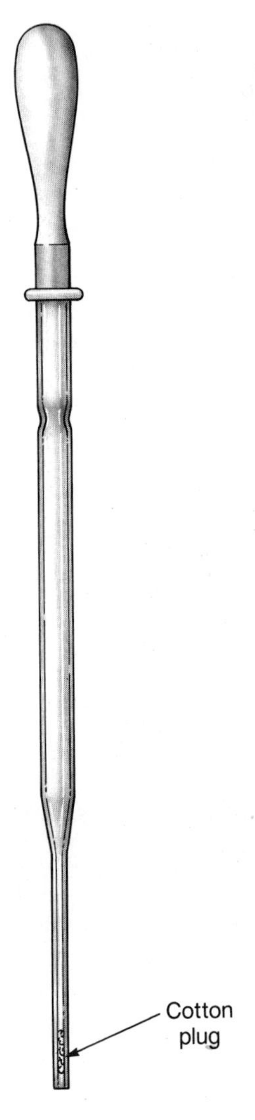

Figure 4.8 Pipet with inserted cotton plug.

To isolate a neutral component, wash the original ether solution with 1 mL of water, dry the ether layer with a drying agent, and after separating the drying agent, remove the ether solvent under vacuum with gentle heating on a steam bath and constant agitation (gentle shaking) to prevent bumping.

QUESTIONS

1. In the back-extraction described on page 47, calculate the amount of A and B remaining in the ether layer after using two 25-mL portions of water instead of one 50-mL wash.

2. In the preparation of acetanilide in Chapter 3, the crude precipitate of the product was recrystallized directly from the aqueous mixture. This procedure depended on the favorable solubility of acetanilide in water and is not generally applicable; in most instances, the crude product would be isolated by extraction. Describe in detail a procedure for isolating the acetanilide based on extraction.
3. In an oxidation experiment, 1 g of a neutral, water-insoluble, ether-soluble ketone was obtained along with some acidic by-products in 10 mL of acetone solution. The ketone was recovered by extraction. Describe all steps in the isolation procedure.
4. Phenol (C_6H_5OH) is a weaker acid than is a carboxylic acid (RCO_2H) and is not converted to an anion with bicarbonate. However, phenol does form the sodium salt with the stronger base hydroxide ion:

$$C_6H_5OH + NaOH \rightarrow C_6H_5O^- + H_2O + Na^+$$

Devise a procedure for the separation of *p*-chlorobenzoic acid, *p*-chlorophenol, and *p*-dichlorobenzene based on an extraction process.
5. In Experiment A, what would be the effect on the values obtained for K_d, and the percentage of acid removed by single versus multiple extractions, if an error were made in weighing the benzoic acid and the solutions were only 0.015 *M*?
6. Suggest the most likely sources of error in Experiment A.
7. If the mixture used in Experiment B contained a water-soluble amine, would it be separated from the other compounds by the procedure used? How would the results differ from those observed with a water-insoluble amine?
8. On the basis of the considerations in Chapter 2 on melting points, would you expect the melting points of the compounds isolated in Experiment B to be generally higher, lower, or the same as the values in Table 4.1?
9. Contrast the merits of filtering through a cotton plug and through filter paper. Why is cotton suggested for removing the drying agent in Experiment B but not for removing charcoal or collecting the acetanilide in Chapter 3?

REFERENCE

Craig, L. C.; Craig, C. In *Techniques of Organic Chemistry*, 2nd ed.; Weissberger, A., Ed.; Interscience: New York, 1956; Vol. III, Part I, Chapter 2.

Name ______________________ Section ______________ Date ________

4 Extraction

PRELABORATORY QUESTIONS

1. If an extraction is being made of an aqueous phase A with a water-immiscible phase B, would one extraction of phase A with 100 mL of B be more efficient than four extractions of phase A with 25-mL portions of B? Explain.

2. Two terms used in conjunction with drying agents are "intensity" and "capacity." How do these terms differ in meaning?

3. An organic acid and a ketone are both water insoluble, but ether soluble. What is the simplest extraction scheme you can think of whereby a mixture of these two substances can be separated?

4. The two immiscible phases in an extraction are water and 1,2-dichloroethane ($ClCH_2CH_2Cl$). Which phase would you expect to be the upper layer? Explain?

5. In a scheme involving multiple extractions of an aqueous solution with an organic solvent, would there be a practical advantage in utilizing an organic solvent that is heavier than water? Explain.

5

Distillation

Distillation is the process of heating a liquid to its boiling point, continuing the heat input to convert liquid to vapor and to force the vapor to another portion of the apparatus, cooling the vapor there to condense it, and collecting the liquid condensate. In this way, a liquid can be recovered from nonvolatile contaminants or mixtures of liquids with different volatilities can be separated. When vaporization occurs from a solid, the process is called **sublimation** (see Chap. 18). The term **evaporation** is used when the residue rather than the distillate is of importance and the volatile liquid is allowed to escape uncondensed into a hood or an aspirator. The term **reflux** means boiling the liquid with return of the condensate to the original flask.

Distillation depends on the fact that at a given temperature, some molecules of a liquid have sufficient kinetic energy to escape from the surface and create a vapor pressure. This tendency for vaporization becomes greater as the kinetic energy is increased by raising the temperature. When a liquid is heated to the temperature at which the vapor pressure equals that of the surrounding atmosphere, the liquid boils, and this temperature, at a given pressure, is called the **boiling point.** Two or more components of a liquid that have different boiling points may be separated by distillation under favorable circumstances. This process is known as **fractional distillation** and involves the distillation of one component (or fraction) at a time.

Consider a mixture of two volatile liquids. The question might be asked, "What is the composition of the vapor that is distilled from this liquid?" The expectation would intuitively be that the vapor would be richer in the more volatile (lower boiling) component; thus, it might be supposed that one could heat the liquid to a temperature between the boiling points of the two components and cleanly distill off the lower boiling component. Such is not the case. For example, it is not possible to heat a mixture of benzene (bp 80°C) and toluene (bp 111°C) to a temperature of 81°C and distill off pure benzene. Consideration of some general relationships will explain why this is so.

In a mixture of two compounds, A and B, the total vapor pressure is the sum of the partial pressures of each component (**Dalton's Law**). Furthermore, the partial pressures are proportional to the amount of each component in the mixture. For an ideal solution of two compounds, in which the molecules do not interact, the vapor pressures P_A and P_B of each compound are defined by **Raoult's Law** as the vapor pressure of the pure compound multiplied by the mole fraction of that compound.

$$P_A = N_A \cdot P_A^\circ \quad \text{and} \quad P_B = N_B \cdot P_B^\circ$$

where

P_A = vapor pressure of component A
N_A = mole fraction of A
P_A° = vapor pressure of pure A at the temperature in question

Combining Dalton's Law and Raoult's Law gives

$$P_{total} = P_A + P_B = N_A \cdot P_A^\circ + N_B \cdot P_B^\circ$$

Consider a mixture of equimolar amounts of A and B where the mole fraction = 0.5 for each and component A is the more volatile, with the higher vapor pressure and the lower boiling point. When the mixture boils, the vapor in equilibrium with the liquid will be richer in the more volatile component A ($N_A \cdot P_A^\circ > N_B \cdot P_B^\circ$).

The distillation of such a two-component mixture is most readily described in terms of a phase diagram in which equilibrium curves for vapor and liquid are plotted with temperature versus composition (Fig. 5.1). The boiling points of pure A and pure B are T_A and T_B respectively. The upper curve represents a boundary above which there is only vapor, and the lower curve is the boundary below which there is only liquid. The lower curve represents the boiling points of various mixtures of components A and B, going from pure A on the left to pure B on the right. At any point on the curve, a horizontal line drawn across to the upper curve shows the composition of the vapor in equilibrium with the liquid.

In an equilibrium mixture of components A and B, corresponding to 0.5 mole fraction A in Figure 5.1, the boiling point of the liquid is T_1. At this temperature, however, the vapor has a composition of 0.8 mole fraction of component A, much richer in A, as seen from Raoult's Law. If a sample of this vapor is removed, the composition of the remaining liquid shifts to the right on the diagram; that is, the relative amount of component B increases and the mixture has a higher boiling point. Continued distillation from the mixture provides distillate progressively enriched in B, but each fraction of the distillate will contain both components. The net result is that a simple distillation of two liquids, whose boiling points differ by less than about 50°C, effects relatively little separation, and the temperature will rise continuously during the distillation (dashed line in Fig. 5.2).

How the separation efficiency of a distillation can be increased can be illustrated by further study of Figure 5.1. If the condensed vapors of 0.8

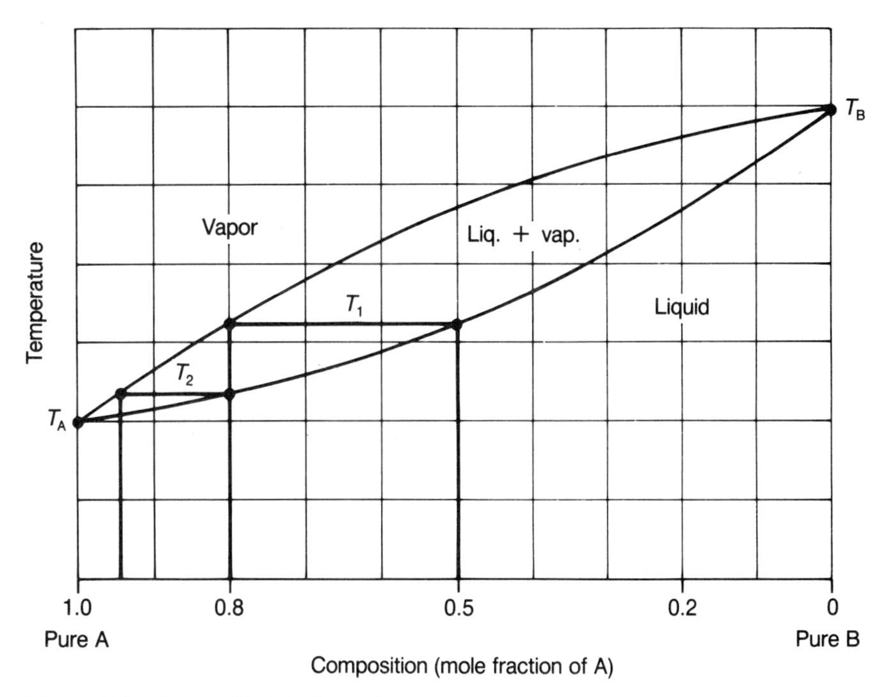

Figure 5.1 Vapor–liquid phase diagram.

mole fraction A are reequilibrated at temperature T_2, the vapors will now have a composition of 0.95 mole fraction A. If the equilibration can be repeated several times, it is possible to obtain a final distillate that is nearly pure A. This process, termed **fractional distillation**, is accomplished by distilling through a column with a large surface area, at which repeated exchange of molecules between the liquid and vapor phases occurs.

Each of the steps seen in Figure 5.1 (from 0.5 to 0.8 to 0.95 mole fraction A) represents in principle a simple distillation step, with one equilibration between vapor and liquid. In fractional distillation, successive distillation steps occur as vapor moves up the column and liquid flows back

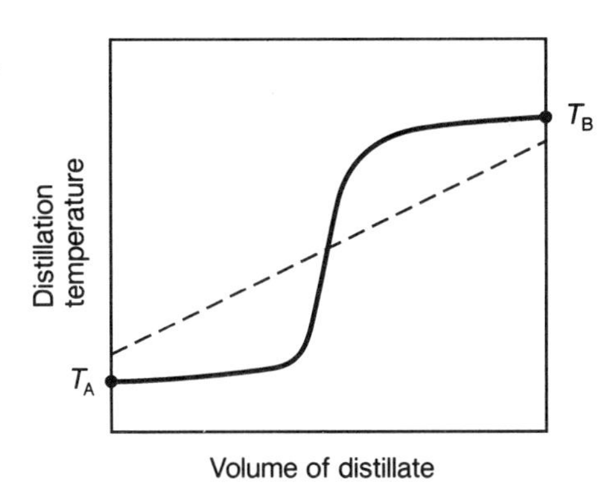

Figure 5.2 Distillation curves. Dashed line equals simple distillation; solid line equals fractional distillation with efficient column.

down. The extent of separation by a column depends on the number of theoretical distillation stages, usually referred to as "theoretical plates." The more efficient the column, the larger is the number of plates in a given length, or the smaller the height equivalent to one theoretical plate (HETP). The HETP is thus an index of the *efficiency* of a column; highly efficient columns have HETP values of less than 1 cm per plate. For a mixture of two compounds with boiling points differing by 10°C, a column of at least 25 theoretical plates is required for a reasonably efficient separation, such as that shown by the solid curve in Figure 5.2.

The essential function of a fractionating column is to allow vapor to condense and revaporize so that the more volatile material continues to progress upward while the less volatile component flows back down to the distilling flask. This is accomplished by providing ample surface for condensation and revaporization as well as sufficient length for the necessary temperature gradient. The column may be packed with glass beads, wire-mesh sponge, glass wool, or other inert material. A problem with a large amount of packing is that there is a rather large "holdup" of material. This term refers to the amount of condensed vapor that remains in the fractionating column when the distillation is completed. One of the most efficient types of column is a so-called spinning band column, in which a spiral of metal or Teflon on a central shaft extending down through the column is rotated at high speed. This action provides rapid exchange of vapor with a continuously flowing film of liquid over the spiral band.

To achieve high efficiency of separation, a column must be heated to a temperature intermediate between that of the distilling flask ("pot temperature") and the still head. Also the distillation must be carried out slowly, with only a small amount of the distillate reaching the condenser being removed; the remainder is allowed to flow back down the column for further equilibration.

AZEOTROPES

In the earlier discussion of fractional distillation, the mixture was assumed to be an ideal solution, that is, it obeyed Raoult's Law. However, this is not always the case. For example, with some dissimilar compounds such as an alcohol and a hydrocarbon, association between molecules of one compound may be greater than association between unlike molecules. The vapor pressure of the mixture at certain compositions is greater than the vapor pressure of the more volatile compound alone, and there is a minimum in the vapor–liquid phase diagram. Such a mixture is called a **minimum-boiling azeotrope** (azeotropes are also called constant boiling mixtures). The mixture at composition X shown in Figure 5.3 distills completely at temperature T_X with no separation of the components. Fractional distillation of a mixture

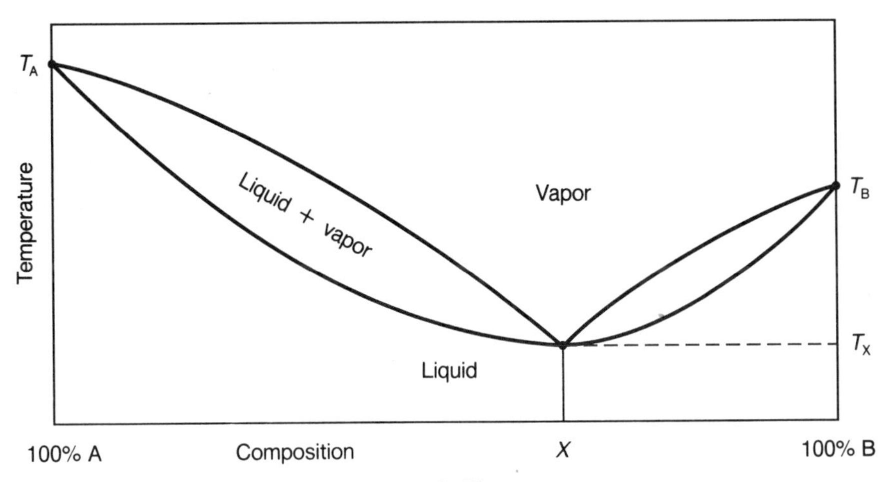

Figure 5.3 Phase diagram for minimum-boiling azeotrope.

at any other composition will provide distillate of composition X until only one component remains in the liquid. **Maximum boiling azeotropes** also occur but are less common. In these cases, fractional distillation of a mixture of two components yields one pure substance until the azeotropic composition is reached, and the remaining material then distills as the azeotrope.

APPARATUS AND TECHNIQUE

The basic elements of a distillation apparatus are a boiling flask, a column through which the vapor rises, a thermometer, a condenser in which the vapor is cooled and liquefied, and a receiver. Standard assemblies for distillation are illustrated in Figures 5.4 and 5.5. The distilling flask is always supported by a heating mantle or ring and wire gauze. The flask with liquid (and a boiling stone) is clamped in place and the column and/or distilling head put in place. The condenser is clamped on a second ring stand with height and angle adjusted so that the joints fit without stress; all clamps should not be tightened until the apparatus is carefully aligned. The rubber condenser tubing is connected with the inlet at the bottom and the outlet at the top. An adapter and suitable receiver (Erlenmeyer flask, round-bottom flask, distillation receiver, or vial) are attached and are supported by a third ring stand if necessary. The receiver should not be propped on a make-shift support. The adapter should be fastened to the condenser with a rubber band or a clamp.

In assembling ground-glass apparatus, it is initially important that the joint is clean, free of grit, and properly aligned. A very thin film of stopcock grease may be used to ensure a snug fit. A common error is to use too much grease; this results in contamination of liquid that comes in contact

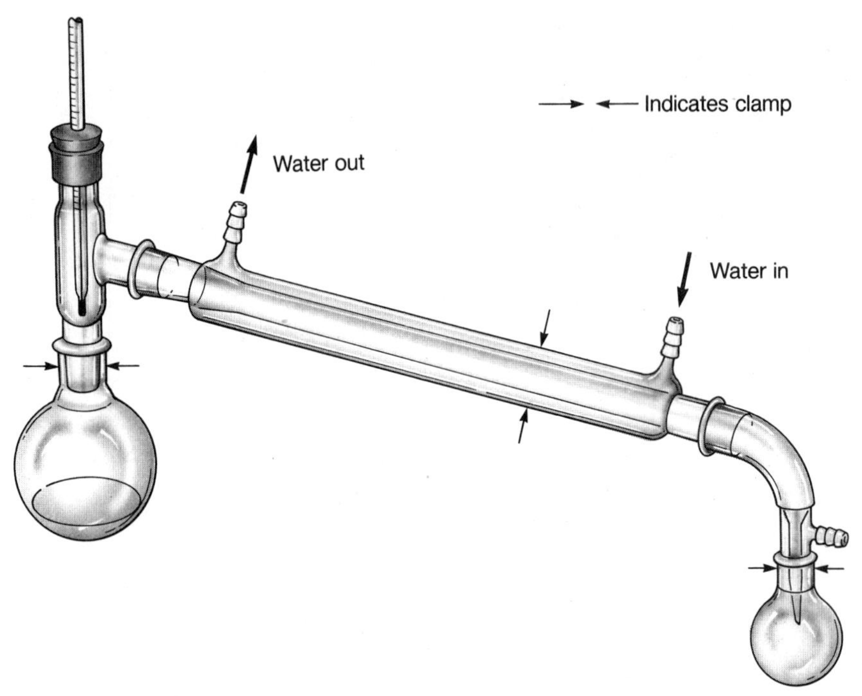

Figure 5.4 Simple distillation setup.

with the joint. Ground joints may be used without grease and will seldom stick if the contents of the apparatus are not basic and if the apparatus is taken apart before it has completely cooled.

It is important to select the proper size flask for a distillation. Too large a flask results in loss due to "holdup" of condensing vapor on the walls. In too small a flask, the liquid will expand and fill the flask to the point where there is insufficient room for boiling to occur without forcing liquid up into the column or distilling head. As a general rule, the flask should be one-third to two-thirds full initially. A boiling stone (small lump of carborundum or other porous material) is essential to maintain constant formation of bubbles (ebullition). Without a boiling stone to provide a steady stream of gas bubbles, superheating and bumping of the liquid can occur. This is particularly true when a liquid on a steam bath is evaporated without a condenser. If the boiling stone is forgotten until the liquid is hot and possibly superheated, the flask must be cooled before the boiling stone is added or a violent and dangerous eruption of the liquid may occur. Since the pores fill with liquid as soon as boiling ceases, a boiling stone cannot be reused and a fresh one must be added if the distillation is interrupted.

In order to reduce loss due to "holdup," a "chaser" may be used. This is a high-boiling, inert liquid that is added to the sample and serves as a heat transfer medium, thus allowing most of the lower boiling sample to be distilled.

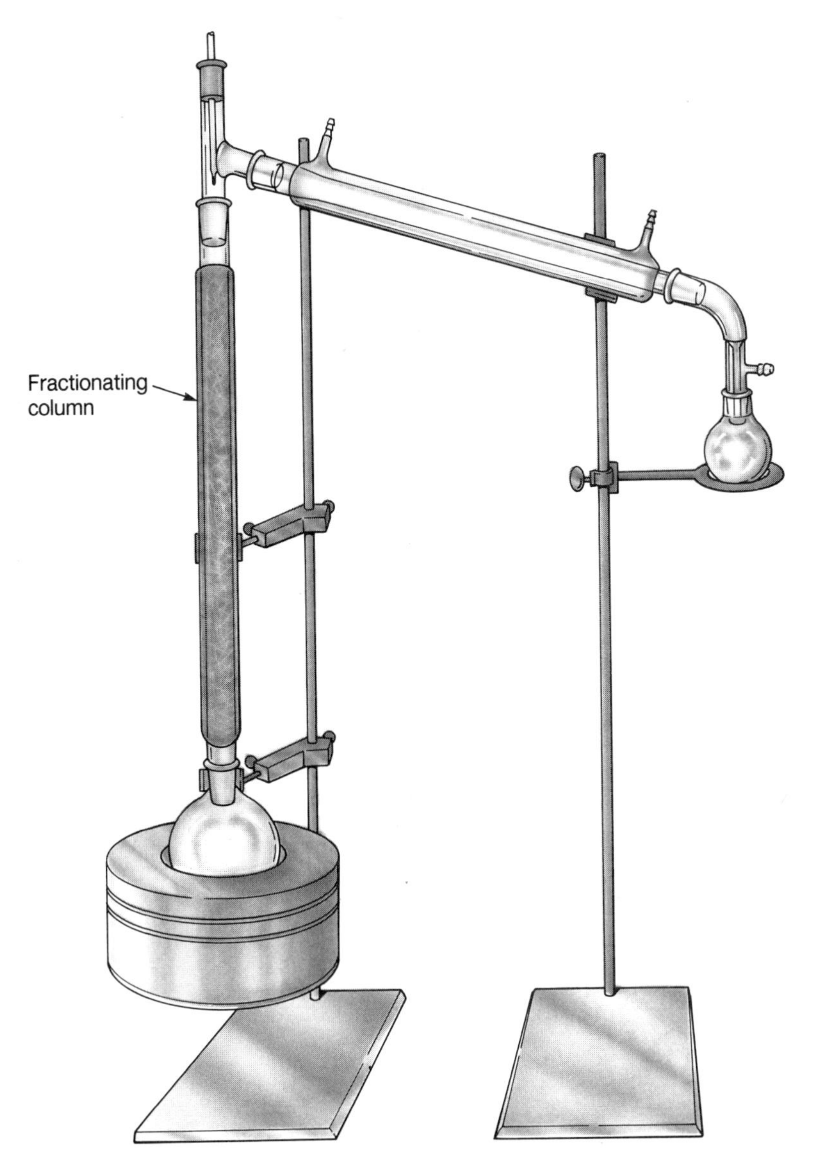

Figure 5.5 Fractional distillation setup.

Distillation assemblies similar to that shown in Figure 5.4 may be used to distill liquid samples as small as 1 to 2 mL in volume by simply using smaller scale apparatus. For samples smaller than this, different apparatus and techniques must be used (see Cheronis, 1954). A Hickman still (Fig. 5.6) may be used for simple distillations of 200 to 900 μL of liquid or for the fractional distillation of similar volumes.

Another method of distillation on a microscale (50 to 500 μL) utilizes the Kugelrohr apparatus (Fig. 5.7). The name comes from the German meaning "bulb tube," which describes the glass distillation tube. The ma-

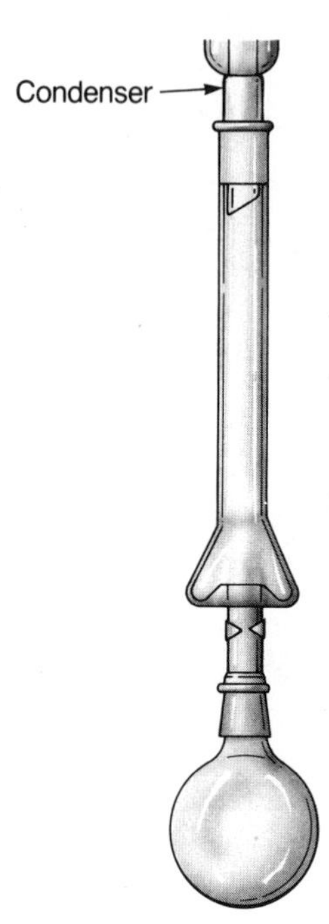

Figure 5.6 Hickman still.

terial to be distilled is placed in the terminal bulb with a pipet, and the tube is then inserted into the glass heating jacket so that all other bulbs are outside the jacket. The apparatus provides for rotation and distillation under vacuum should these features be desired. As the material distills, it condenses in the cooler bulb outside the jacket. By inserting this receiver bulb into the jacket, the material can be distilled again (at the same or different

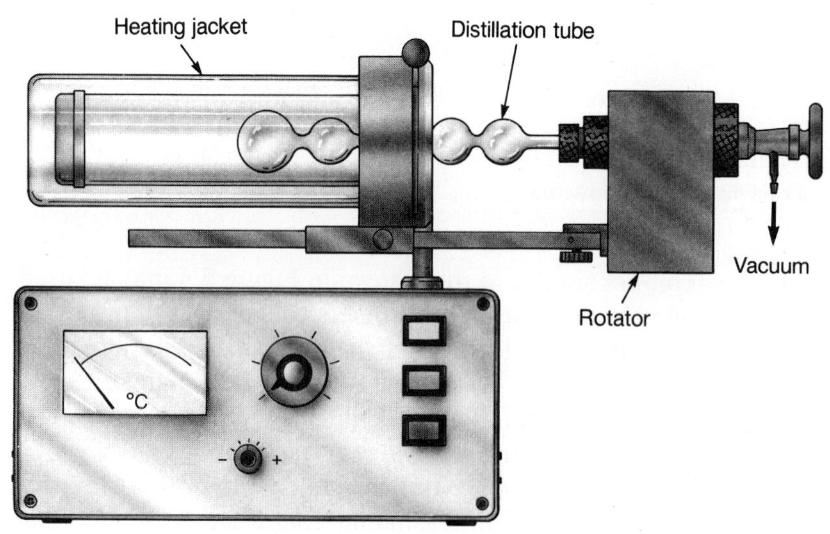

Figure 5.7 Kugelrohr distillation apparatus.

temperature) to the next bulb. In this manner, a number of distillations can be carried out without removing the material from the tube.

SAFETY NOTE A DISTILLATION MUST *ALWAYS* BE STOPPED BEFORE THE FLASK BECOMES COMPLETELY DRY. WITHOUT THE ABSORPTION OF HEAT DUE TO VAPORIZATION, THE FLASK TEMPERATURE CAN RISE VERY RAPIDLY. MANY LIQUIDS, PARTICULARLY ALKENES AND ETHERS, MAY CONTAIN PEROXIDES THAT BECOME CONCENTRATED IN HIGHLY EXPLOSIVE RESIDUES (SEE PAGE 2).

Experiments

A. DISTILLATION OF A PURE COMPOUND (Macroscale)

Assemble a simple distillation apparatus (see Fig. 5.4) using a 50-mL boiling flask that contains 20 mL of 1-propanol (or other liquid as designated by your instructor) and a boiling stone. Be sure the thermometer is correctly positioned with the top of the thermometer bulb just below the bottom of the distilling head side arm, the water flow is up the condenser, and the assembly is protected from drafts by glass wool or aluminum foil. Heat the flask gently with a mantle or other source until the liquid begins to boil, and then adjust the heat to distill the liquid at a rate of about 2 mL per minute (2 drops every 3 seconds) using a graduated cylinder or distillation receiver as a receptacle. Record the temperature when the first drops of distillate are collected in the receptacle and after each 2 mL of distillate is obtained. Always stop a distillation before the distilling flask becomes dry; in this case you should be able to collect 15 to 17 mL of distillate. Make a plot of temperature versus volume of distillate.

B. SIMPLE DISTILLATION OF A MIXTURE (Macroscale)

Place 20 mL of a mixture provided by your instructor in a 50-mL flask with a boiling stone. Assemble the flask in a simple distillation apparatus and conduct a distillation as described in Experiment A, again noting the temperature as every 2 mL of distillate is collected. Make a plot of temperature versus volume of distillate. Save the total 20 mL of liquid for use in Experiment C.

C. FRACTIONAL DISTILLATION (Macroscale)

Place 20 mL of the mixture from Experiment B or provided by your instructor in a 50-mL flask with a boiling stone. Assemble a distillation apparatus with a fractionating column (large-diameter condenser packed with glass beads, glass wool, or wire-mesh sponge) between the distillation flask and the still head (see Fig. 5.5). Use a 10-mL graduated cylinder or a

Table 5.1 Unknown Components

COMPOUND	BP
hexane	69.0°C
cyclohexane	80.7
toluene	110.6
p-xylene	138.3

distillation receiver as a receptacle and have four clean dry test tubes available (label these fractions 1, 2, 3, and 4). Distill the mixture at the rate of about 1 or 2 mL per minute and record the temperature at both the beginning and end of each fraction collected. Collect an initial fraction of 2 mL and quickly empty it into the first labeled test tube. Then collect separate fractions during the distillation of the low-boiling component, the period of rapid temperature rise, and the distillation of the high-boiling component.

Plot the boiling point versus the volume of distillate. The fractions may be examined by gas chromatography (Chap. 6) or refractive index to determine the relative amounts of the two compounds present in each fraction.

The mixture provided by your instructor will contain two components listed in Table 5.1 that differ in boiling point by at least 25°C. From your data, determine the boiling points of the two components, their identities, and their relative ratio in small, whole numbers such as 1:1, 1:2, 2:1, and so on.

D. DISTILLATION OF AN UNKNOWN COMPOUND (Microscale)

Obtain 3 to 5 mL of an unknown liquid from your instructor and carry out a simple distillation as described in Experiment A, but use a 5-mL distilling flask. Report the boiling range for the main fraction of distillate and save it for use in the micro boiling point determination (Experiment F).

E. FRACTIONAL DISTILLATION (Microscale)

Obtain an unknown mixture from your instructor that consists of two compounds from Table 5.1 (see last paragraph of Experiment C). Place 3.5 mL of this mixture in a 5-mL flask and conduct a fractional distillation using a fractionating column with a low holdup (2 to 3 in. of wire mesh sponge). Do not use a cooling condenser, but connect the still head directly to the take off adaptor. Carefully distill the mixture at not more than 2 to 4 drops per minute into a 10-mL graduated cylinder or distillation receiver cooled in an ice bath.

Plot the boiling point versus the volume of distillate to determine the boiling point of the two components, their identities, and their relative ratio.

F. MICRO BOILING POINT DETERMINATION

Seal one end of a piece of 7- or 8-mm glass tubing using a Fisher burner. Allow the glass to cool and then cut the tube using a triangular file to make a "boiler tube" about 3 to 4 inches long.

Place the liquid whose boiling point is to be determined into the boiler tube so that the level of the liquid is approximately 1 cm from the bottom. Break off a melting point capillary 1 inch from the closed end and drop it in the boiler tube with the closed end up. Attach the boiler tube to a thermometer using a small rubber band cut from 8- or 10-mm rubber tubing, and place the thermometer with attached boiler tube in a Thiele–Dennis tube (Fig. 5.8). Slowly heat the Thiele–Dennis tube until a rapid stream of bubbles comes out of the capillary. Remove the heat and allow the apparatus to cool. When the bubbles just cease and liquid is drawn up into the capillary tube, observe the temperature. This is the boiling point of the liquid. Carry out a micro boiling point determination with your unknown (Experiment D above) or with 1-propanol if you were not assigned an unknown. It may be helpful to practice this technique by carrying out a micro boiling point determination using water.

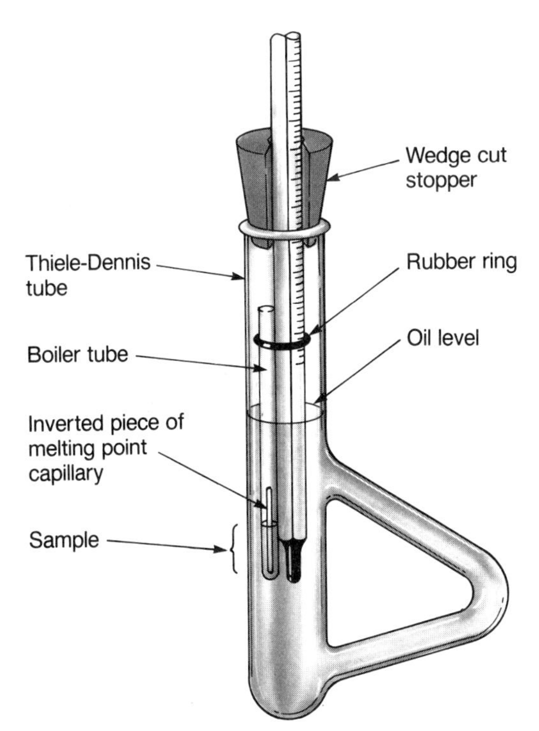

Figure 5.8 Micro boiling point setup.

QUESTIONS

1. The normal boiling point of benzene is 80°C. What is the vapor pressure of benzene at 80°C?
2. What effect would a decrease in barometric pressure have on the boiling point? An increase?
3. Why should a distilling flask be filled not less than one-third full nor more than two-thirds full?
4. Compound A has a vapor pressure of 400 mm Hg at 50°C and compound B has a vapor pressure of 480 mm Hg at 50°C. If A and B are miscible liquids, what is the vapor pressure at 50°C of a mixture of one mole of A and three moles of B?
5. If a fractional distillation mixture of A and B (see Fig. 5.1) is carried out with a column in which there are four equilibrium stages (plates), what will be the composition (±1%) of the distillate when the mixture in the distilling flask is 50% B; 95% B?
6. In contrast to the system illustrated in Figure 5.3, some azeotropes boil at a higher temperature than either of the components. The boiling points of acetone and chloroform are 56.5°C and 61.2°C, respectively. A 1 to 4 mixture of these compounds has a constant boiling point of 64.7°C. Sketch a phase diagram that reflects these data.

REFERENCES

Cheronis, N. D. In *Techniques of Organic Chemistry*; Weissberger, A., Ed.; Interscience: New York, 1954; Vol. VI.

Weissberger, A., Ed.; *Techniques of Organic Chemistry*; 2nd ed.; Interscience: New York, 1965; Vol. IV.

Name ______________________ Section ____________ Date ________

5 Distillation

PRELABORATORY QUESTIONS

1. Define the boiling point of a liquid.

2. Why is it necessary to position the thermometer bulb carefully when carrying out a distillation?

3. What is the purpose of placing a boiling stone in the distillation flask prior to beginning a distillation?

4. Why not distill a liquid all the way to dryness, that is, to get as much distillate as possible?

5. What is the effect of the following on the boiling temperature of a liquid?
   a. an insoluble solid, such as glass or clay

   b. a soluble, nonvolatile impurity

6

Gas Chromatography

The term **chromatography** refers to a general method of separation in which a mixture is partitioned between a stationary phase and a moving phase. The moving phase may be a vapor or a liquid; the stationary phase is a solid or a liquid film coated onto a solid. As the mobile phase flows over the stationary phase, compounds in the mixture are continuously equilibrated between the two phases according to their distribution coefficients. The greater the affinity of a compound for the moving phase, the higher its concentration will be in that phase, and therefore the faster it will be transported along the stationary phase. As a result, individual compounds become separated into zones or bands as the molecules are carried along in the moving phase. Eventually, if the resolution is sufficient, these bands emerge from the chromatograph as discrete fractions.

In **gas chromatography** (GC), sometimes called vapor phase chromatography (VPC), the mobile phase is a stream of inert gas and the stationary phase is a liquid film supported on a solid and packed in a heated column. A sample of the mixture is injected with a syringe into a vaporization chamber and from there it enters the gas stream. The sample then passes through the column, where the partitioning process and separation take place. From the column, the components travel past a detector that delivers a signal to an electronic recorder. If desired, the compounds can be collected in a cold trap. A schematic diagram of the apparatus is shown in Figure 6.1.

The separation process in gas chromatography is illustrated in Figure 6.2 for a mixture of two compounds, A and B. As the sample is carried along the column by the gas flow, the faster-moving compound B becomes separated from A and passes the detector first. The recorder chart displays a series of peaks resulting from the detector response to each component in the mixture. In a typical gas chromatogram (Fig. 6.3), whereas the earlier peaks are quite narrow; the width becomes progressively broader for later peaks. The position of each peak, expressed as the time required for the compound to pass through the column, is called the **retention time**, $\boldsymbol{R_t}$.

For a given compound in the same column and exactly the same op-

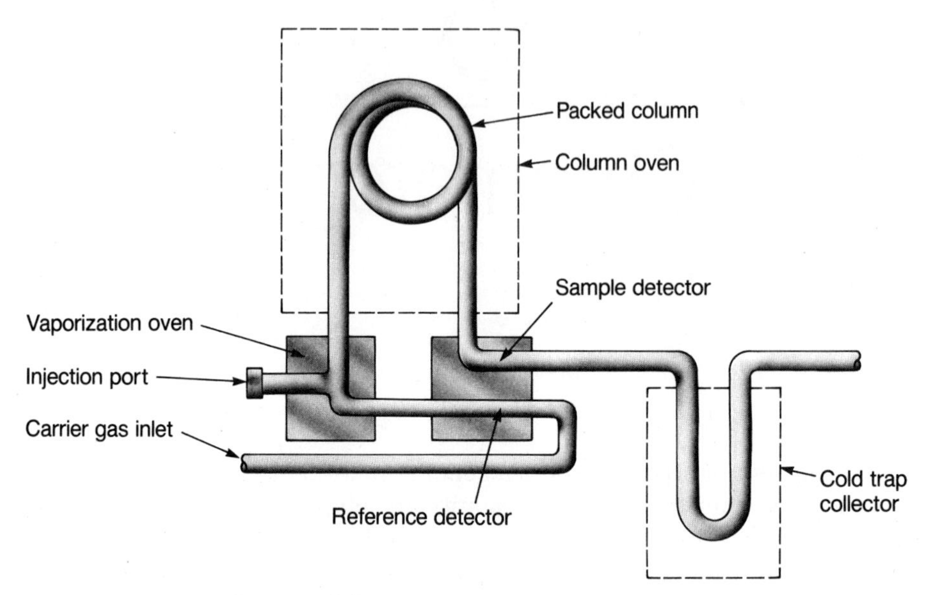

Figure 6.1 Schematic diagram for gas chromatograph.

erating conditions, the retention time is a characteristic value and is independent of the presence of other compounds that may be present in the mixture. For qualitative analysis, if a peak in a gas chromatogram of a mixture is suspected to arise from a certain compound, a sample of the mixture can be "spiked" with an authentic sample of the compound in question. The absence of an additional peak in the chromatogram of the spiked sample is strong evidence that the compounds are identical. For positive identification, the compounds should be compared using several columns packed with different stationary phases. (In general, it is not acceptable to make an identification of any substance solely on the basis of

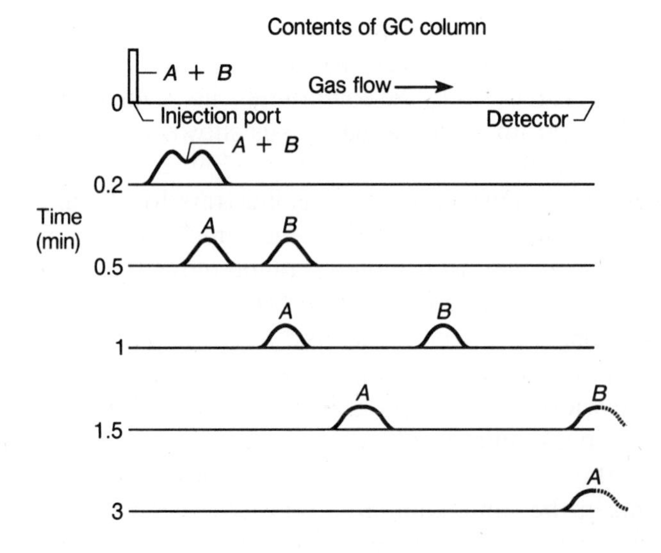

Figure 6.2 Separation of compounds by gas chromatography (GC).

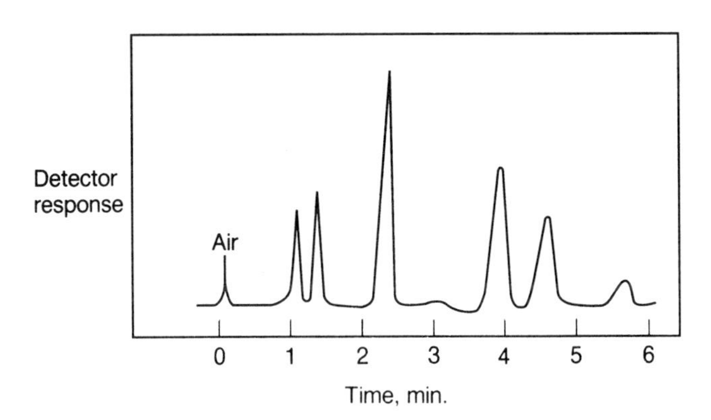

Figure 6.3 Typical gas chromatogram.

one physical measurement, whether it be only melting point, refractive index, density, or one retention time.)

The power of the method of gas chromatography lies in its high resolution and sensitivity, permitting a quantitative analysis of complex mixtures with less than a milligram of sample. The area of a peak in the chromatogram trace is proportional to the amount of the compound present in the sample, and as described below, the composition of the mixture can be determined by very simple measurement.

The resolution in a gas chromatogram can be expressed, as in the case of distillation, in terms of the number of theoretical plates (p. 70). A simple formula can be used to derive the number of plates from the position and width of a peak, as shown in Figure 6.4. With the values d and w expressed in the same units (millimeters or minutes), the number of plates n is given by

$$n = 16 \left(\frac{d}{w}\right)^2$$

In a typical gas chromatograph with a 2-meter column, the number of plates is usually in the range of 2000 to 2500, more than an order of magnitude higher than the best distillation apparatus.

For most analytical work the column is a metal tube bent into a compact coil or spiral. The length is usually 2 to 3 meters (5 to 10 feet) and the diameter 2 to 5 mm. In general, the resolution of peaks increases with increasing length and decreases with increasing diameter. For difficult analytical separations in which very high resolution is needed, much longer columns of very small diameter (e.g., 0.2 mm) can be used.

Gas chromatography also can be used for preparative scale separations and purification and is one of the most powerful tools available for this purpose. Small amounts of material can be separated by repeated injections and collection on an analytical-type column. In general, the maximum amount per injection without overloading the column is less than 20 μL. For larger samples, special columns with a diameter of 2 cm or larger are used. With these "preparative" columns, samples up to 1 mL in size can be applied.

The column packing used for the stationary liquid phase is a crucial

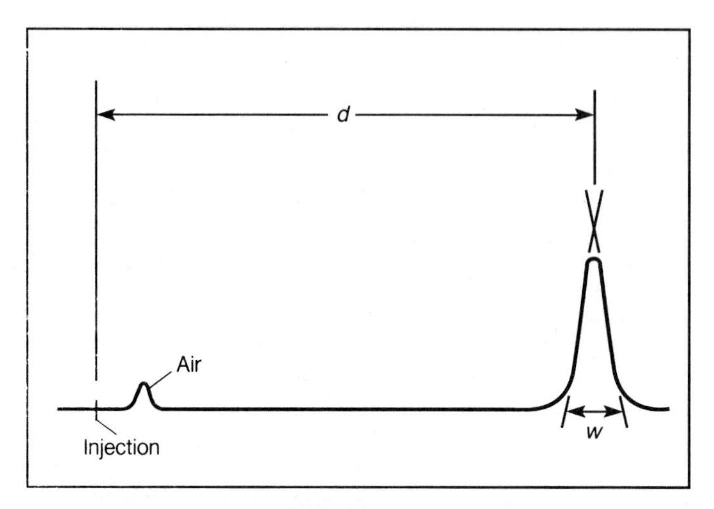

Figure 6.4 Calculation of number of theoretical plates.

factor in gas chromatography. Separation depends on the distribution of compounds between the vapor and liquid and is therefore affected by both the vapor pressures of the compounds to be separated and their solubilities in the liquid phase. The vapor pressure is controlled by the temperature of the oven surrounding the column, which is usually set at or a little below the average boiling point of the mixture. To obtain adequate solubilities in the liquid phase, it is chosen to be similar in polarity to the compounds in the mixture.

Several types of liquid, with numerous minor variations, are used for most gas chromatography. An important requirement is that the liquid phase be nonvolatile, so that it is not swept along and out of the column during use at elevated temperatures. For this reason the liquid is usually a polymeric substance. For nonpolar mixtures, various long-chain hydrocarbon greases (e.g., squalane or Apiezon) or silicone oils (e.g., SE-30, DC-200, or OV-101) are used. More polar liquid phases are polyethylene glycol (e.g., Carbowaxes, Durabond-Wax, or UCON) and polyesters such as poly-diethyleneglycol succinate (DEGS).

$$HO\!-\!(CH_2CH_2O)_n\!-\!H$$

polyethylene glycol

$$\left[\begin{array}{c}CH_3 \quad\ CH_3\\ | \qquad\quad |\\ -Si-O-Si-O-\\ | \qquad\quad |\\ CH_3 \quad\ CH_3\end{array}\right]_n$$

silicone oil

$$-\!\left(CH_2CH_2OCH_2CH_2O\overset{O}{\overset{\|}{C}}CH_2CH_2\overset{O}{\overset{\|}{C}}-O\right)_n\!-$$

diethyleneglycol succinate

These liquids are supported on finely divided, uniform particles of crushed firebrick or diatomaceous earth, usually in amounts of 5 to 15%. To provide an even coating, the stationary phase is prepared by evaporating a solution of the liquid dissolved in a low-boiling solvent in the presence of the support. Most commercial gas chromatographs come equipped with several standard columns. For most of these columns, the maximum operating temperature is 200 to 275°C. At higher temperatures the liquid phase will slowly "bleed off" and damage the detector.

Detectors used in gas chromatography are of two general types: thermal conductivity and flame ionization. **Thermal conductivity detectors**, which are normally used in undergradute laboratory equipment, contain heated filaments or thermistor beads whose electrical resistance varies with temperature. One of these elements is placed in the gas stream coming out of the column and another in a reference stream of the pure carrier gas. The gas used with a thermal conductivity detector is helium, which has a very high thermal conductivity because of its high diffusion rate. When the stream containing a compound from the sample passes the detector, the rate of cooling by the gas stream decreases because of the lower thermal conductivity of the organic compound. This change in temperature results in a change in resistance and voltage, which is detected by the recorder.

In a **flame ionization detector**, the stream from the column passes through a hydrogen flame, and ions formed by the combustion of the compound are captured by a grid to produce a current. Flame ionization is much more sensitive than thermal conductivity, and nitrogen, which is less expensive than helium, can be used as a carrier gas.

Since the thermal conductivity or extent of ionization varies somewhat from one compound to another, particularly with compounds of different types, the response of the detector per milligram or micromole of sample differs. For accurate work, corrections or response factors must be applied to take account of these differences. Factors for a few compounds are given in Table 6.1 for thermal conductivity detectors, relative to benzene, which is taken as 1.00. These factors can be expressed on a weight basis or on a molar basis by multiplying by the ratio of molecular weights. For closely

Table 6.1 Response Factors for Thermal Conductivity Detectors

COMPOUND	WEIGHT FACTOR (W_F)	MOLAR FACTOR (M_F)
Benzene	1.00	1.00
Toluene	1.02	0.86
Heptane	0.90	0.79
Ethyl acetate	1.01	0.90
Ethanol	0.82	1.39

similar compounds, such as isomeric alkanes or alkyl halides, the response factors are usually very nearly the same.

The measured areas in a chromatogram are multiplied by the factors shown in Table 6.1 to give the actual amounts of the compounds. For example, in a chromatogram of a benzene–ethanol mixture, let us say the peak areas are benzene, 50 mm^2, and ethanol, 70 mm^2. The composition on a weight basis is as follows:

benzene: $50\ mm^2 \times 1.00 = 50$; ethanol: $70\ mm^2 \times 0.82 = 57.4$

total corrected area = 107.4

% benzene = $50/107.4 \times 100 = 46.6\%$

% ethanol = $57.4/107.4 \times 100 = 53.4\%$

OPERATING TECHNIQUE

The major variables in carrying out gas chromatography are column size and packing, column temperature, flow rate of carrier gas, sample size, and an instrument setting called the attenuation. The selection of a column packing depends on the factors discussed earlier; in undergraduate laboratory work the choice of column is usually limited to either a polar or nonpolar stationary phase.

Temperature and flow rate are interdependent, and both affect the resolution and time required for the chromatogram. The higher the temperature or the flow rate, the faster the mixture will pass through the column and, thus, the smaller the differences in retention times of the components. If either temperature or flow rate is too high, resolution may fall off to the point that separation is incomplete. If the temperature is too low, on the other hand, the retention times will become very long, and also, the peaks in the chromatogram will be broadened. The temperature is usually set at or a little below the boiling point of the sample, and the flow is then adjusted (around 20 to 30 mL/minute) to give the optimum performance.

The amount of sample depends on the column size and is usually 1 μL or less. Too large a sample overloads the column, and resolution is lost. The **attenuation** setting controls the sensitivity of the instrument and the amplitude of the peaks on the chart (higher attenuation leads to lower sensitivity and lower amplitude). This is set so that the largest peak in the sample registers full scale. If the sample contains a low-boiling solvent, the solvent peak often goes off scale.

The sample is injected by means of a syringe with very small bore. The sample, usually about 1 μL or less, is drawn up into the barrel of the syringe and then 1 to 2 μL of air is drawn in to empty the needle. This is done to avoid vaporization of sample from the needle before injection of the main sample. The air also provides a peak on the chromatogram, with very short R_t that can be used as a marker to measure retention times of other peaks. To make the injection, the needle is inserted through a soft rubber septum

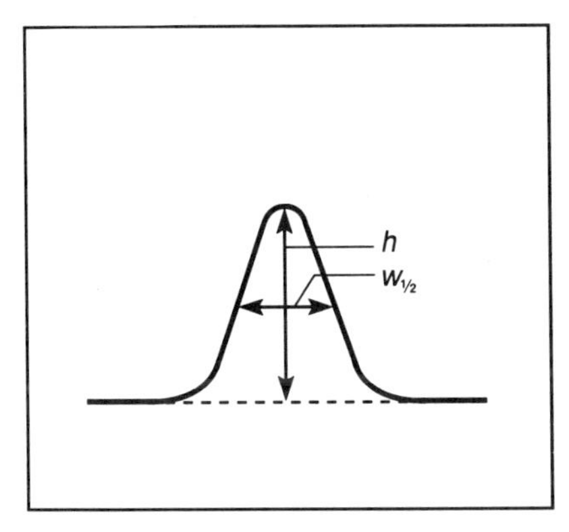

Figure 6.5 Measurement of peak area.

on the injection port to a depth of 2 to 3 cm. The plunger on the syringe is then pushed in with a quick, smooth motion to place all the sample on the column in the shortest time possible.

For measurement of the peak areas in the chromatogram, some instruments have **integrators**, which record the instrument response for each peak. If an integrator is not available, the areas can be obtained by multiplying the height of each peak (h) by the width of the peak at half the height ($w_{1/2}$) as shown in Figure 6.5. Alternatively, if the peak is asymmetic or very broad, the peaks can be cut out of the chart and weighed on an analytical balance. The area under the recorded curve is proportional to the amount of material causing the peak. As mentioned earlier, detector response differs from one compound to another; therefore, for accurate work the area–quantity relationship should be determined by calibration of the chromatograph. With calibration, the area under the curve can be used in the quantitative analysis of mixtures. For most routine analyses, the approximate ratios of components can be estimated reasonably well using only the chromatogram of the analysis sample.

SAFETY NOTE THE LIQUIDS USED IN THIS EXPERIMENT ARE RELATIVELY HARMLESS; HOWEVER, AVOID EXCESSIVE INHALATION OF THE VAPORS. EXERCISE CARE WITH THE MICROSYRINGE AS YOU WOULD ANY SHARP OBJECT.

Experiments

Your laboratory instructor will demonstrate the use of the gas chromatograph and discuss retention times, sample size, peak areas, and other factors affecting this experiment.

Procedure

1. You will be given an unknown mixture made up of four liquids (these may be a series of hydrocarbons or alcohols; see Table 6.2). Samples of

Table 6.2 Unknown Components

	compound	*bp (°C)*
alcohols	methanol	65
	1-propanol	97
	1-butanol	118
	1-pentanol	137
alkanes	pentane	36
	hexane	69
	heptane	98
	octane	126

the pure liquids will be available for your use. Determine the retention times of each of the four liquids and run a gas chromatogram of the unknown. By comparison of the retention times of the known compounds with the unknown, determine which liquids are in your unknown. (Record temperature, chart speed, gas flow rate, and column used.)

2. Determine the approximate percentage composition of your unknown by comparison of the areas under the peaks. (With a thermal conductivity detector, assume the peak areas are directly proportional to moles of sample present.)
3. Prepare a mixture that corresponds to the estimated percentages of your unknown, determine its chromatogram, and compare it with the chromatogram of your unknown.
4. Label your chromatograms and affix them in your notebook.

As an alternate introductory experiment in gas chromatography, fractions from the fractional distillation experiment, Chapter 5, can be examined to evaluate the separation efficiency of the distillation.

QUESTIONS

1. What would be the effect on the retention time of raising the column temperature? Of increasing the carrier gas flow rate?
2. A mixture of heptane (bp 98°C) and 1-propanol (bp 97°C) is not fully resolved on a silicone column but is easily separated on a Carbowax column. Explain why, and predict which compound would have the shorter R_t on Carbowax.
3. Using the formula on page 83 and the peak as seen in Figure 6.4, calculate the number of theoretical plates in the column shown.
4. Using a chromatogram from your experiment, calculate the number of theoretical plates for the column you used. An alternative formula is $n = 5.8(d/w_{1/2})^2$, where $w_{1/2}$ is the width of the peak at half its height

(see Fig. 6.5). This latter formula is theoretically less accurate than the one given on page 83, but it is often easier to obtain an accurate value of $w_{1/2}$ than w. Explain why this might be so.

REFERENCES

Crippen, R. C. *Identification of Organic Compounds with the Aid of Gas Chromatography*; McGraw-Hill: New York, 1973.

Perry, J.A. *Introduction to Analytical Gas Chromatography: History, Principles, and Practice*; Dekker: New York, 1981.

Schupp, O. E., III. In *Techniques of Organic Chemistry*; Weissberger, A., Ed.; Interscience: New York, 1968; Vol. XIII.

Name ______________________ Section ______________ Date ________

6 Gas Chromatography

PRELABORATORY QUESTIONS

1. What volume of sample is appropriate in the running of a gas chromatograph?

2. What is meant by the "retention time" in gas chromatography?

3. Which would you expect to have the shorter retention time, hexane (bp 69°C) or toluene (bp 111°C)? Explain.

4. How is an air peak often useful in a gas chromatogram?

5. Would gas chromatography be a good way to analyze the following mixtures? Explain your answers.

a. ethanol and 1-butanol

b. pentane and heptane

c. sugar and water

7

Liquid Phase Chromatography

In liquid phase chromatography, compounds are separated by partitioning them between a moving liquid and a finely divided solid phase. The process is closely analogous to gas chromatography (see Chapter 6), and the same basic principle applies. A mixture is distributed between the two phases according to differences in affinity of the components for liquid and solid. As the mixture moves over the solid phase, the compounds move apart into discrete zones. The solid phase can be in the form of a thin layer on some kind of backing or packed into a vertical column. In either case, the sample to be separated is placed at one end of the layer or column, and solvent is then allowed to flow over the sample and through the solid.

In most chromatography involving moving liquid and stationary solid phases, molecules are **adsorbed** on the solid, which has a very high surface area. The most common solids for adsorption chromatography are alumina and silica gel; a few other materials, including magnesium silicate, charcoal, and calcium carbonate, also can be used. In some cases, such as chromatography on paper or cellulose pulp, the partitioning actually occurs between the moving solvent and water molecules that are adsorbed on the solid surface.

The adsorbents used in chromatography vary considerably in activity, or adsorptive power. A major determinant of activity is the degree of hydration of the surface. Alumina can be made progressively less active by adding small amounts of water. In most adsorbents, the surface provides a highly polar environment, and compounds with polar groups (see p. 30) will be adsorbed more strongly than less polar ones. Hydrogen bonding to the surface is an important effect and causes a compound with an OH group, for example, cholesterol, to be adsorbed more strongly than similar compounds lacking the OH group, such as the ketone, Δ^4 = cholestenone.

HO O

cholesterol Δ^4-cholestenone

Separation in adsorption chromatography depends on the competition between adsorption on the solid surface and desorption by the solvent needed to elute it. As a corollary, it follows that the more polar the solvent, the more rapidly the compounds will move with the liquid phase. A list of a few common chromatographic solvents, in order of increasing polarity, is given in Table 7.1. Relatively nonpolar solvents, such as hydrocarbons and chlorinated hydrocarbons, are most commonly used. If a solvent such as an alcohol is required for elution of the compounds because of very strong adsorption, a combination of less-powerful adsorbent and less-polar solvent would probably provide a more successful separation.

I. Thin Layer Chromatography

Thin layer chromatography (TLC) is primarily a tool for rapid qualitative analysis and is extremely effective and convenient for this purpose. The chromatography is carried out on a small plate, such as a microscope slide, that is covered on one side with a thin coating of the adsorbent. Very small amounts of samples are applied to the adsorbent surface as small spots in a row near one end of the plate. The plate is then placed, sample-end down, in a widemouth jar containing a shallow pool of solvent. As the solvent rises over the adsorbent layer by capillary action, the compounds in the samples move to varying heights on the plate. The individual compounds can then be detected as separate spots on the plate.

The chief uses of TLC are to determine the number of components in a sample, to detect a given compound or compounds in a very crude mixture,

Table 7.1 Chromatographic Solvents

Least polar	Alkanes (pentane, hexane)
↓	Aromatic hydrocarbons (toluene)
	Chlorinated alkanes (dichloromethane, chloroform)
	Ether
	Ethyl acetate or acetone
Most polar	Alcohols (methanol, ethanol)

and to serve as a preliminary trial in finding conditions before running a column chromatogram. Since tiny amounts of material are exposed on an open surface, the use of TLC is limited to relatively nonvolatile substances; in this respect it is complementary to gas chromatography. However, TLC is not as easily adapted to quantitative determinations. Neither the sensitivity nor the resolving power, in general, is as high as in vapor phase chromatography (VPC), and the detection methods do not lend themselves readily to quantitative determinations. The TLC method can be scaled up by using larger plates (20 × 20 cm or larger) and thicker coatings of adsorbent, so that gram quantities of a mixture can be separated and the components recovered from the plate. A relatively recent preparative modification involves a type of radial chromatography where the separation is performed on a thin layer of adsorbent on a circular plate about 24 cm in diameter which is rotated at constant speed. The substances are placed near the center of the plate and are driven toward the outer edge by centrifugal force. Depending on the adsorbent thickness, up to 1.5 g sample loads can be separated.

To detect the compounds in a mixture, the developed TLC plate is treated with a general reagent, such as iodine vapor. Nearly all compounds absorb iodine or react with it to form violet or brown spots on the slide. However, the relative intensity of the spots is not an accurate indication of the amounts of the compounds present, since the extent of reaction varies. Other widely used general reagents are phosphomolybdic acid and vanillin.

Another method for visualizing spots is illumination of the plate with an ultraviolet lamp. Many substances, particularly aromatic compounds, will show bright fluorescence that may have a characteristic color. A similar technique uses an adsorbent layer that contains a trace of fluorescent dye. Compounds that are fluorescent will show up as bright spots on a light background; any others will appear as dark spots, since they absorb the ultraviolet light and prevent fluorescence of the dye.

The distance that a compound travels on the TLC plate is expressed as the **R_f value**, which is the ratio of the distances from the starting line to the compound and to the solvent front. Under exactly the same experimental conditions, the R_f for a compound is a characteristic value. However, R_f may be affected by small changes in the coating thickness, solvent, temperature, the amount of sample applied, and the presence of other compounds. In practice, it is usually not possible to control all of these conditions completely. For this reason, a comparison of two samples suspected of being the same compound should always be carried out by spotting both samples on the same plate. An even better way to compare the samples is to dissolve them together and apply this mixture, along with the individual samples, or apply solutions of each compound to the same lane on the plate.

A typical example of the use of TLC is the examination of a reaction mixture as illustrated in Figure 7.1. A sample of the crude mixture before

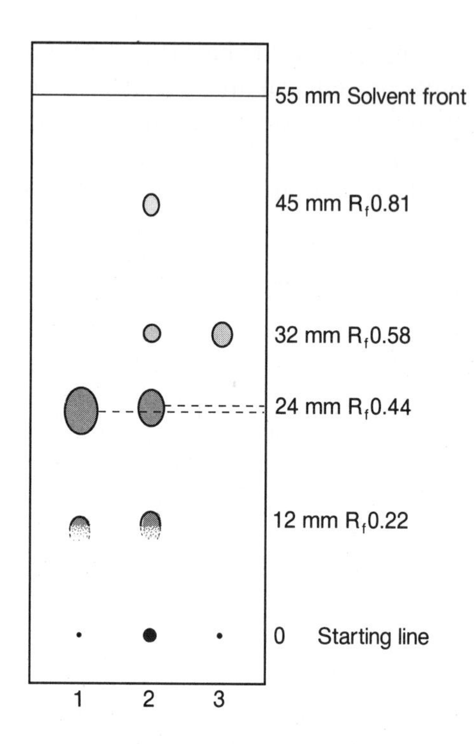

Figure 7.1 Thin layer chromatogram of a reaction mixture. 1. Isolated product. 2. Crude reaction mixture. 3. Authentic starting material.

the isolation of products, spotted in the center lane, shows the presence of one main component and three other compounds, with R_f values from 0.22 to 0.81. A small dark spot at the starting line indicates some polymeric tar. Comparison with the right-hand lane, in which the starting material of the reaction is spotted, shows that the spot at R_f 0.58 in the product mixture is unreacted starting material. The main product isolated from the reaction is spotted in the left lane. It contains a trace of the slower-moving compound but neither of the faster-moving ones.

PREPARATION OF TLC PLATES

The most commonly used adsorbent for routine TLC work is silica gel containing a small amount of gypsum ($CaSO_4$) or other material to bind the coating on the plate. TLC sheets with a very uniform coating of silica gel or alumina on plastic (Mylar) are available commercially. These plates can be obtained with or without a fluorescent dye. An organic polymer is used as a binder for the coating, and these layers are flexible and durable. The sheets are 20 × 20 cm and can be cut with scissors into strips. Twenty-four 2.5 × 6.6-cm, six 6.6 × 10-cm, or four 6.6 × 14-cm rectangles can be obtained from one sheet.

For routine qualitative TLC work, plates can be prepared inexpensively by applying silica gel to glass microscope sides. Two methods for coating TLC slides are most often used: (a) dipping the slides into a slurry of the adsorbent in organic solvent or (b) spreading an aqueous suspension of the adsorbent over the slides. The dipping procedure is simple, and slides can be prepared in a few seconds. Spreading is more time-consuming, and the plates must then be dried at room temperature for several hours or briefly in an oven. The spreading method produces a more uniform and reproducible coating and is the only practical way to prepare larger plates. The adhesive action of the binder is more fully utilized in the aqueous preparation, and thicker coatings can be prepared.

SAFETY NOTE THIS PROCEDURE SHOULD BE CARRIED OUT IN A HOOD; AVOID BREATHING SOLVENT VAPORS.

Dipping Procedure

The slurry is prepared by using a ratio of 1 g silica gel in 3 ml of methylene chloride containing 1% methanol. A sufficient amount must be made up to fill the container to a depth of approximately 8 cm. The level must be maintained by refilling periodically, and solvent must be added as necessary to replace evaporation losses.

Shake the slurry to mix it thoroughly. Hold two clean slides together by one end, immerse them in the slurry to approximately 5 mm from the top, and withdraw them with a slow steady motion (Fig. 7.2). Touch a bottom corner to the lip of the bottle to remove the excess coating and

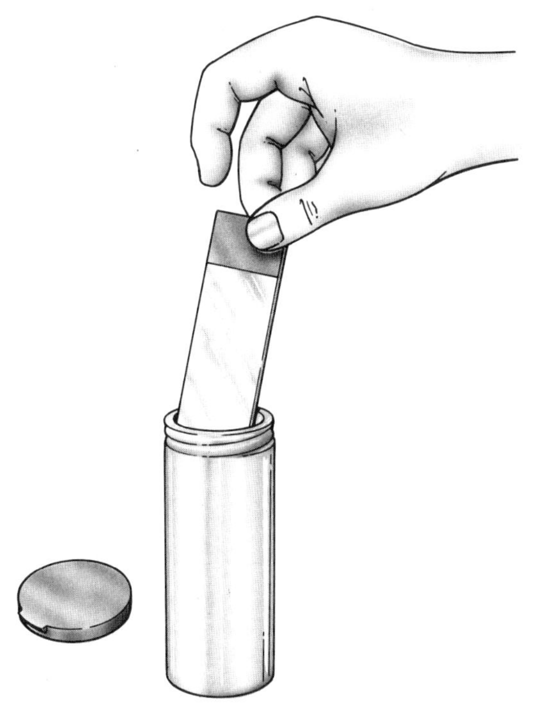

Figure 7.2 Dipping method for coating slides.

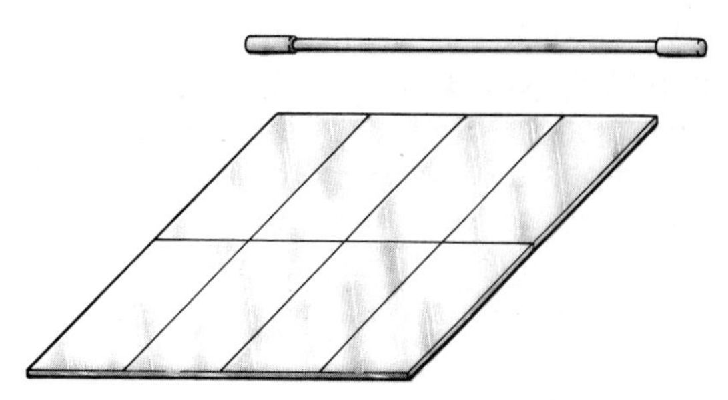

Figure 7.3 Spreading method for coating slides.

securely recap the bottle of slurry. Separate the slides and lay them on a bench, coated side up. The plates should be dried by briefly heating them in a 50°C oven or on a hot plate to remove all methanol. Streaked or unevenly coated plates result from incomplete mixing of the slurry. If the layers are thin and grainy, the slurry is too dilute. Poorly coated plates, of course, can be redipped.

Spreading Procedure

Wipe clean a section of bench top and lay out eight slides in a block four slides wide and two high, with edges touching (Fig. 7.3). Prepare a roller by placing short lengths of rubber tubing at each end of a piece of 7-mm glass rod. Weigh out 4.0 g of silica gel in a 7-dram plastic cap vial. When these preparations have been made, add 8 ml of water to the silica gel and shake vigorously for 30 seconds. The aqueous slurry will thicken in a few minutes and must be spread without delay. Pour the slurry evenly across one end of the block of slides and spread it across the slides by rolling the rod across and back with a smooth motion. Excessive rolling will cause ridges at the edges of the slides.

When the surface of the slides has become dull (1 to 2 minutes), push them apart with a spatula tip and slide them onto a piece of cardboard; handle them carefully to avoid damaging the edge of the coating. Allow the slides to dry in air for 10 to 15 minutes. Drying can be completed in an oven at 80 to 100°C for 30 minutes or at room temperature in a desiccator overnight.

APPLICATION OF SAMPLE

The sample is applied with a *very fine* capillary. This should be about one-tenth the diameter of a melting-point capillary and can be prepared conveniently by softening a 1-cm section in the middle of an open-ended melting-point capillary with a microburner and drawing it out to about 4 to 5 cm. The thin portion is broken to make two TLC capillaries. A larger quantity of capillaries can be made from a soft-glass transfer pipet; the pipet

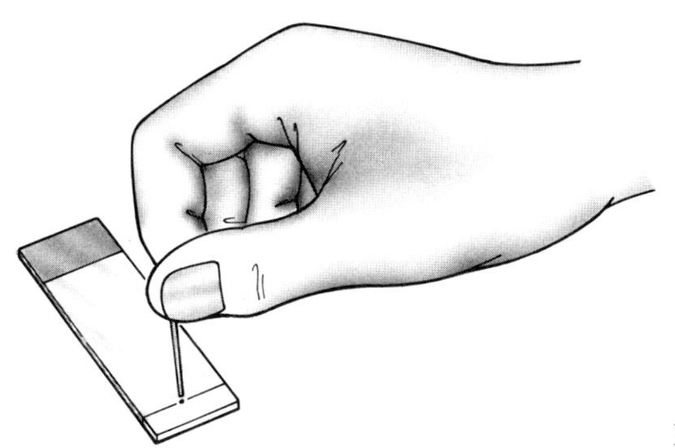

Figure 7.4 Spotting TLC plates.

is drawn out to the appropriate thickness (about a 2-foot length) and the capillary section is broken into 3-inch pieces.

The sample to be applied can be dissolved in any volatile solvent; acetone or methylene chloride is convenient. Make a roughly 5 to 10% solution of the sample in a 10 × 75-mm test tube or 10-mL Erlenmeyer flask. With a sharp pencil, mark a light, straight line approximately 1 cm from one edge of the plate. Dip a capillary into the solution and then touch the end very lightly at a spot on the pencil line (Fig. 7.4). It is a good idea to practice sample spotting first on a scrap plate to check your technique and make sure that the capillary does not deliver too large a spot. The spot should spread to a diameter of no more than 1 mm. If it is necessary to apply a larger amount, let the first spot dry completely and then touch the capillary again at the same place. When working with ultraviolet (UV) fluorescent substances or adsorbent, a useful technique is to carry out the spotting under a UV lamp. This makes it considerably easier to see the size of the spot and the amount of sample applied. It is important that no crystals from the sample be transferred to the plate or form when the spot is applied, since this causes streaking during development.

Two common errors are allowing the spot to become too broad and applying too much sample. Commercial TLC sheets and slides prepared by dipping usually have a thinner coating than those made by spreading, and the former therefore require smaller amounts of sample. For a preliminary investigation, it is useful to make several spots with one, two, or three applications to determine the proper volume of solution to apply. Three spots can easily be placed on a microscope slide. Four spots is the maximum, since the zones spread as the plate is developed.

DEVELOPMENT

The choice of solvent for development depends on the nature of the compounds to be separated. For relatively nonpolar compounds, the solvents shown in the upper part of Table 7.1 are appropriate. To obtain the proper polarity, a few percent of a polar solvent such as an alcohol can be added to a less polar one, such as ether. A third or even fourth component is sometimes used to further adjust the polarity and R_f values. It must be kept in mind that the outcome of the chromatogram depends on the quality and activity of the TLC plates as well as the solvent. A certain solvent mixture that works well for a particular mixture on one batch of plates may not be optimal with other plates.

The best separation of closely similar compounds by TLC is usually achieved when the R_f values are 0.4 to 0.6, since the spots become more diffuse as they move farther up the plate. Chloroform is a generally useful TLC solvent for a fairly wide range of compounds, but it is a suspected carcinogen and should be used only in a hood. Methylene chloride often can replace chloroform.

Widemouthed screw-cap or snap-cap bottles are useful chambers for small-scale TLC work. To maintain equilibrium conditions during the development of the chromatogram, it is important that the space in the developing chamber above the solvent remains saturated with solvent vapor. This is accomplished by fitting a piece of filter paper around the inner wall of the jar to act as a wick; the wick must not be in contact with the TLC plate during development and must be placed in such a way so at least the upper part of the plate will be visible. Solvent is added to the jar to a depth of about 0.5 cm, saturating the wick; the solvent level must not be above the sample spots on the plate. Wait a few minutes for equilibration and then lower the plate, spotted end down, into the chamber and cap the jar.

When the solvent has risen to about 0.5 cm from the top of the slide, remove the plate, immediately mark the solvent front with a sharp pencil, and allow the excess solvent to evaporate. Visualize the chromatogram by viewing with a UV light or placing the plate for a few minutes in a jar containing a few crystals of iodine. Prick the outline of the spots with a sharp pencil (spots visualized with I_2 often fade in a few minutes).

If the R_f values are too low, that is, if the compounds have moved only a short distance, the chromatogram can be developed a second (and third) time, allowing it to dry between each development. Another possibility is to repeat the chromatogram with a freshly spotted plate and using a more polar solvent.

Experiments

A. TLC OF NITROANILINES

This experiment is a simple demonstration of TLC to provide practice in the basic techniques of the method.

SAFETY NOTE NITROANILINES ARE TOXIC; AVOID CONTACT WITH THE SKIN. AVOID PROLONGED BREATHING OF METHYLENE CHLORIDE VAPORS.

Procedure

Prepare eight TLC plates by the method designated by your instructor, and prepare six to eight TLC capillaries. Obtain, in labeled 10 × 75-mm test tubes or vials, very small samples of *ortho*-, *meta*-, and *para*-nitroaniline and 2,6- and 2,4-dinitroaniline (about 5 mg, or just enough to cover the bottom of the test tube, is sufficient). Dissolve the samples in approximately 0.5 mL of acetone; the entire sample must be in solution.

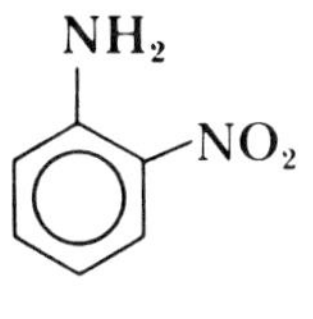

o-nitroaniline 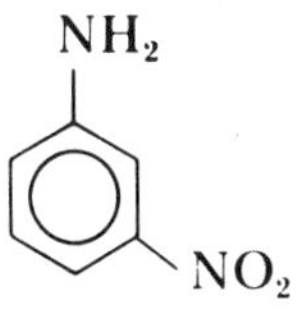 *m*-nitroaniline 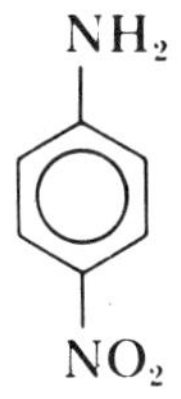 *p*-nitroaniline

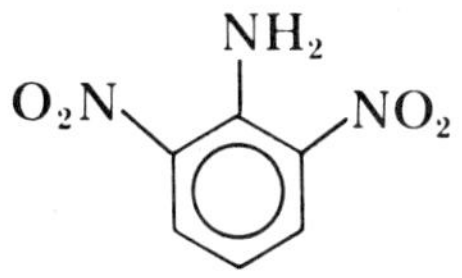

2,6-dinitroaniline 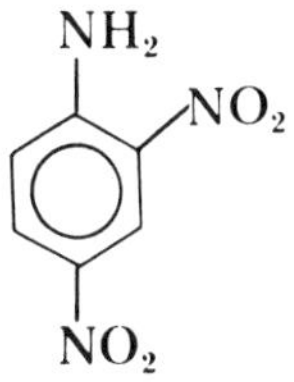 2,4-dinitroaniline

On one plate, spot samples of *o*- and *m*-nitroaniline and 2,6-dinitroaniline; on a second plate, spot samples of *m*- and *p*-nitroaniline and 2,4-dinitroaniline. Develop the plates with methylene chloride. If the spots are not distinct or are too large or have run together, repeat the chromatogram. Sketch the appearance of the developed plates in your notebook, and then develop again. If you feel it may be useful, develop a third time. The spots should be visible on the dry plate, but placing them in an iodine vapor jar or viewing them under UV light may bring them out more distinctly. (It is not practical to redevelop a plate that has been exposed to iodine, even if the color disappears, since some chemical change may have occurred.)

On other plates, spot mixtures of two or three of the compounds in various combinations in the same lane; apply the second solution right over the spot of the first, taking care to avoid overloading. Spot the individual compounds in other lanes on the same plates. Develop and record the appearance of each plate.

When satisfactory results have been obtained with single compounds and known mixtures, obtain samples of unknowns from your instructor and determine which compound(s) are present in each.

B. TLC ANALYSIS OF ANALGESIC DRUGS

Analgesics (compounds that relieve pain) range from aspirin, which is consumed at the rate of many millions of pounds per year, to morphine and related narcotics. In addition to aspirin, several other chemically similar compounds are widely used in nonprescription or proprietary analgesic tablets. Among these are phenacetin, salicylamide, and acetaminophen. Caffeine is sometimes added to these formulations to overcome drowsiness. A few other compounds such as *N*-cinnamylephedrine (cinnamedrine) are included for other therapeutic effects, such as antispasmodic or slight sedative action. In addition to the active ingredients, the tablets of these drugs contain starch, lactose, and other substances that act as binders and permit rapid solution. Inorganic bases are sometimes also included.

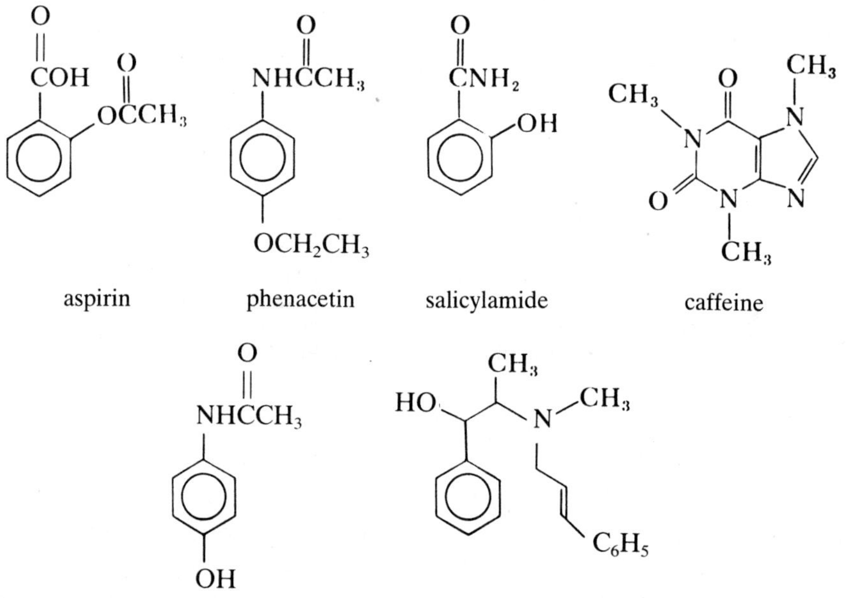

aspirin phenacetin salicylamide caffeine

acetaminophen cinnamedrin

In this experiment, you will obtain as an unknown a proprietary analgesic drug. The objective is to identify the unknown drug by TLC comparison with several known compounds. The unknown will be one of those listed in Table 7.2. The amounts of ingredients in some cases are given in *grains* per tablet [grain (gr) is an apothecary unit; 1 gr = 64 mg].

Pure samples of the compounds found in the analgesics are used as reference standards. These are aspirin (acetylsalicylic acid), acetaminophen (4-acetamidophenol), caffeine, cinnamedrin (cinnamylephedrine), and phenacetin (*p*-acetophenetide).

Several solvent systems have been recommended for the separation of these analgesic drugs, and the relative merits of these solvents have been

Table 7.2 Common Analgesic Drugs

DRUG (BRAND NAME)	INGREDIENTS
Anacin	Aspirin and caffeine
Empirin Compound	Phenacetin (2.5 gr), aspirin (3.5 gr), caffeine (0.5 gr)
Midol	Aspirin (7 gr), cinnamedrin (0.23 gr), caffeine (0.5 gr)
Tylenol	Acetaminophen (325 mg)
Vanquish	Aspirin (227 mg), acetaminophen (194 mg), caffeine (33 mg)

compared (see Roswell, D. F. and Zaczek, N. M.). The solvents include the following:

butanol–ethyl acetate–methyl isobutyl ketone (2:9:9)

ethyl acetate

ethyl acetate–ethanol–acetic acid (25:1:1)

ethylene dichloride–acetic acid (17:4)

methylene chloride–ether–acetic acid (40:20:2)

toluene–ether–acetic acid–methanol (20:10:3:0.2)

Your instructor will recommend one of these or some other solvent. The effect of slight changes in polarity can be observed by comparing ethyl acetate alone and with added ethanol and acetic acid, using a combination of the reference compounds and drug samples.

Since the proper choice of solvent system depends somewhat on the plates and the conditions used for the experiment, it is best to experiment in advance with several systems to find the one that is most effective.

Procedure

Place approximately 10 mg of the five reference compounds in labeled vials or 10 × 75-mm test tubes and dissolve the samples in a few drops of methanol. In a sixth tube place a quarter of an analgesic tablet and add 1 mL of methanol. Crush the tablet with a rod, stir well, and allow the insoluble material to settle.

Obtain a 7 × 14-cm piece of fluorescent TLC sheet and spot samples of the six solutions 1 cm from one end and 1 cm apart. The unknown sample should be in the center lane. When the spots are dry, place the sheet in a widemouthed jar with a solvent pool about 5 to 6 mm deep. Cap the jar and allow the chromatogram to develop; approximately 40 minutes are required for the solvent to rise to within 1 to 2 cm of the top.

Remove the sheet, recap the jar, and mark the solvent boundary with a pencil mark or small scratch. Examine the chromatogram under a UV lamp and sketch the appearance of the plate in your notebook, indicating the location and approximate size of the spots and any distinctive colors. After this examination, place the sheet in a jar of iodine vapor for approximately 30 seconds, remove, and again record the appearance. Identify the spots in the chromatogram, including as many of the spots in the unknown lane as possible. From the identity and number of the spots in the unknown and the composition of the possible unknowns, deduce the identity of your unknown.

QUESTIONS

1. What will be the result of the following errors in TLC technique?
   a. too much sample applied
   b. solvent of too high polarity
   c. solvent pool in developing jar too deep
   d. forgetting to remove plate when solvent has reached the top of the plate
2. Explain why the presence of crystals of the sample at the starting point causes streaking of the plate during development.
3. Arrange the nitroanilines in order of decreasing polarity as observed by TLC. Correlate the polarities with the structures and suggest an explanation for the polarity sequence.
4. From the chromatogram of the analgesics, what conclusions can be drawn about the functional groups responsible for ultraviolet fluorescence?

REFERENCES

Bobbitt, J. M. *Thin Layer Chromatography*; Van Nostrand Reinhold: New York, 1963.

Kirchner, J. G. Thin Layer Chromatography, In *Techniques of Chemistry*; Perry, E. S.; Weissberger, A., Eds.; Wiley-Interscience: New York, 1978; Vol. XIV.

Lien, V. T. *J. Chem. Education*. **1971**, *48*, 478.

Randerath, K. *Thin-Layer Chromatography*; Academic Press: New York, 1966.

Roswell, D. F.; Zaczek, N. M. *J. Chem. Education*. **1979**, *56*, 834.

Thin Layer Chromatography, A Laboratory Handbook, 2nd ed.; Stahl, E., Ed.; Springer Verlag: New York, 1969.

Touchstone, J.; Dobbins, M. F. *Practice of Thin Layer Chromatography*; John Wiley and Sons: New York, 1978.

Name ______________________ Section ______________ Date ________

7 I. Thin Layer Chromatography

PRELABORATORY QUESTIONS

Circle the *most* correct answer for Questions 1 to 5 below.

1. A common adsorbent used in TLC is **a.** alumina, **b.** sodium bicarbonate, **c.** iron powder, **d.** nylon powder, **e.** iodine.

2. After development, the detection of colorless spots can be accomplished by use of **a.** ionizing detector, **b.** ultraviolet light, **c.** infrared light, **d.** heat, **e.** sugar solution.

3. The R_f may be affected by small changes in **a.** coating thickness, **b.** solvent, **c.** temperature, **d.** amount of applied sample, **e.** all of the above.

4. The text discusses the student preparation of TLC plates using **a.** window pane glass, **b.** microscope slides, **c.** Mylar film, **d.** starch adsorbent, **e.** none of these.

5. In TLC analysis, the sample is applied with a(n) **a.** capillary, **b.** eye dropper, **c.** pipet, **d.** syringe, **e.** glass rod.

6. Place the following common organic solvents in order of *increasing* polarity: **a.** methyl ethyl ketone, **b.** cyclohexane, **c.** isopropyl alcohol, **d.** acetic acid, **e.** 1,1,2-trichloroethane.

7. Draw the structures for aspirin and caffeine.

II. Column Chromatography

Chromatography on a column of adsorbent is a way to separate a mixture and to isolate the components in larger amounts than is possible by TLC. In this method, the sample is applied in a very narrow band at the top of the column, and solvent is allowed to flow through the adsorbent. In this process, the chromatogram develops into separate zones containing the individual compounds. The zones (or bands as they are called) can then be eluted (washed off) in sequence with further solvent and collected in separate fractions.

In the standard method for liquid chromatography on a column, the adsorbent is first saturated with the least polar solvent to be used in the separation. This is often hexane, or, for more polar compounds, ether or methylene chloride. The chromatogram is then developed with this same solvent. More polar solvents (lower in the series as seen in Table 7.1) may be needed to elute the compounds from the column. The second (or third) solvent should be blended gradually into the eluting liquid to avoid disrupting the bands that have been developed. In practice, the development and elution steps are often not entirely distinct, since the fastest-moving band may begin to emerge from the column before separation of the slower-moving bands is complete.

A typical chromatography column is shown in Figure 7.5. The amount of adsorbent, such as alumina or silica gel, is usually about 20 to 30 times the amount of sample. The dimensions of the actual adsorbent column should be in a height-to-diameter ratio of about 10:1 for most work. Larger ratios, with longer columns, may be needed for difficult separations, but the chromatography will be slower. It is necessary to have some free volume in the tube above the adsorbent for a supply of solvent. A stoppered separatory funnel mounted above the column with the stopcock open and the tip below the solvent level can be used as a convenient constant-head solvent reservoir (Fig. 7.6).

The column is prepared in the following manner:

1. A wad of glass wool or cotton is placed in the bottom of the tube and a layer of sand is poured over this. The sand retains fine particles and also provides a flat horizontal base for the adsorbent column.
2. The tube is filled with the first solvent to be used and then the dry adsorbent is added in a fine stream. The tube is tapped and shaken to dislodge air bubbles and the solvent is drained out at the bottom to make room as needed but the solvent level must always be kept above the adsorbent.
3. When the adsorbent has settled into a compact column, another layer of sand is added at the top of the adsorbent to prevent disturbance of the surface when solvent is added.

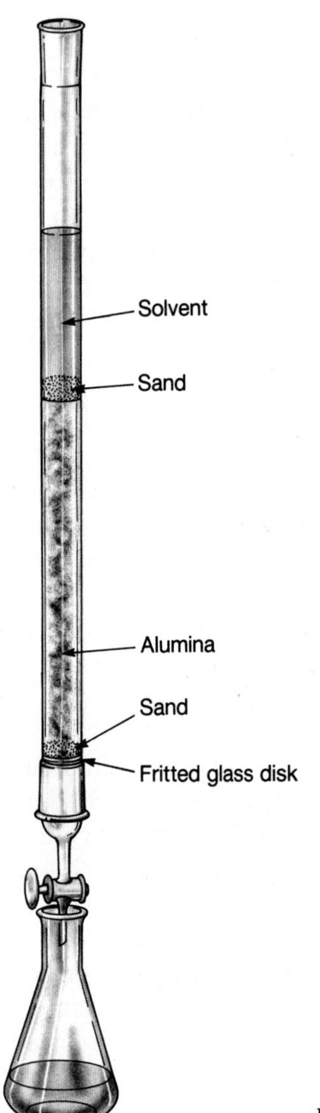

Figure 7.5 Typical chromatography column.

4. The solvent is permitted to drain down to just above the top sand layer. *The adsorbent column must be kept covered with solvent during the chromatography; otherwise, channels and cracks will develop.*
5. The sample is dissolved in a minimum volume of solvent and the solution is added to the column with a pipet and bulb. The solution is allowed to drain onto the column and more solvent is added immediately.

To carry out the separation, solvent is allowed to pass through the column as fast as the flow rate permits. When the compounds to be separated are colored, the progress of the development and elution can be followed by the appearance and movement of bands on the column, and

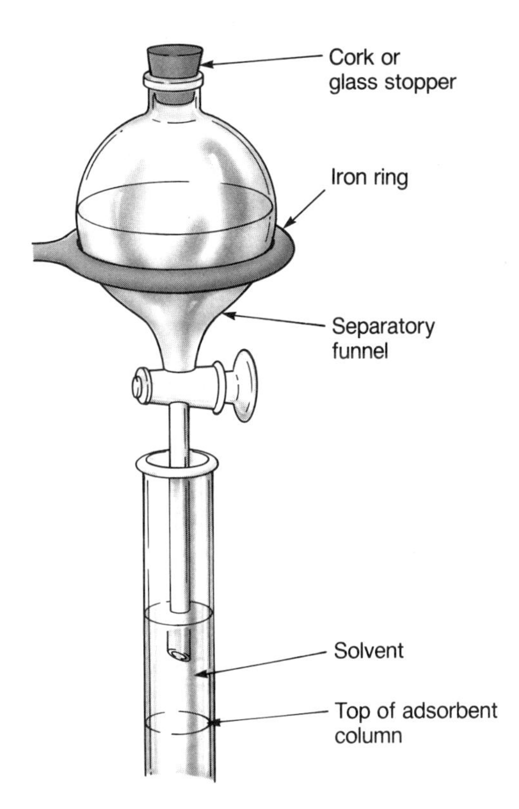

Figure 7.6 Constant-head solvent reservoir.

fractions are collected as each band is eluted. If the mixture is colorless and there are no visible clues such as fluorescence under a UV lamp, fractions of the eluate are collected at regular intervals and samples are examined by TLC or other means. If the movement of compounds on the column slows or stops, the solvent polarity is increased by adding 5 to 10% of a more polar solvent, followed by switching to the pure polar solvent.

In an alternative technique called **dry column chromatography**, the sample is placed at the *top* of the column of the dry adsorbent, usually contained in nylon tubing. Solvent is allowed to pass through the column to develop the chromatogram. When the solvent reaches the bottom of the tube, the flow is stopped. The adsorbent column is then extruded from the tubing and the bands are cut apart. Dry column chromatography closely parallels TLC and, in effect, extends the TLC method to larger quantities.

Liquid column chromatography can be adapted to function as a high-resolution analytical tool by use of a very small column through which the solvent is pumped at high pressure. These so-called high performance liquid chromatography (HPLC) instruments are analogous to their vapor–liquid counterparts in operation and extend analytical chromatography to non-volatile compounds. The sample is injected into the solvent stream and the separated compounds are detected in the eluate by very sensitive methods, usually by UV or refractive index. The signals from these detectors are then transmitted electronically to a recorder.

SEPARATION OF LEAF PIGMENTS

The chloroplast cells of leaves and other plant tissue in which photosynthesis occurs contain chlorophylls and a number of other pigments that can be extracted when the cells are ruptured. The extracts contain several major pigments and a number of minor ones, plus fats and many other colorless compounds. Most of the pigments are of two structural types—the green chlorins and the yellow-orange carotenoids, with a number of variations in functional groups. The structures of four common pigments are shown in the following illustration.

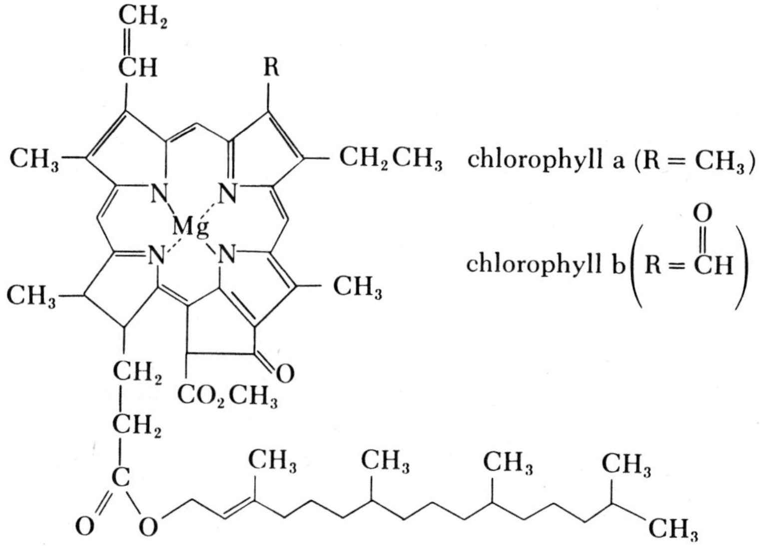

chlorophyll a ($R = CH_3$)

chlorophyll b $\left(R = \overset{O}{\overset{\|}{C}}H\right)$

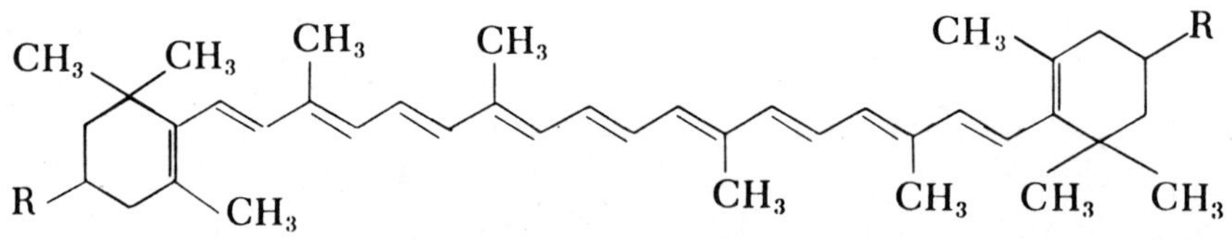

β-carotene ($R = H$)
zeaxanthin ($R = OH$)

The complex mixtures of pigments from plants present a formidable challenge in separation, and the earliest use of chromatography dealt with this problem. Many different chromatographic methods and adsorbents have been investigated. Since the compounds include a range of chemical types, complete separation of all the pigments in an extract cannot be accomplished in a single chromatogram. Different adsorbents are required for separating the closely related compounds of any one type. The chlorophylls, for example, can be separated on powdered sugar, but separation of the less polar carotenes requires a more active adsorbent. In this experiment, column chromatography is illustrated with a leaf extract, using alumina as the adsorbent. Under the conditions used, some changes of the more sensitive pigments occur during the extraction and chromatography, and only partial separation is achieved, but the principles of the method can be seen.

Experiments

C. EXTRACTION OF PIGMENTS

The chloroplasts of all seed-bearing plants contain approximately the same mixture of pigments. The exact composition depends on the age of the plant and conditions of growth, and the amounts of the individual pigments may vary considerably. Any dark green leaves can be used; spinach is a particularly good source.

Since only part of the final extract will be used for chromatography, the usual precautions and rinsing to effect complete transfers need not be observed in this extraction.

Procedure

Remove the stems from about 4 g of spinach leaves and chop the leaves into small pieces with scissors. Place the leaf tissue in a mortar, add one-half teaspoonful of sand and 5 mL of methanol, and grind with a pestle to a coarse mash for about 5 to 10 minutes.* Transfer the mash to a centrifuge tube, add 10 mL of hexane, and stir thoroughly. Pour off the hexane layer into a second centrifuge tube, squeezing the leaf residue with a spatula. To the hexane solution add an equal volume of water, stopper the centrifuge tube, and shake thoroughly to extract the methanol. Remove the lower layer (containing some emulsified upper layer) with a pipet and repeat the extraction of the hexane layer with a second portion of water.

Place a pellet of cotton in a small funnel, add one-half teaspoonful of sodium sulfate, and filter the hexane solution through the drying agent into a clean centrifuge or test tube. Evaporate the dried solution to approximately 0.5-mL volume on the steam bath.

*An alternate procedure that works well is to use a Waring blender in place of the mortar, pestle, and sand. This method allows the preparation of the mash for many students at one time.

D. CHROMATOGRAPHY (Microscale)

To conserve time and solvent and permit use of a simple column, the chromatogram is carried out on a very small scale. The column is a 15-cm length of 8-mm glass tubing, drawn out at one end and sealed off in a fine tip. Two of these columns can be prepared by softening the center of a 30-cm length of tubing in a flame and drawing the ends apart. After the column is filled, the tip is broken to start the chromatogram. Alternatively, a Pasteur pipet with a previously sealed tip can be used.

The results of the chromatography depend on a number of variables. Among these are the source of pigments, the efficiency of the extraction (and thus the total concentration of pigment in the extract), the activity of the alumina, and the volumes of each solvent used in development. Very small changes in some of these factors can affect the separation and can even cause changes in the relative positions of individual pigments in the chromatogram.

Since it is impossible to specify all of these variables, one procedure for carrying out the chromatogram, no matter how detailed, will not give optimal results in every case. In the procedure given, a general sequence of solvent changes and collection of fractions is suggested, but further details will depend on the individual chromatogram. The overall objective is to obtain as many individual pigments as possible in separate collection containers.

Since each chromatogram requires only a short time, several can be carried out, using the experience gained in previous runs to refine and improve the separation.

Procedure

Clamp the column upright, half-fill it with hexane, and, with a thin rod, insert a very small wad of cotton or glass wool and tamp it gently into the constriction. Add enough sand to give about a 3-mm layer. Place 2 g of alumina (1.5 g if a sealed-tip Pasteur pipet is used) on a piece of glassine paper and pour it into the column in a slow stream. When the alumina piles up above the liquid, dislodge it by shaking or stirring it with a thin rod. Tap the column to settle the adsorbent and shake loose any air bubbles. Again add sand to give a 3-mm layer above the adsorbent; the free space above the column is approximately 2 mL (somewhat less if a pipet is used).

Before proceeding further, read through the procedure and obtain all the supplies needed; when the chromatogram is started you will have to proceed without stopping. In separate labeled test tubes place 10 mL each of hexane, toluene, methylene chloride, and ethyl acetate. Four empty 18 × 150-mm test tubes and several transfer pipets and 2-mL bulbs will be needed. Have a sample of leaf extract ready in one pipet and bulb and 2 mL of hexane in another pipet.

With a file, score the tip of the column and snip it off. Allow the solvent to drain into a test tube, and when the solvent level in the column reaches the sand, place the concentrated leaf extract onto the column. Hold a pipet filled with hexane above the column, and *as soon as the pigment solution has just drained into the sand*, rinse with a few drops of hexane and then fill the space above the column.

While the hexane is draining into the column, add 1 mL of toluene to the test tube containing the remaining hexane. This will be the first of several solvent mixtures of increasing polarity used to elute the pigments. Before the hexane level reaches the sand in the column, add 1 to 2 mL of the hexane–toluene mixture. Continue replenishing the solvent as it drains out, noting any changes in the appearance of the column.

As bands begin to separate and move down the column, place empty test tubes under the column to collect the separate pigment solutions as they are eluted. If there is little or no movement of pigments on the column, start using pure toluene as the eluting solvent. Continue developing and eluting the pigments with solvents and mixtures of increasing polarity (toluene + 10% methylene chloride, methylene chloride, methylene chloride + 10% ethyl acetate, ethyl acetate), collecting each band in a separate test tube as it elutes. The solvent used for elution should be changed only when there is no appreciable further movement or separation of bands. For the final elution, a mixture of ethyl acetate plus 1 or 2% of methanol is sufficient; pure methanol is not necessary for any of these relatively non-polar compounds.

Summarize in your notebook the course of the development and elution and the volume and color of each of the fractions collected. To reuse the column, tap the contents out of the tube into a waste receptacle and rinse the tube with a little acetone. Pull air through with the aspirator to dry the tube, and reseal the tip in a flame. Tubes prepared from Pasteur pipets can be discarded.

QUESTIONS

1. What would be the effect on the results of a column chromatogram if

   a. an excessive amount of solvent were used in adding the sample to the column?

   b. solvent is not maintained above the adsorbent, and a crack develops in the column?

   c. methanol were not completely removed from the leaf extract before chromatography?

2. Of the following mixtures, predict which compound would be eluted first on adsorption chromatography:

a. 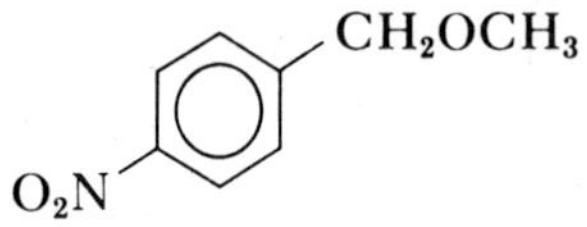 CH_2OCH_3, O_2N and CH_2OH, O_2N, CH_3

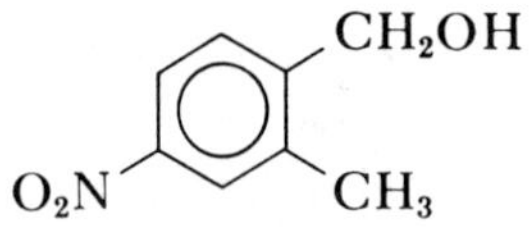

b. 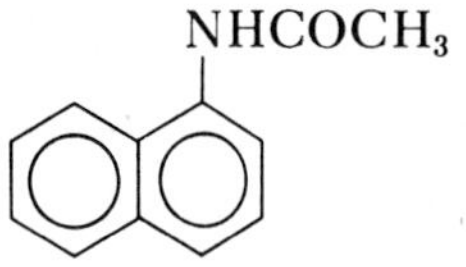 $NHCOCH_3$ and C_2H_5, $N—COCH_3$

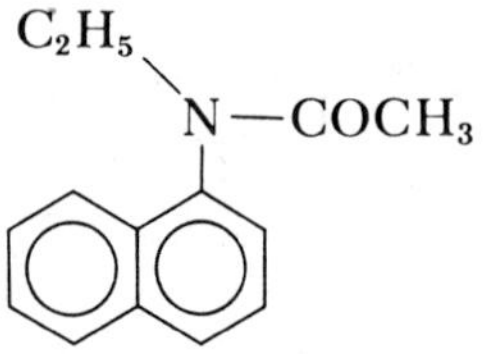

c. 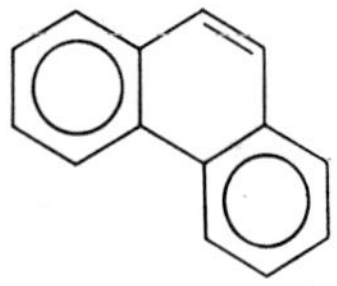 and

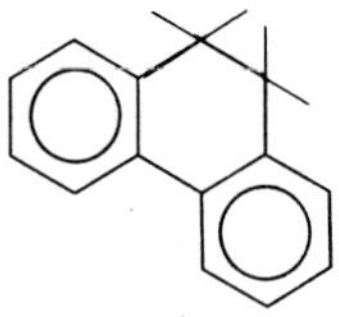

3. In a typical chromatogram of a leaf extract, four separate colored bands were eluted in the following sequence: (1) yellow-orange, (2) green, (3) yellow, (4) green. Assuming that these bands represent the four pigments shown on page 110, suggest which compounds correspond to bands 1 to 4.

REFERENCES

Bobbitt, J. M. *Introduction to Chromatography*; Van Nostrand Reinhold: New York, 1968.

Chromatography: A Laboratory Handbook of Chromatographic and Electrophoretic Methods; Heftmann, E., Ed.; Van Nostrand Reinhold: New York, 1975.

Perry, S. G.; Amos, R.; Brewer, P. I. *Practical Liquid Chromatography*; Plenum Press: New York, 1972.

Snyder, L. R.; Kirkland, J. J. *Introduction to Modern Liquid Chromatography*, 2nd ed.; Wiley-Interscience: New York, 1979.

Name ______________________ Section ______________ Date ________

7 II. Column Chromatography

PRELABORATORY QUESTIONS

1. As a rough "rule of thumb," the height of the adsorbent in a column in relation to its diameter should be what ratio?

2. The ratio of the weight of adsorbent to the weight of sample being separated should be approximately what?

3. If movement of substances down the column ceases, what is the next step?

4. Name the two most common adsorbents used in column chromatography.

5. What is the metal found in chlorophyll?

8

Steam Distillation of Essential Oils

In Chapter 5 it was seen that distillation of a mixture (more precisely, a solution) of two miscible liquids depends on the vapor pressure and mole fraction of each of the components. There is a different situation in the distillation of a mixture of two compounds that are not mutually soluble. In this case, the vapor pressure above the mixture is the sum of the partial pressures of the components, $P_T = P_A + P_B$; that is, each exerts its own vapor pressure independently of the other (Dalton's Law). Since the vapor pressures are additive, the boiling point of the mixture (i.e., the temperature at which $P_T = 1$ atm) is lower than the boiling point of either of the components (Fig. 8.1).

As long as separate phases are present in the liquid, the mixture will have a constant boiling point. In addition, the distillate will have a constant composition, which is determined by the ratio of the vapor pressures (Eq. 8.1). Since the number of moles of a compound (n) is equal to its weight divided by its molecular weight (mw), Equation 8.2 can be derived for the weight ratio of compounds in the distillate.

$$\frac{n_A}{n_B} = \frac{P_A}{P_B} \qquad [8.1]$$

$$\frac{\text{wt. A}}{\text{wt. B}} = \frac{P_A}{P_B} \times \frac{\text{mw A}}{\text{mw B}} \qquad [8.2]$$

Thus, if the molecular weights and partial pressures of the two compounds are known, the weight ratio of the components in the distillate can be calculated. Conversely, if the vapor pressure and molecular weight of one component and the total pressure during the distillation are known, the molecular weight of the second component can be calculated from the weight ratio.

For example, assume that one component is water, the other is an unknown organic compound, X, and the distillation occurs at 99.4° at a

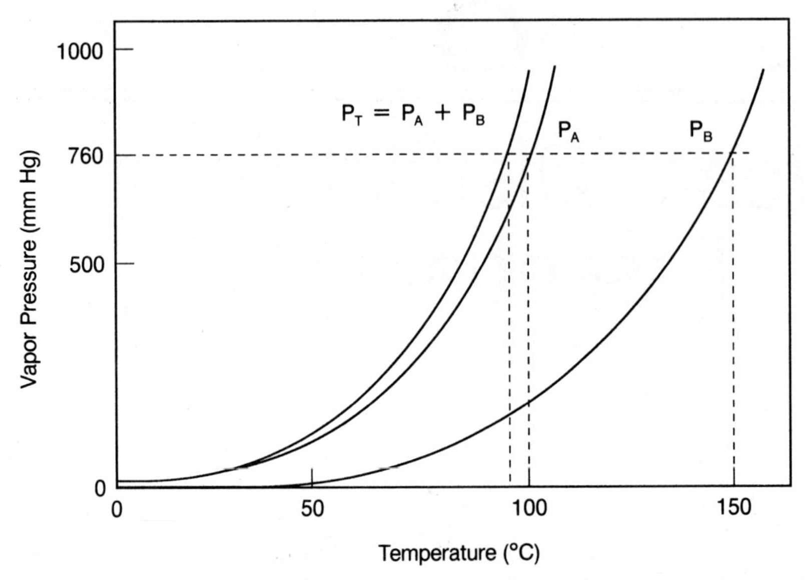

Figure 8.1 Vapor pressure versus temperature for a mixture of two immiscible compounds with boiling points of 100 and 150°C.

barometric pressure of 750 mm to give a mixture of water and X in a ratio of 9 g/1 g. The partial pressure of X can be calculated from the total pressure and the partial pressure of H_2O at 99.4°C, which from handbook tables is 744 mm.

$$\begin{aligned} P_x &= P_T - P_{H_2O} \\ &= 750 - 744 \text{ mm} \\ &= 6 \text{ mm} \end{aligned}$$

Then

$$\frac{\text{mw X}}{\text{mw H}_2\text{O}} = \frac{\text{wt.X}}{\text{wt. H}_2\text{O}} \times \frac{P_{H_2O}}{P_x}$$

$$\frac{\text{mw X}}{18} = \frac{1}{9} \times \frac{744}{6}$$

$$\text{mw X} = 250$$

When one of the two components is water, this process is termed **steam distillation**, and it provides a means of distilling a slightly volatile compound at a temperature far below its atmospheric boiling point. Steam distillation is useful in the separation of a water-insoluble organic compound that is present in a large amount of nonvolatile material. In this case, simple distillation, even in a vacuum, is ruled out because destructively high temperatures would be needed to distill the relatively small amount of volatile material and because mechanical entrapment will prevent its complete removal. Steam distillation is a useful alternative to extraction for the isolation of volatile organic compounds, such as essential oils from plant material. Extraction with solvents removes gums and fats as well as the volatile oils; the latter are separated selectively by steam distillation.

APPARATUS AND TECHNIQUE

Steam distillation on a macroscale is most conveniently carried out in a two- or three-neck round-bottom flask, or a one-neck flask fitted with a Claisen adapter. A typical setup is illustrated in Figure 8.2. The top of the distilling head and the extra neck are sealed off with a cork or stopper; the boiling point is usually of no interest. Steam is provided by the addition of water and sufficient heat from a heating mantle or other suitable heat source. Alternatively, steam can be supplied from an external boiling flask or a steam line on the desk; this is necessary to provide agitation when steam distilling a tarry or very bulky mass. When using an external source of steam, a trap must be included between the source of steam and the distilling flask since a drop in steam pressure can cause the contents of the flask to be sucked back through the steam inlet. Steam distillation on a microscale can be carried out similarly, except in a smaller apparatus.

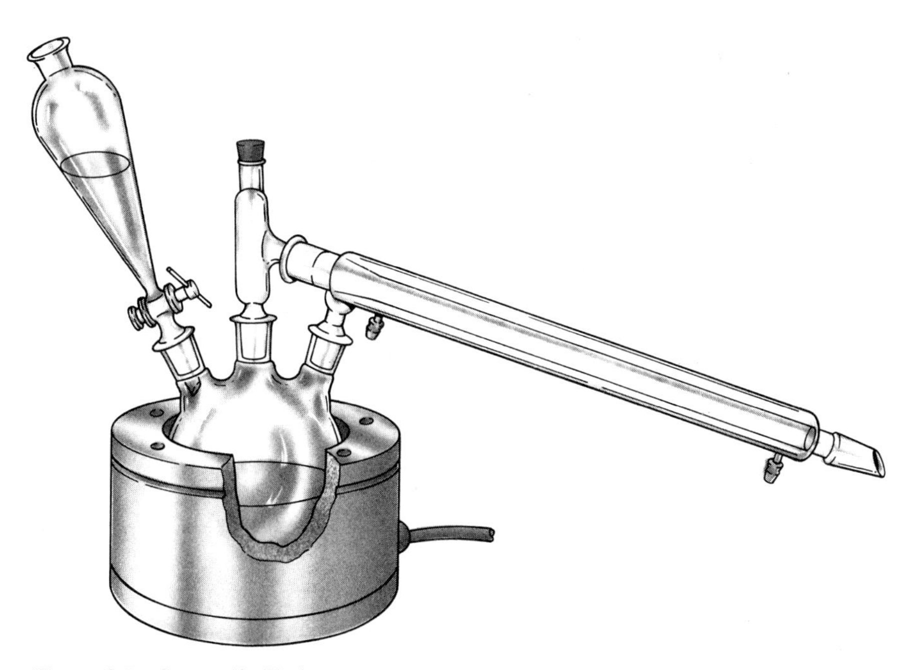

Figure 8.2 Steam distillation setup.

ESSENTIAL OILS

The characteristic aromas of plants are due to the volatile or essential oils,* which have been used since antiquity as a source of fragrances and flavorings. These oils occur in all living parts of the plant; they are often concentrated in twigs, flowers, and seeds. Essential oils are generally complex mixtures of hydrocarbons, alcohols, and carbonyl compounds, mostly belonging to the broad group of plant products known as terpenes.

Some terpenes, such as the hydrocarbon limonene, are found in many different plant species. Certain other compounds, particularly aromatic aldehydes and phenols, occur as a major constituent in only one or a few plants. Thus, the characteristic aromas of clove and cinnamon oils, for example, are due mainly to one compound. However, the full flavor and bouquet of a spice or flavoring usually depend on a blend of several substances and cannot be reproduced by a single compound.

These essential oils are isolated from the plant tissue by steam distillation, sometimes using a very primitive apparatus in remote parts of the world. Individual compounds can be separated from the essential oil by fractional distillation at reduced pressure, crystallization, or chromatography. In this experiment, the essential oil will be steam-distilled on a macroscale from one of two widely used spices—caraway and clove—and the major constituent will be characterized as a derivative. A further experiment describes the isolation of the essential oils of clove on a microscale.

Experiments

A. ISOLATION OF THE ESSENTIAL OILS OF CARAWAY OR CLOVE (Macroscale)

SAFETY NOTE SINCE STEAM IS BEING GENERATED, CARE SHOULD BE TAKEN TO AVOID BURNS.

Place 10 g of the ground spice (either caraway or clove) in a 500-mL two- or three-neck flask,† add 150 mL of water, and mark the water level with a grease pencil. Assemble the setup as in Figure 8.2 with a 125-mL Erlenmeyer flask as the receiver. Fill the dropping funnel with water. Add a boiling stone and heat the mixture until steady distillation begins. Then, add water from the funnel at a rate to maintain the original level. Collect

*The term "essential" here does not mean "necessary," but derives from the word "essence."

†If a two- or three-neck flask is not available, a one-neck flask fitted with a Claisen adapter works equally as well.

the distillate until no further droplets of oil can be seen; a minimum of 100 mL should be distilled and collected.

Pour the distillate into a separatory funnel, rinse the receiver with 5 mL of methylene chloride, and add this to the funnel. Shake the funnel, allow the layers to separate completely, and drain the lower organic layer into a tared 25 × 100-mL test tube or 25-mL Erlenmeyer flask. Extract the aqueous layer again, using 3 to 4 mL of methylene chloride, and drain the lower layer carefully into the test tube or flask. If the layer separations have been made cleanly, drying of the solution is unnecessary. If there is a significant amount of water present, add a small amount of sodium sulfate, swirl, and transfer the solution with a pipet and bulb through a cotton wad in a funnel into another tared test tube or flask. Evaporate the methylene chloride on the steam bath in a hood until the solution has been concentrated to an oily residue; then, tilt the warm tube or flask with the open end down and rotate to spread the oil on the wall and allow the heavy solvent vapor to flow out.

Weigh the oil and calculate the percent yield based on the weight of plant material. Continue with either Part C or Part D.

B. ISOLATION OF THE ESSENTIAL OIL OF CARAWAY (Microscale)

Place 2.0 g of ground caraway in a 100-mL two-neck flask* and add 50 mL of water. Assemble the apparatus as shown in Figure 8.2 with a 50-mL Erlenmeyer flask as the receiver. Fill the dropping funnel with water and mark the water level in the two-neck flask with a grease pencil, as you will want to maintain the level roughly constant. Heat the mixture until steady distillation begins, then add water at a rate to maintain the original level. Collect the distillate until the distillate is no longer cloudy; a minimum of 50 mL should be collected.

Pour the distillate into a separatory funnel, rinse the receiver with 2 mL of methylene chloride, and add this to the funnel. Shake the funnel, allow the layers to separate completely, and drain the lower organic layer into a tared 25 × 100-mL test tube or 10-mL Erlenmeyer flask. Extract the aqueous layer again, using 2 mL of methylene chloride, and drain the lower layer carefully into the same test tube or flask. If the layer separations have been made cleanly, drying of the solution is unnecessary. Should there be a significant amount of water present, add a small amount of sodium sulfate, swirl, and filter the solution with a pipet and bulb through another

*If a two- or three-neck flask is not available, a one-neck flask fitted with a Claisen adapter works equally as well.

pipet containing a cotton wad into another tared test tube or flask. Carefully, evaporate the methylene chloride on the steam bath in a hood until the solution has been concentrated to an oily residue; then, tilt the warm tube or flask with the open end down and rotate, to spread the oil on the wall and allow the heavy solvent vapor to flow out.

Weigh the oil and calculate the percent yield based on the weight of plant material. Continue with the preparation of the hydrazone, which is described in Part C, Microscale.

C. CARAWAY OIL

Caraway seed, which is actually a dried fruit, is obtained from the biennial herb *Carum carvi*, grown mainly in Holland. The essential oil of caraway, varying from 1 to 3% of the seed weight, contains two principal compounds, carvone and limonene. Both compounds contain the 1-methyl-4-isopropyl-cyclohexane skeleton, which is the most common structural unit among monoterpenes. Carvone is the major constituent in good-quality caraway oil and is the dominant note in the aroma of seed oil. The carvone can be isolated almost quantitatively as the 2,4-dinitrophenylhydrazone, and the carvone content of the essential oil can be estimated from the weight of the derivative.

carvone + 2,4-dinitrophenylhydrazine → carvone 2,4-dinitrophenylhydrazone, mp 193–194°C

limonene

SAFETY NOTE CARE SHOULD BE TAKEN TO AVOID CONTACT OF THE 2,4-DINITROPHENYLHYDRAZINE REAGENT WITH THE SKIN. IF SKIN CONTACT SHOULD OCCUR, IMMEDIATELY WASH WITH COPIOUS AMOUNTS OF WATER.

(Macroscale)

To prepare the hydrazone, dilute the oil with 2 mL of ethanol and add 1 mL of 2,4-dinitrophenylhydrazine (DNPH) reagent.* Collect the solid on a Hirsch funnel, wash it with a small amount of ethanol, dry the crystals, and determine the melting point. If the yield of oil was insufficient to permit accurate weighing of the oil, weigh the hydrazone and, assuming 100% conversion of carvone to the derivative, calculate the weight and percentage of carvone in the essential oil.

(Microscale)

To prepare the hydrazone, dilute the oil with 0.5 mL of ethanol and add 0.25 mL of 2,4-dinitrophenylhydrazine reagent. Collect the solid on a Hirsch funnel, wash it with a small amount of ethanol, dry the product, and determine the melting point. As in the macroscale experiment, if accurate weighing of the oil was not possible, weigh the hydrazone, assume 100% conversion of the carvone to the derivative, and calculate the weight and percentage of carvone in the essential oil.

D. CLOVE OIL (Macroscale)

Oil of cloves (*Eugenia caryophyllata*) is rich in 4-allyl-2-methoxyphenol (eugenol); other compounds, including eugenol acetate and the sesquiterpene caryophyllene, are present in very small amounts. For characterization, the eugenol can be converted to the benzoate ester.

SAFETY NOTE AVOID BREATHING BENZOYL CHLORIDE VAPORS AND AVOID SKIN CONTACT. SHOULD THE LATTER OCCUR, WASH WITH COPIOUS AMOUNTS OF WATER. THE ETHER USED IN THE EXTRACTION IS HIGHLY FLAMMABLE.

To prepare the benzoate, place about 0.2 mL of clove oil in a centrifuge tube or small Erlenmeyer flask and add 1 mL of water. Add 1 *M* KOH or NaOH solution drop by drop until the oily layer dissolves; a completely clear solution will not result, but there should be no oily droplets. To this solution, add 4 to 5 drops of benzoyl chloride (use a hood); an excess of the acid chloride should be avoided. Stopper and shake the mixture and then warm *unstoppered* on the steam bath for approximately 5 minutes. Cool, add 2 to 3 mL of ether, and mix the layers by shaking. With a transfer pipet and bulb, remove the lower aqueous layer from the tube or flask. Add a small amount of sodium sulfate to dry the ether solution, and transfer

---

*2,4-Dinitrophenylhydrazine reagent is prepared by dissolving 1.0 g of 2,4-dinitrophenylhydrazine in 5 mL of conc H_2SO_4 and adding this solution to a mixture of 8 mL of water and 25 mL of ethanol.

the solution through a pipet containing a cotton plug to another test tube or flask. Add 1 mL of methanol, evaporate the solution to about 0.5-mL volume, cool, and scratch to crystallize the ester. Collect and dry the crystals and determine the melting point.

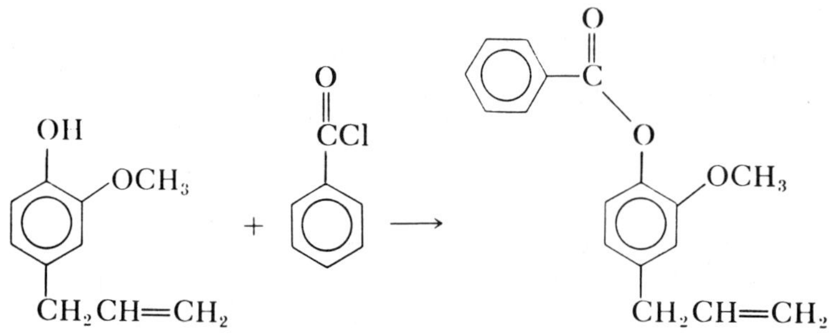

OH
OCH_3
$CH_2CH{=}CH_2$
eugenol

+

O
CCl
benzoyl chloride

⟶

O
C
O
OCH_3
$CH_2CH{=}CH_2$
eugenol benzoate
mp 70°C

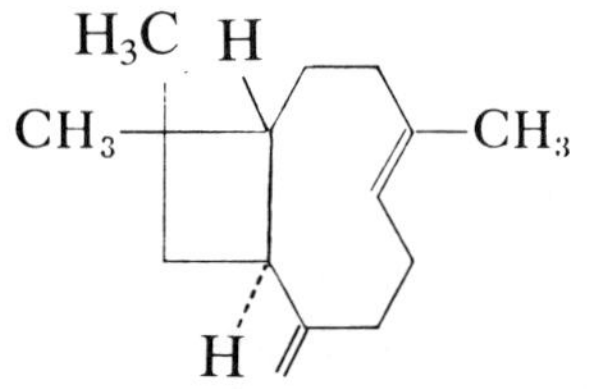

H_3C H
CH_3
CH_3
H
caryophyllene

QUESTIONS

1. Could methylene chloride (or a higher-boiling organic compound with similar solvent properties) be used instead of water for codistillation of essential oils from ground seeds? Explain your answer.
2. The vapor pressures of the essential oils described in this chapter are about 10 to 15 mm in the temperature range of 96 to 99°C. Assume a value of 12 mm for the vapor pressure of the essential oil and atmospheric pressure of 750 mm, and then calculate the amount of water theoretically needed to steam-distill 1 g of the essential oil that you isolated. Assume for purposes of calculation that each oil contains only the compound whose derivative you prepared.

REFERENCE

Guenther, E. *The Essential Oils*; Van Nostrand: New York, 1949–1952; 6 vols.

Name ______________________ Section ____________ Date ________

8 Steam Distillation of Essential Oils

PRELABORATORY QUESTIONS

1. Define the term "essential oil."

2. When one has a mixture of *non*–mutually soluble components, what is the relationship of the total pressure (P_T) in relation to the partial vapor pressures (P_A and P_B) of the components?

3. When is it desirable to separate components by steam distillation rather than by normal distillation?

4. Is the boiling point of a steam distillation involving two immiscible components always less than 100°C? Explain.

5. Isoprene has the structure, $CH_2{=}C(CH_3){-}CH{=}CH_2$, the skeletal structure of which forms the building block for many naturally occurring substances called terpenes. This is called the "isoprene rule" and carvone is such a terpene.

a. Identify the location of the skeletal isoprene units in carvone by circling the appropriate carbon atoms associated with each unit.

b. Does caryophyllene conform to the isoprene rule? Explain.

Steam Distillation of Essential Oils

PRELABORATORY QUESTIONS

9

Infrared Absorption Spectroscopy

Infrared absorption spectroscopy is the measurement of the amount of radiation absorbed by compounds within the infrared region of the electromagnetic spectrum. The infrared region consists of radiation with wavelengths between 2.5 and 15 μm (1 μm = 1 micrometer or micron = 10^{-6} m), which corresponds to frequencies between 4000 and 650 cm^{-1} (cm^{-1} = "reciprocal centimeters" or wave numbers). An infrared spectrometer measures the amount of radiation absorbed as a function of the frequency of the radiation and provides absorption spectra such as that shown in Figures 9.2 through 9.8. Spectrometers can be constructed with a chart drive that is linear in either frequency or wavelength; both types are in common use. The wavelength and frequency scales have a reciprocal relationship, and conversion from one to the other is very simple:

$$\bar{\nu}(\text{cm}^{-1}) = \frac{1}{\lambda(\text{cm})} = \frac{10{,}000}{\lambda(\mu\text{m})}$$

and

$$\lambda(\mu\text{m}) = \frac{10{,}000}{\bar{\nu}(\text{cm}^{-1})}$$

When electromagnetic radiation in the infrared region is absorbed by molecules, the molecules are excited to a higher energy state. Infrared radiation absorption causes energy changes on the order of 2 to 10 kcal/mole, and the excited state is one involving a greater amplitude of molecular vibration. Each absorption band or peak in the spectrum corresponds to the excitation of a different type (mode) of vibration of the atoms, and the positions of these peaks (measured in μm or cm^{-1}) provide useful information about the structure of the molecule.

For simple molecules containing few atoms, the number and position of peaks in the infrared spectrum can be calculated, and the spectrum can

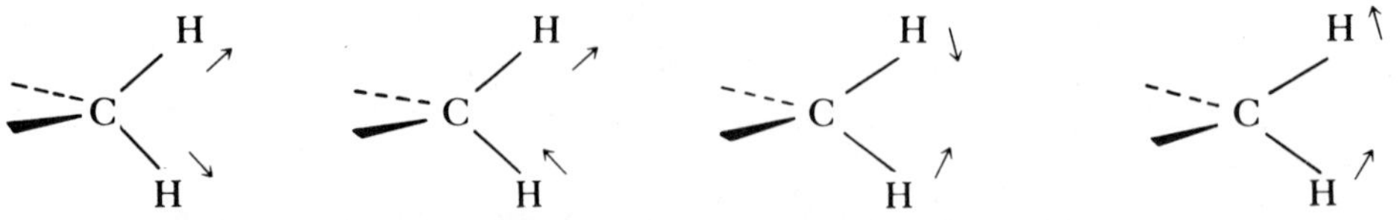

Symmetric stretching ($\sim$2850 cm^{-1}) | Asymmetric stretching ($\sim$2925 cm^{-1}) | Scissor bending ($\sim$1450 cm^{-1}) | Rocking motion ($\sim$750 cm^{-1})

Figure 9.1 Normal modes of CH_2 vibration.

be completely analyzed. The major use of infrared spectra in organic chemistry, however, depends on empirical correlations of band positions with structural units. These correlations have been derived from spectra of a large number of compounds of known structure. By the use of this information, the presence or absence of certain functional groups or other structural features of a new compound can be determined.

Two types of vibrations, stretching and bending, are responsible for most of the peaks of importance in the identification of organic compounds. A few of the types of vibrations observed are shown in Figure 9.1, using CH_2 as a typical group. Also shown are the approximate frequencies of the radiation absorbed in exciting each type of vibration. When expressed in cycles per second (Hz), these are also the vibrational frequencies of the indicated motions. The vibrational frequency, and thus the frequency of radiation absorbed, is determined by the force constant for the deformation (i.e., the rigidity or strength of the bond) and the masses of the atoms that are involved in the vibration; specifically, the stronger the bond and the lighter the atoms, the higher is the vibrational frequency. Bending motions are easier than stretching motions, so the former absorb at lower frequencies. These trends are illustrated in Table 9.1, which lists the types of vibrations that appear in various regions of the infrared spectrum.

Peaks in the first four regions listed in Table 9.1 are largely due to vibrations of specific types of bonds, and these are by far the most useful in compound classification. A few of the peaks appearing below 1500 cm^{-1} are characteristic of certain functional groups, but most of the absorption bands in this region are associated with vibrations of larger groups or the molecule as a whole. An infrared spectrum can be divided into two portions for examination. The region 4000 to 1500 cm^{-1} is useful for the identification of various functional groups. The region from 1500 to 600 cm^{-1}, sometimes called the "fingerprint region," is quite complex yet represents a unique pattern for each organic compound and, so, is useful for comparing two compounds for identification.

Fourier transform infrared spectroscopy (FTIR) is now being widely used. In this method, the electromagnetic radiation is split into two beams and one is made to travel a longer path than the other. Recombination of

Table 9.1 Absorption Frequencies and Types of Vibrations

FREQUENCY (CM^{-1})	VIBRATION
3600–2700	X—H single bond stretching: O—H, N—H, C—H
3300–2500	Hydrogen-bonded O—H--X stretching
	Ammonium ion $\diagdown\overset{+}{N}\text{—H}$ stretching (with —N and $\diagup$ bonds)
2400–2000	Triple bond and cumulated double bond stretching: C≡C, C≡N, N=C=O, C=C=O, N=C=N
1850–1550	Double bond stretching: C=O, C=N, C=C
1600–650	Single bond bending: NH_2, CH_3, C—C—C
	Single bond stretching: C—C, C—O, C—N

the two beams creates an interference pattern or **interferogram**. Mathematical manipulation (Fourier transformation) of this interference pattern by a computer converts it into the usual infrared (IR) spectrum. The advantages of FTIR are that the whole spectrum is measured in a few seconds and high resolution is obtained.

STRUCTURAL GROUP ANALYSIS

Extensive correlations exist between absorption peak positions and structural units of organic molecules; the most useful of these are summarized in Table 9.2. In spectra to be presented later, the vertical scale is percent transmittance, 100% being the top of the spectrum and 0% the bottom. Thus, absorption of radiation at a certain frequency results in a decrease in transmittance and appears as a dip in the curve (Figs. 9.2 through 9.8).

For the more common structural units shown in Table 9.2, more specific assignments and deductions may often be made. Some of these are discussed in the following sections. For even more specific correlations and for similar data on the other functional groups, any of several books on this subject can be consulted (see references).

ALKANES

The most prominent peaks in infrared spectra of saturated hydrocarbons and from saturated portions of more complicated organic compounds are those due to C—H stretching and bending (see Fig. 9.2). The stretching frequencies are in the regions 3000 to 2800 cm^{-1} and are usually rather weak. Since most organic compounds contain several CH_3, CH_2, and/or

Table 9.2 Structural Units and Absorption Frequencies

BOND		TYPE OF COMPOUND	FREQUENCY (CM^{-1})
—C—H (with three single bonds on C)	(stretch)	Alkane	2800–3000
=C—H	(stretch)	Alkenes, aromatics	3000–3100
≡C—H	(stretch)	Alkynes	3300
—O—H	(stretch)	Alcohols, phenols	3600–3650 (free) 3200–3500 (H-bonded) (broad)
—OH	(stretch)	Carboxylic acids	2500–3300
—N—H	(stretch)	Amines	3300–3500 (doublet for NH_2)
O=C—H (aldehyde C–H)	(stretch)	Aldehyde	2720 and 2820
—C=C—	(stretch)	Alkenes	1600–1680
—C=C	(stretch)	Aromatic	1500 and 1600
—C≡C—	(stretch)	Alkynes	2100–2270
—C(=O)—	(stretch)	Aldehydes, ketones	1680–1740
—C≡N	(stretch)	Nitriles	2220–2260
C—N	(stretch)	Amines	1180–1360
—C—H	(bending)	Alkane	1375 (methyl)
—C—H	(bending)	Alkane	1460 (methyl and methylene)
—C—H	(bending)	Alkane	1370 and 1385 (isopropyl split)
—C—H	(bending)	R—CH=CH_2	1000–960 and 940–900
—C—H	(bending)	R_2C=CH_2	915–870
—C—H	(bending)	*cis* RCH=CHR	790–650
—C—H	(bending)	*trans* RCH=CHR	990–940
—C—H	(out of plane bending)	*mono* subst. benzene	770–730 and 710–690
—C—H	(out of plane bending)	*o*-subst. benzene	770–735
—C—H	(out of plane bending)	*m*-subst. benzene	810–750 and 710–690
—C—H	(out of plane bending)	*p*-subst. benzene	860–800
—C—O	(stretch)	Primary alcohol	1050
—C—O	(stretch)	Secondary alcohol	1100
—C—O	(stretch)	Tertiary alcohol	1150
—C—O	(stretch)	Phenol	1200

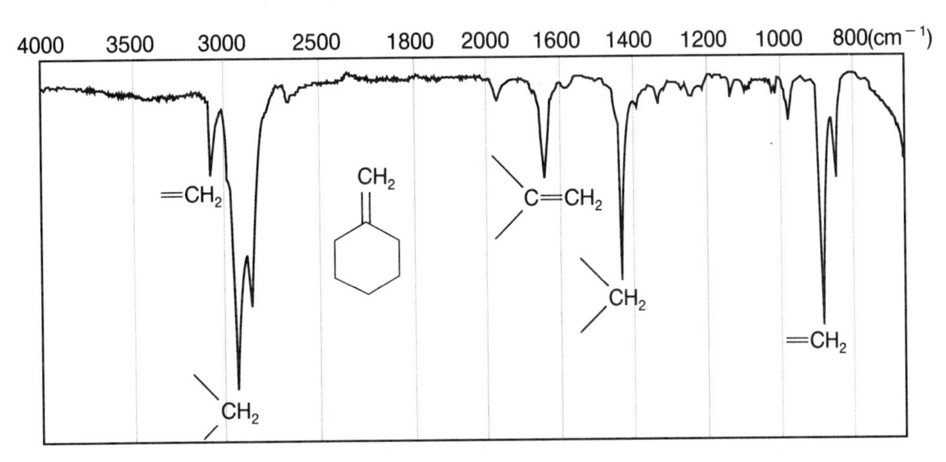

Figure 9.2 Infrared spectrum of an alkene.

CH groupings, the presence or absence of peaks in this region is simply taken to indicate the presence or absence of aliphatic C—H bonds in the molecule. The major bending modes of CH_2 and CH_3 groups appear in the 1470 to 1420 cm^{-1} and 1340 to 1380 cm^{-1} regions, but interpretation is usually complicated by the presence of several bands of this type or additional bands from other sources. The usually large number of different C—C bonds in an organic molecule makes the C—C stretching vibrations in the 1300 to 1100 cm^{-1} region uninterpretable in most cases.

ALKENES

Olefinic C—H stretching peaks generally appear in the region 3100 to 3000 cm^{-1}, thus differentiating between saturated and unsaturated hydrocarbons. Of the C—H bending modes, the out-of-plane vibrations in the 1000 to 650 cm^{-1} region are often useful in predicting the substitution pattern of the double bond. This is illustrated in Table 9.2. The C═C stretching frequency in the 1600 to 1675 cm^{-1} region also varies with substitution, but to a lesser degree. The out-of-plane bending and C═C stretching absorptions of an alkene are illustrated in Figure 9.2.

ALKYNES

The C—H stretching vibration of terminal acetylenes generally appears at about 3300 cm^{-1} as a strong sharp band. The C≡C stretching band is found in the region 2150 to 2100 cm^{-1} if the alkyne is monosubstituted (Fig. 9.3) and at 2270 to 2150 cm^{-1} if disubstituted. The latter are usually quite weak absorptions (the partial peak at 1601 cm^{-1} is a frequency marker).

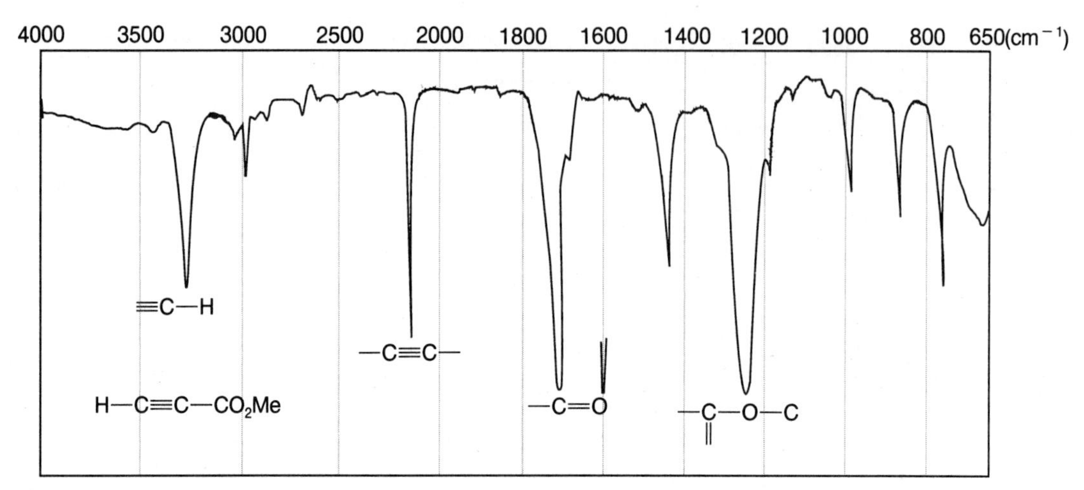

Figure 9.3 Infrared spectrum of an alkyne.

AROMATIC RINGS

Aromatic C—H stretching absorption appears in the region 3100 to 3000 cm^{-1}. This fact, taken with the corresponding frequencies of aliphatic and olefinic C—H stretching bands, allows a reliable determination of the types of carbon-bound hydrogen in the molecule. Aromatic C—H out-of-piane bending bands in the 900 to 690 cm^{-1} region are reasonably well determined by the substitution pattern of the benzene ring, as indicated in Table 9.2. In the absence of other interfering absorptions, such as those from nitro groups, these strong, usually sharp bands can be used in distinguishing positional isomers of substituted benzenes. Sharp peaks at approximately 1600 and approximately 1500 cm^{-1} are very characteristic of all benzenoid compounds; a band at 1580 cm^{-1} appears when the ring is conjugated with a substituent (Fig. 9.4).

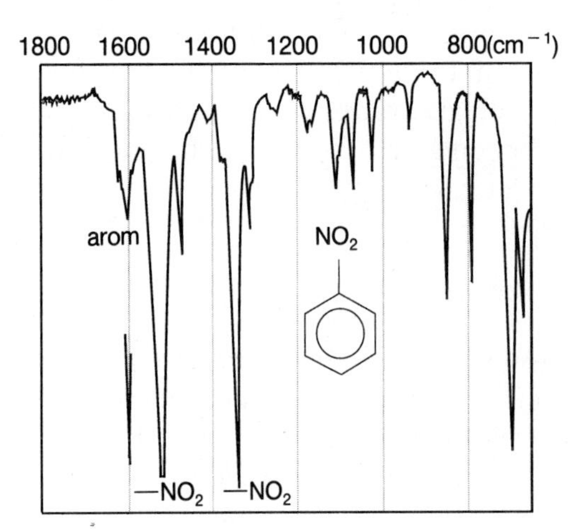

Figure 9.4 Infrared spectrum of an aromatic compound.

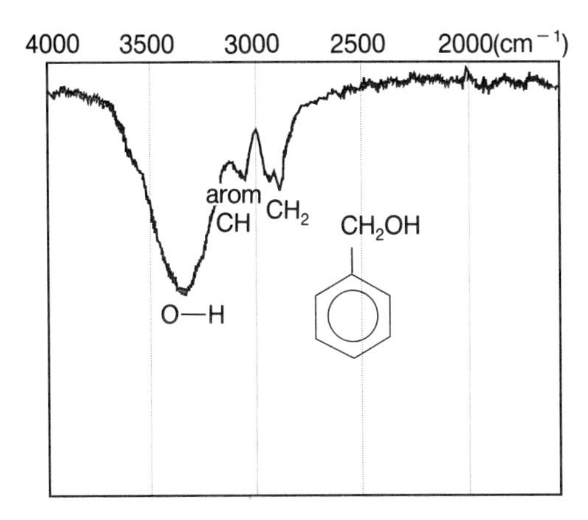

Figure 9.5 Infrared spectrum of an alcohol.

ALCOHOLS, PHENOLS, AND ENOLS

The very characteristic infrared band due to O—H stretching appears at 3650 to 3600 cm^{-1} in dilute solutions. In spectra of neat (undiluted) liquids or solids, intermolecular hydrogen bonding broadens the band and shifts its position to lower frequency (3500 to 3200 cm^{-1}) (Fig. 9.5). Intramolecular hydrogen bonding (to C=O, —NO_2 groups), as in enols, lowers the frequency and broadens the absorption even more. Strong bands due to O—H bending and C—O stretching are observed at 1500 to 1300 cm^{-1} and 1220 to 1000 cm^{-1}, respectively. In simple alcohols and phenols, the exact position of the latter is useful in classification of the hydroxyl group (see Table 9.2). Within the range given, typical frequencies observed are phenols>tertiary>secondary>primary. Further branching and unsaturation also affect the frequency somewhat.

ALDEHYDES AND KETONES

The C=O stretching frequencies of saturated aldehydes and acyclic ketones are observed at 1735 to 1710 cm^{-1} and 1720 to 1700 cm^{-1}, respectively. Adjacent unsaturation lowers the frequency by 25 to 50 cm^{-1}. Thus, aryl aldehydes generally absorb at 1700 to 1690 cm^{-1} and diaryl ketones at 1670 to 1660 cm^{-1}. Intramolecular hydrogen bonding to the carbonyl oxygen also lowers the frequency by 25 to 50 cm^{-1}. Aldehydes are also recognizable by the C—H stretching vibration, which appears as two peaks in the 2850 to 2700 cm^{-1} region (Fig. 9.6). The former may be obscured by aliphatic C—H stretching, but the latter is usually quite prominent. Cyclic ketones (four- or five-membered rings) absorb at higher frequencies (1780 and 1745 cm^{-1}, respectively).

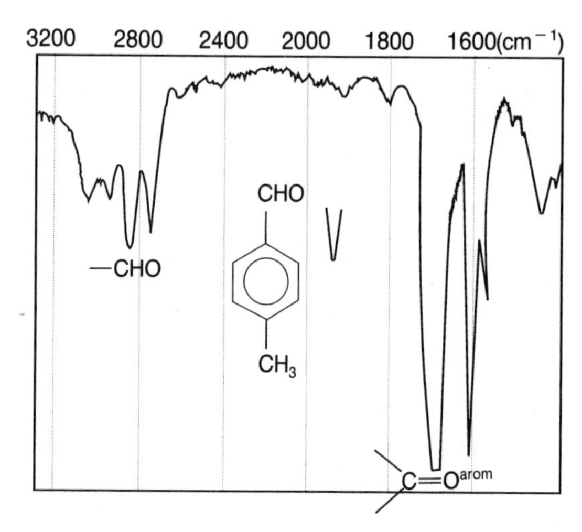

Figure 9.6 Infrared spectrum of an aldehyde.

CARBOXYLIC ACIDS

The most characteristic absorption of carboxylic acids is a broad peak extending from 3300 to 2500 cm^{-1} due largely to hydrogen-bonded O—H stretching (Fig. 9.7). The C—H stretching vibrations appear as small peaks on top of this band. The carbonyl group of aliphatic acids appears at 1730 to 1700 cm^{-1} and is shifted to 1720 to 1680 cm^{-1} by adjacent unsaturation.

CARBOXYLIC ESTERS AND LACTONES

Saturated ester carbonyl stretching is observed at 1740 to 1720 cm^{-1}. Unsaturation adjacent to the carbonyl group lowers the frequency by 10 to 15 cm^{-1} (see Fig. 9.3), whereas unsaturation adjacent to the oxygen (enol and phenol esters) increases the frequency by 20 to 30 cm^{-1}. The C—O—C stretching of esters appears as two bands in the 1280 to 1050 cm^{-1} region. The asymmetric stretching peak at 1280 to 1150 cm^{-1} is usually strong and

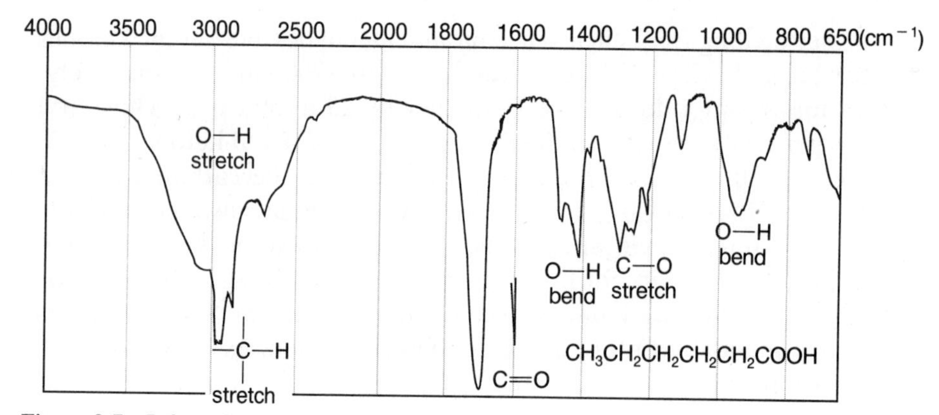

Figure 9.7 Infrared spectrum of a carboxylic acid.

varies with substitution, much as does the corresponding ether band. Cyclic esters (lactones), like cyclic ketones, absorb at higher frequencies as the ring size decreases.

ANHYDRIDES

Acid anhydrides are readily recognized by the presence of two high-frequency (1830 to 1800 cm^{-1} and 1775 to 1740 cm^{-1}) carbonyl absorptions. As with other carbonyl stretching vibrations, the frequency is increased by incorporating the group in a ring and decreased by adjacent unsaturation. Cyclic anhydrides differ from acyclic anhydrides also in that the lower-frequency band is stronger in the former, while the reverse is true of the latter.

AMIDES AND LACTAMS

Amide carbonyl stretching is observed in the 1670 to 1640 cm^{-1} region. In contrast to other carbonyl groups, both adjacent unsaturation and ring formation (lactams) cause the absorption to shift to higher frequencies. Primary and secondary amides also show N—H stretching at 3500 to 3100 cm^{-1} (see discussion of amines), and N—H bending at 1640 to 1550 cm^{-1}.

ETHERS

The asymmetric C—O stretching absorption of ethers appears in the region 1280 to 1050 cm^{-1}. As in alcohols, the exact position of this strong peak is dependent on the nature of the attached groups. Phenol and enol ethers generally absorb at 1275 to 1200 cm^{-1} and dialkyl ethers at 1150 to 1050 cm^{-1}. Epoxides have three characteristic absorptions in the 1270 to 1240, 950 to 810, and 850 to 750 cm^{-1} regions of the spectrum.

AMINES

Primary and secondary amines show N—H stretching vibrations in the 3500 to 3300 cm^{-1} region (Fig. 9.8). Primary amines generally have two bands approximately 70 cm^{-1} apart due to asymmetric and symmetric stretching modes. Secondary amines show only one band. Intermolecular or intramolecular hydrogen bonding broadens the absorptions and lowers the frequency. In general, the intensities of N—H bands are less than of O—H bands. The N—H bending and C—N stretching absorptions are not as strong as the corresponding alcohol bands and occur at approximately 100 cm^{-1} higher frequencies. In addition, NH_2 groups give an additional broad band at 900 to 700 cm^{-1} caused by out-of-plane bending.

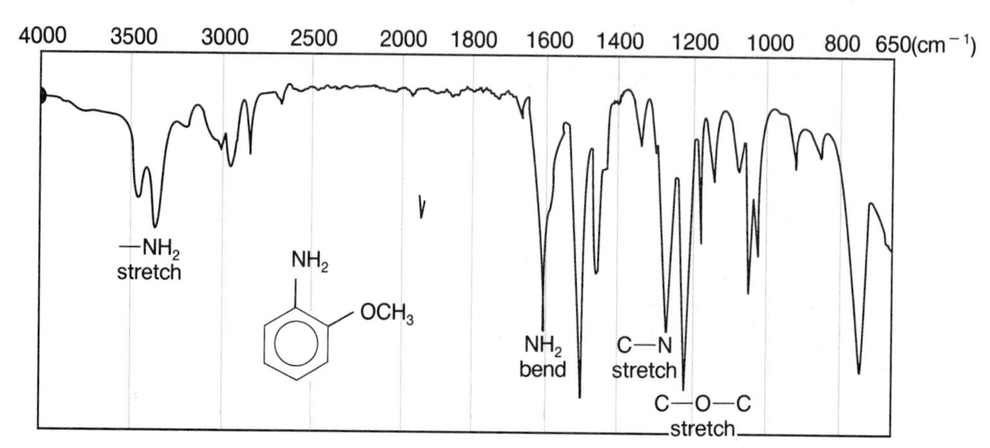

Figure 9.8 Infrared spectrum of an amine.

NITRO GROUPS

Due to the high polarity of the N—O bonds in nitro compounds, the absorptions at the N—O stretching frequencies are very strong. In nitroaromatics, the asymmetric stretching band appears at 1560 to 1520 cm^{-1}, and the symmetric stretching at 1360 to 1320 cm^{-1} (see Fig. 9.4). Since peaks due to several groups other than NO_2 appear in these regions, the presence or absence of a nitro group often cannot be determined with certainty from infrared data alone.

NITRILES

A sharp, usually strong absorption at 2260 to 2220 cm^{-1} due to C≡N stretching is very characteristic of nitriles.

SAMPLE PREPARATION

FILMS

Infrared spectroscopy is quite convenient for the identification of films such as cellophane, Saran wrap, and so forth. To do this, cut a hole (approximately 2 × 3 cm) in a 3- × 5-inch file card and tape the film over the hole in the card, taking care that none of the tape overlaps the hole. The sample is then ready for insertion in the sample beam.

LIQUIDS

The simplest method for mounting a liquid sample consists of placing a thin film of the liquid between two transparent windows. The most common material used for the windows is NaCl, which is transparent throughout the

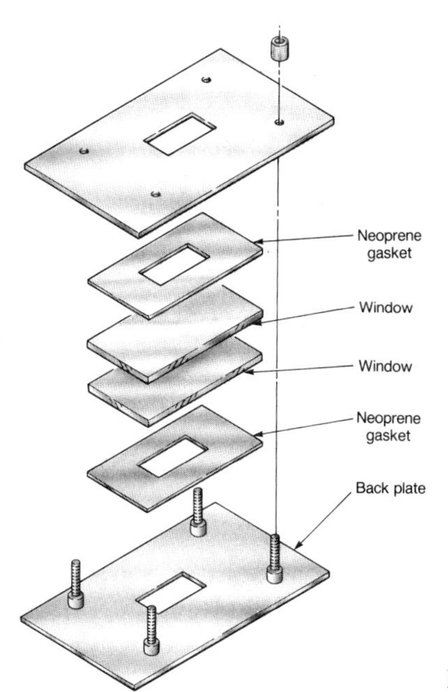

Figure 9.9 Infrared salt plate assembly.

normally used region of the infrared spectrum (10,000 to 650 cm^{-1}). Large polished single crystals are used, and it is important to remember when handling them that NaCl is water soluble. They should be picked up *only* by the edges, preferably with gloves, to avoid marring the polished surface with moisture from your fingers.

For mounting the salt plates with the liquid sample between them, the holder illustrated in Figure 9.9 may be used. The bottom metal plate is placed on a flat surface, and one of the rubber gaskets is placed around the opening in the plate. This serves to cushion the relatively fragile salt plate, which is placed on top of it. A drop of the liquid compound is placed in the center of the lower window, and the second salt plate is placed carefully on top, spreading the drop into a thin film. The other rubber gasket and the face plate are then added to the top of the "sandwich," and the entire stack is held together by the thumb nuts on the threaded studs as shown. All four nuts should be tightened firmly but not excessively. The assembly is placed in the holder provided on the instrument.

After obtaining the spectrum, disassemble the cell and rinse the windows well with a dry, volatile solvent (CH_2Cl_2, hexane, or the like). Let the solvent evaporate, and store them in a desiccator to protect them from atmospheric moisture.

As alternatives to the cell illustrated in Figure 9.9, smaller circular salt plates may be used and held together in threaded holders that simply screw together or, if the liquid is somewhat viscous, the two plates will stick together by themselves and can be set in the sample beam without mounting brackets.

SOLIDS

Solutions

Sealed cells are available for volatile samples or solutions of compounds in volatile solvents. Solution spectra in nonpolar liquids are useful in minimizing intermolecular association of polar groups in the molecule.

An approximately 10% solution of the compound is prepared using a solvent such as chloroform, and the solution is placed in the sealed cell through an injection port using a syringe. Another cell is filled with pure solvent and placed in the reference beam. In this manner the absorption of the solvent in the reference beam and in the sample will approximately cancel each other.

KBr Pellet

As an alternative to measuring the spectrum of a liquid solution of a solid compound, a solid solution or dispersion in KBr is usually more convenient. Approximately 1 to 2 mg of the compound is mixed with 50 to 100 mg of dry KBr powder. The mixing must be complete and vigorous enough to reduce the particle size below which it will diffract or scatter the light (less than 2 microns). This can be accomplished by grinding the mixture with an agate mortar and pestle or by using the pounding action of a miniature ball mill. The latter technique will be described here.

Weigh the specified quantities of material into the plastic vial provided and add a glass ball that fits loosely into the vial. Place the stoppered vial in the shaker and mix for 30 seconds at full speed. Remove the ball from the vial and loosen the powder from the sides of the vial by tapping or by scraping with a spatula.

The powder is formed into a transparent pellet by pressure in a small die. One type of die that may be used consists of a threaded metal block and two bolts with polished end surfaces (Fig. 9.10). To use the die, screw one of the bolts in five or six turns, so that one or two threads remain showing. Pour the powder into the open end and distribute it evenly over the end of the bolt by tapping gently. Screw the other bolt down on top of the sample and tighten the bolts securely with the wrenches provided (use of a torque wrench is recommended). Let the die stand for about 1 minute, during which time the KBr flows to fill the empty spaces between the original particles.

Loosen the bolts with the wrenches and remove them, leaving the KBr as a window in the middle of the die. If the pellet is very cloudy, either the compound was not ground well or the bolts were not tightened enough, and a poor spectrum will result due to light scattering. Slight scattering due to a translucent pellet can be partially compensated for with an attenuating device in the reference beam of the instrument. In the absence of a com-

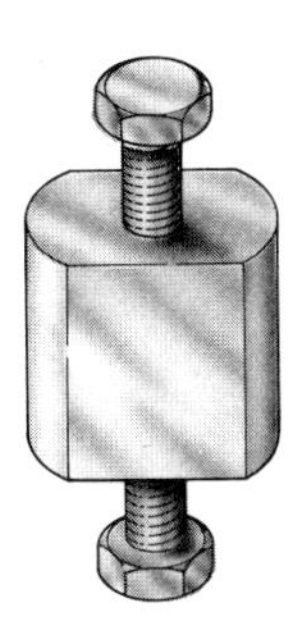

Figure 9.10 Pellet press.

mercial attenuator, a piece of wire screen works adequately to balance the lowered transmittance of the pellet.

Nujol Mull

The infrared spectra of solid compounds can also be measured in a Nujol mull, prepared by grinding 5 mg of the solid to a fine dispersion in a drop of Nujol, which is a purified grade of mineral oil. The mixture is then placed between salt plates and the spectrum of the thin film is measured. Since Nujol consists of a mixture of saturated hydrocarbons, absorption bands will be present at 2850 to 3000, 1460, and 1375 cm^{-1} in all Nujol mull spectra.

Experiments

Precautions

1. No aqueous solution or other material that may dissolve the sodium chloride optics should be brought near the instrument.
2. The sample holders (except for film holders) are sodium chloride crystals and should be treated with extreme care! Avoid using water or hydroxylic solvents because these dissolve sodium chloride, and avoid mechanical shock because the salt crystals break easily.
3. Do not operate the instrument without prior detailed instructions. You may inadvertently cause considerable damage.
4. Keep the instrument clean.

SAFETY NOTE NOTHING IN THIS EXPERIMENT IS PARTICULARLY HAZARDOUS; HOWEVER, USUAL PRECAUTIONS SHOULD BE TAKEN TO AVOID EXCESSIVE EXPOSURE TO VAPORS OF SOLVENTS AND/OR SAMPLES. CHLOROFORM AND CARBON TETRACHLORIDE, SOMETIMES USED AS SOLVENTS, ARE SUSPECTED CARCINOGENS.

Procedure

1. A demonstration of the use of IR spectrophotometers will be given. Students are reminded to sign the log book when using instruments and to promptly label each spectrum (date, sample identity, instrument number, etc.) for future use.
   - **A.** Determine the infrared spectrum of polystyrene film. The 1601 cm^{-1} band is often used as a frequency marker (e.g., see Figs. 9.3 and 9.7); note particularly its location on your spectrum (it is important to correctly position the chart paper before recording a spectrum). Label this spectrum and secure it in your laboratory notebook.
   - **B.** Determine the infrared spectrum of paraffin oil (Nujol). Note the simplicity of the spectrum and the location of the absorption bands near 3000 cm^{-1} (above or below 3000 cm^{-1}?). Use this spectrum for future reference should you have occasion to measure a spectrum in a Nujol mull.
   - **C.** Determine the infrared spectrum of cyclohexanol. Examine this spectrum and identify the absorption bands due to the O—H stretch, the C—O stretch, and the C—H stretching.
   - **D.** Determine the spectrum of ethyl benzoate,

$$CH_3CH_2—O—\overset{\overset{\displaystyle O}{\|}}{C}—C_6H_5.$$

     Examine this spectrum and note the location of the absorption band due to the

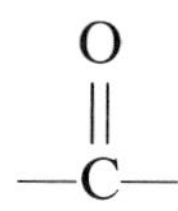

     stretch. Does absorption occur above and below 3000 cm^{-1}? Note the absence of absorption above 3100 cm^{-1} (O—H stretch). Attempt to identify the C—O—C stretching bands and as many of the aromatic absorption bands as you can. Label the spectrum and secure it in your laboratory notebook.
   - **E.** You will be given an unknown to characterize. It will be either an alkane, aromatic hydrocarbon, alcohol, or ketone. Assign the major peaks in your spectrum and conclude from your data the nature of your sample.

QUESTIONS

1. Show the following conversions to the units requested.

2.5 microns = ____cm^{-1}

1700 cm^{-1} = ____μm

2. Give absorption frequency ranges (in cm^{-1}) for each of the following.

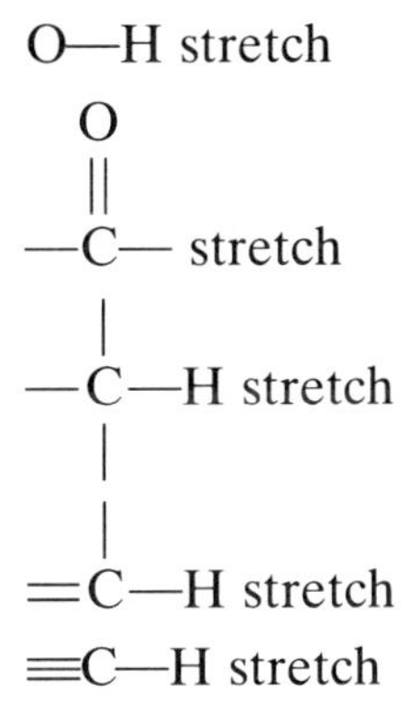

3. How could you distinguish between cyclohexane and cyclohexene using IR?
4. In general, how could you identify a compound as an alkane, alkene, alkyne, or arene using IR?
5. An unknown has the following physical data.
   Elemental analysis: 60.0%C, 13.3%H, 26.7%O
   Molecular weight = 60
   IR: 3400, 2950, 1460, 1385, 1365, 1100 cm^{-1}

   Draw three structures consistent with the analysis and molecular weight data. Which one is consistent with the IR data? Assign absorptions to support your answer.
6. Figures 9.11 to 9.13 are the infrared spectra of three compounds: A($C_{10}H_{12}O$), B($C_{10}H_{12}O_2$), and C($C_{10}H_{12}O$). Interpret the spectra for functional groups, and if possible suggest the structure of each compound. The NMR spectra of these compounds are given in the Questions at the end of Chapter 10 (see Figs. 10.12–10.14).

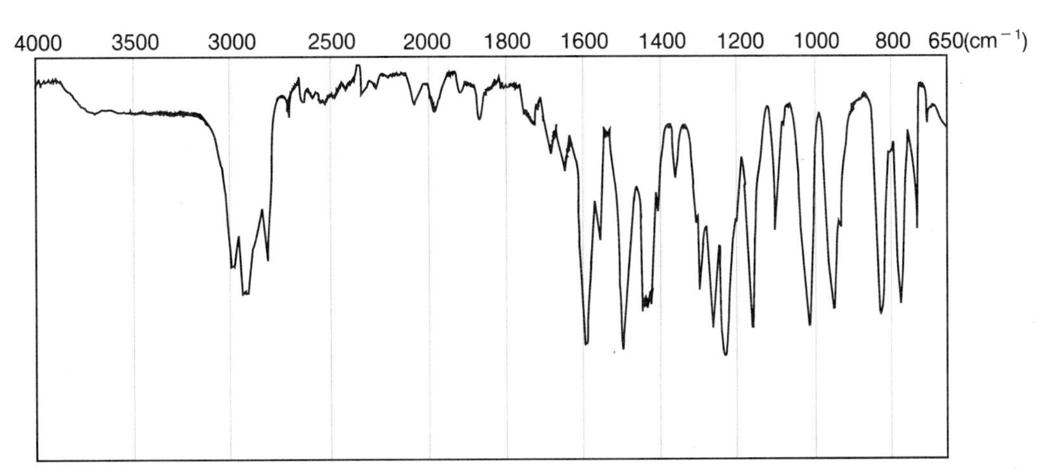

Figure 9.11 Compound A.

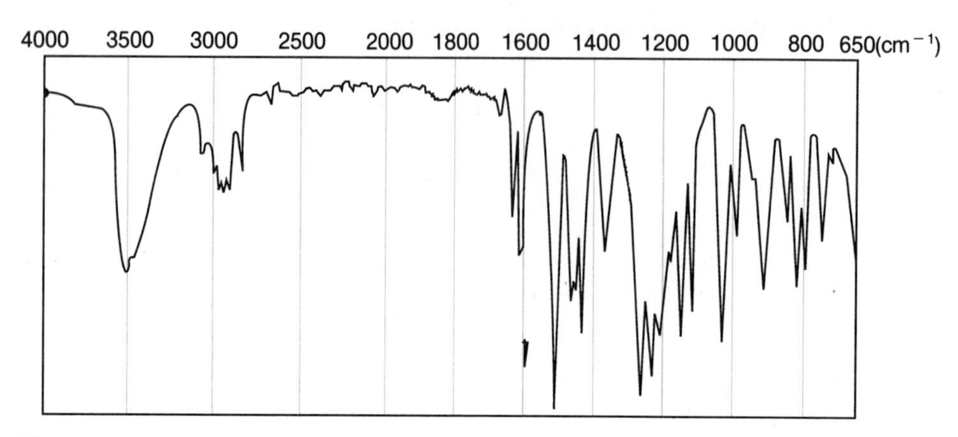

Figure 9.12 Compound B.

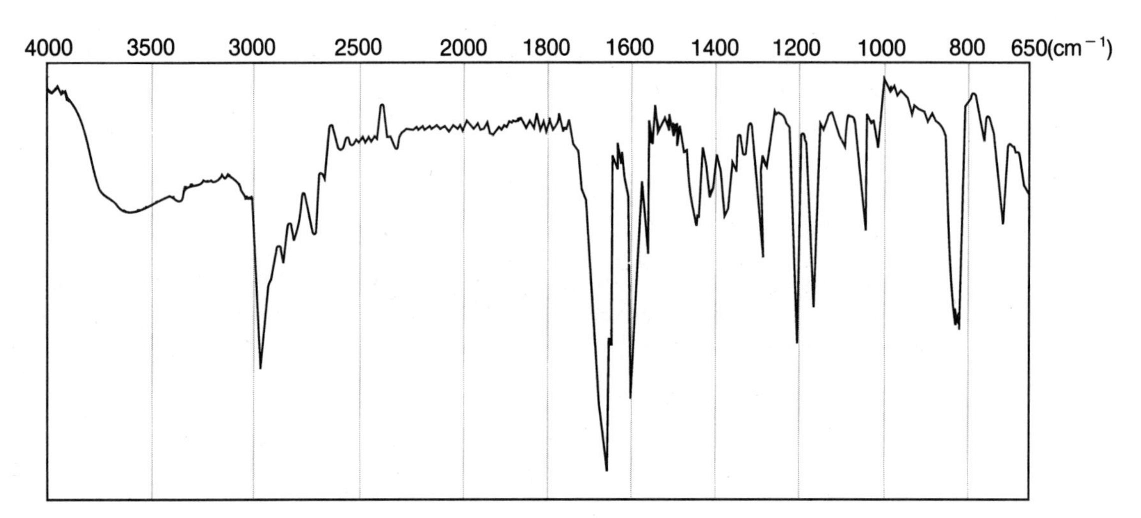

Figure 9.13 Compound C.

REFERENCES

Anderson, D. H.; Woodall, N. B.; West, W. *Techniques of Organic Chemistry*, 3rd ed.; Weissberger, A., Ed.; Interscience: New York, 1959; Vol. I, pt. III.

Cooper, J. W. *Spectroscopic Techniques for Organic Chemists*; Wiley: New York, 1980.

Nicolet IR Spectral Search and Retrieval Program for Aldrich Library of FT-IR Spectra; Nicolet Instrument Corp.: Madison, WI, 1985.

Pouchert, C. J.; *The Aldrich Library of FT-IR Spectra*, 1st ed.; Aldrich Chemical Co., Inc.: Milwaukee, WI, 1985.

Silverstein, R. M.; Bassler, G. C.; Morrill, T. C. *Spectrometric Identification of Organic Compounds*, 4th ed.; Wiley: New York, 1981.

Williams, D. H.; Fleming, I. *Spectroscopic Methods in Organic Chemistry*, 4th ed.; McGraw-Hill: London, 1987.

Name ______________________ Section ______________ Date ________

9 Infrared Absorption Spectroscopy

PRELABORATORY QUESTIONS

1. Why should aqueous solutions not be used in infrared cells?

2. Describe the procedure to prepare a liquid sample for infrared examination.

3. What is Nujol? What is a Nujol mull?

4. What is a "KBr pellet"? Why not just dissolve a solid sample in a suitable solvent to take its IR spectrum?

5. How could you tell if an unknown is an alcohol by examining its IR spectrum?

10

Nuclear Magnetic Resonance Spectroscopy

Many atomic nuclei have the property of nuclear spin. When placed between the poles of a magnet, the axis of rotation of the spinning nuclei precess around the axis of the field (Fig. 10.1). This precession can be detected by irradiation with energy in the radiofrequency region of the spectrum. When the precession and irradiation frequencies are equal, the system is said to be in **resonance**. The resonance frequency is a function of the magnetic field strength and of the magnetic moment of the specific nucleus. For protons, resonance is routinely measured at 60 or 90 megahertz (MHz) in a magnetic field of 1.4092 tesla (14.092 kilogauss) and 2.1138 tesla, respectively; other combinations of frequency and field are also used (e.g., 100, 300 and even 600 MHz). Listed in Table 10.1 are some of the more commonly observed nuclei and the corresponding frequencies required for resonance. The most useful nuclei for organic chemistry are, quite reasonably, hydrogen and carbon.

The value of nuclear magnetic resonance (NMR) spectroscopy lies not in simple qualitative or quantitative observation of the presence or absence

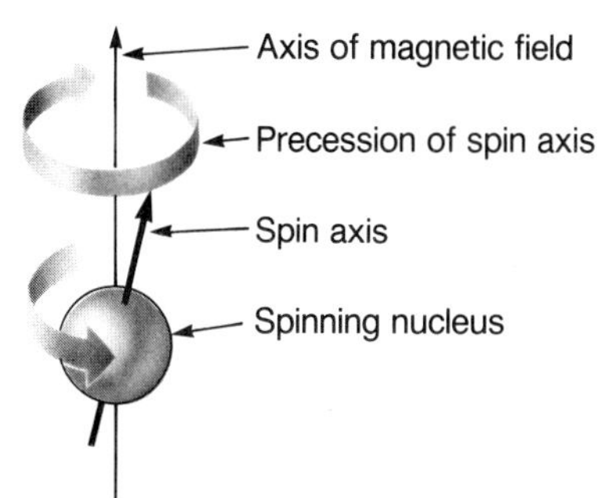

Figure 10.1 Behavior of a nucleus with spin in a magnetic field.

Table 10.1 Nuclei Commonly Observed by NMR Spectroscopy

NUCLEUS	RESONANCE FREQUENCY (MHz) *at 1.41 T*	*at 2.35 T*
^{1}H	60.0	100.0
^{13}C	15.1	25.2
^{15}N	6.1	10.1
^{19}F	56.5	94.0
^{31}P	24.2	40.5

of a specific element, but in the fact that the *effective* magnetic field at a given nucleus in a compound depends on the electron distribution around it and thus the chemical environment of the nucleus. Therefore, when an organic compound is placed in an NMR spectrometer in which the magnetic field strength or the irradiating frequency can be varied, a spectrum of resonance signals is observed, one for each different type of hydrogen in the molecule. A similar sweep at approximately one fourth the proton frequency will give a spectrum of signals, one for each type of carbon atom in the molecule.

Because of the significant differences in the interpretation of proton and carbon NMR spectra, they will be discussed separately, but each complements the other in determining the structure of an organic compound.

PROTON NMR SPECTROSCOPY (PMR)

Protons are by far the most commonly observed nuclei for several reasons. Among these are the relative ease of observation and the fact that the overwhelming majority of chemical compounds contain hydrogen. Equally important from the point of view of utility is that the proton spectrum of a compound provides three basic kinds of information useful in characterizing or identifying the compound:

1. The signal strength, or peak area, as measured by electronic integration, is directly proportional to the number of identical protons in the sample that produce the signal.
2. The position of the peak in the spectrum, or chemical shift, is determined by the atom or structural grouping to which the proton is bonded.
3. Because of coupling with the magnetic fields due to spins of neighboring protons, a signal may be split into several peaks. The number (multiplicity) and separation of these peaks is characteristic of the number and steric relationship of these nearby protons.

These three factors can be seen in the spectrum of ethyl phenylacetate

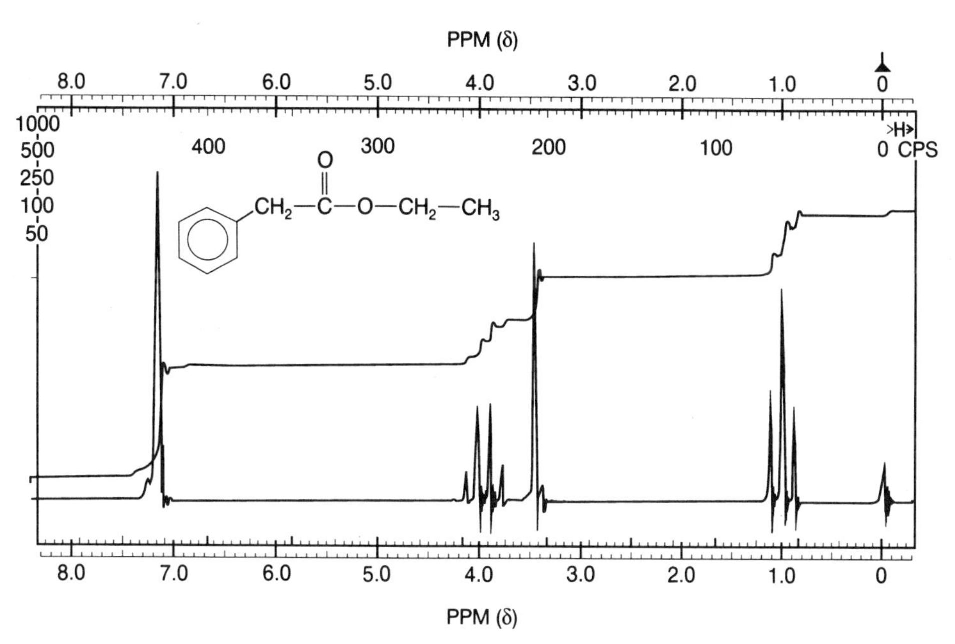

Figure 10.2 NMR spectrum of ethyl phenylacetate.

(Fig. 10.2). The four main steps in the integration line, from left to right, have relative heights of 5:2:2:3, indicating the number of protons of each type. The positions of the four signals, from left to right, correspond to C_6H_5, —OCH_2—, Ph—CH_2—CO, and —CH_3 protons. Finally, the splitting of the —OCH_2— and —CH_3 signals into a quartet (four peaks) and a triplet (three peaks) indicates that each group is adjacent to the other in the molecule. Each of these three properties of a proton NMR spectrum is described in more detail below.

PEAK AREAS

As previously indicated, the relative peak areas are given by the changes in height of the integration curve. These may be obtained by counting lines if the spectrum is recorded on ruled paper or by measuring with a ruler. From the measured heights, derive the smallest set of integers that are in the same ratio. Thus the integration steps seen in Figure 10.2 are, from left to right, 60, 24, 24, and 36 mm on the original spectrum. Dividing each by 12 provides the integer ratios 5:2:2:3, as mentioned. The error in electronic integration is often 5 to 10%, so ratios such as 1.1:2 and 2.8:5 should be rounded off to 1:2 and 3:5, respectively. With multiplets or overlapping signals, the area of the entire grouping is taken together.

If several signals are present, it is often possible to make a fairly good estimate of the total number of protons in the compound, as was done in Figure 10.2. It should be recognized, however, that any multiple of a ratio may be correct; that is, two signals in a ratio of 1:2 may correspond to a total of 3, 6, 9, . . . protons in the compound.

CHEMICAL SHIFTS

The chemical shift of a peak (or "line") in a spectrum is related to the environment of the proton(s) causing the signal. If the proton is strongly **shielded** by electrons around it, screening it from the external magnetic field, it appears at **high field** (to the right side of the spectrum). Conversely, if adjacent atoms withdraw electrons from around the proton, it is **deshielded**, and its signal appears at lower field. Because of special magnetic effects (**anisotropy**), protons attached to unsaturated atoms are additionally deshielded and also appear at lower fields.

The chemical shift of a signal is expressed in terms of its position relative to the peak due to tetramethylsilane [TMS; $(CH_3)_4Si$]. A small amount of TMS is usually added to the sample as an internal standard when the spectrum is run. The TMS protons appear at very high field, well separated from the signals of the protons in common compounds. On the scale that we will use, the TMS protons are assigned a value of $\delta = 0$, and chemical shifts of other protons are expressed in parts per million (ppm) downfield from the TMS signal. The chemical shift can also be expressed in cycles per second (cps or Hz); for a 60-MHz instrument, the value in cps is given by $\delta(\text{ppm}) \times 60$. Both scales generally appear on the chart paper used for recording NMR spectra. For chemical shift positions, the ppm scale is virtually always used because this parameter is independent of the frequency and so is always the same regardless of the instrument being used. On the other hand, coupling constants are always given in Hz because they do not change with instruments of different frequency.

Aliphatic Protons

Listed in Table 10.2 are typical values of the chemical shift for aliphatic protons in various environments. Several useful generalizations can be drawn from the data in these tables:

1. The protons in methyl groups (RCH_3) appear at higher field (lower δ value, further to the right in the spectrum) than those in methylene groups (R_2CH_2); methine protons (R_3CH) appear at still lower field. For example, the three different types of protons in 2,4-dimethylpentane have the following chemical shift values:

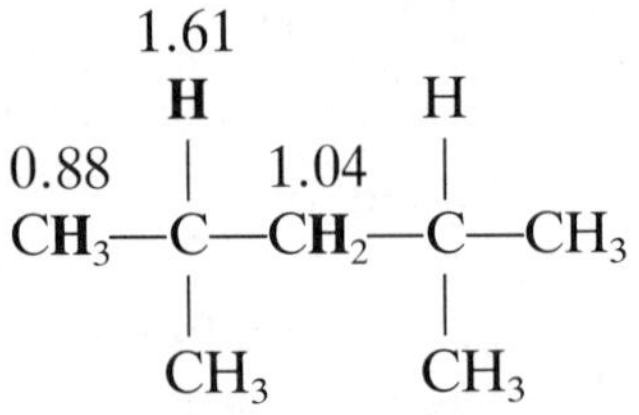

```
        1.61
         H        H
  0.88   |  1.04  |
  CH3—C—CH2—C—CH3
         |        |
        CH3      CH3
```

2. For protons on carbon attached to an electronegative atom or group X, the chemical shift (δ value) increases with the electronegativity of X.

Table 10.2 Chemical Shifts of Aliphatic Protons*

PROTON	δ(PPM)	PROTON	δ(PPM)
$\mathbf{CH}_4$	0.23		
Methyl Protons		*Methylene Protons*	
$R{-}\mathbf{CH}_3$	0.8–1.2	$R\mathbf{CH}_2R$	1.0–1.5
$R'CH{=}CH\mathbf{CH}_3$	1.6–1.9	$Ar\mathbf{CH}_2R$	2.5–2.9
$Ar\mathbf{CH}_3$	2.2–2.5	$R\overset{\overset{O}{\|}}{C}\mathbf{CH}_2R$	2.3–2.7
$R\overset{\overset{O}{\|}}{C}\mathbf{CH}_3$	2.1–2.4	$R\mathbf{CH}_2OR'$	3.4–3.8
$Ar\overset{\overset{O}{\|}}{C}\mathbf{CH}_3$	2.4–2.6	$R\mathbf{CH}_2OAr$	3.9–4.3
$R'O\overset{\overset{O}{\|}}{C}\mathbf{CH}_3$	1.9–2.2	$R\mathbf{CH}_2O\overset{\overset{O}{\|}}{C}R$	3.7–4.1
$ArO\overset{\overset{O}{\|}}{C}\mathbf{CH}_3$	2.0–2.5	$R{-}\mathbf{CH}_2NR'_2$	2.5–2.8
		$R{-}\mathbf{CH}_2Br$	3.4–3.6
$R''_2N\overset{\overset{O}{\|}}{C}\mathbf{CH}_3$	1.8–2.2		
		$R{-}\mathbf{CH}_2Cl$	3.6–3.8
$R'_2N{-}\mathbf{CH}_3$	2.2–2.6		
		$\mathbf{CH}_2Cl_2$	5.33
$ArNH{-}\mathbf{CH}_3$	2.8–3.1		
		Methine Protons	
$R''\overset{\overset{O}{\|}}{C}NH\mathbf{CH}_3$	2.8–3.0	$R_3\mathbf{CH}$	1.4–2.0
		$Ar\mathbf{CH}R_2$	2.8–3.2
$ArO\mathbf{CH}_3$	3.7–4.0	$R_2\mathbf{CH}OH$	3.5–4.0
		$\mathbf{CH}Cl_3$	7.27
$R''CO_2\mathbf{CH}_3$	3.6–4.0		
		Miscellaneous	
$R'O\mathbf{CH}_3$	3.2–3.5	$\mathbf{CH}_3CN$	2.0
		$\mathbf{CH}_3NO_2$	4.3
$R''OCH_2\mathbf{CH}_3$	1.2–1.4		
		$\mathbf{CH}_3I$	2.15
$R''OCH_2CH_2\mathbf{CH}_3$	0.9–1.1	$\mathbf{CH}_3Br$	2.7
		$\mathbf{CH}_3Cl$	3.0
$R''OCH_2CH_2CH_2\mathbf{CH}_3$	0.9–1.1	$\mathbf{CH}_3F$	4.3

*In this table R = alkyl; R′ = alkyl or H; Ar = aryl; R″ = alkyl, aryl, or H.

Table 10.3 Effective Shielding Constants in X—CH_2—Y*

GROUP	Δδ	GROUP	Δδ
CH_3—	0.65†	HO—	2.56
$R_2C{=}CR$—	1.32	RO—	2.36
C_6H_5—	1.85	ArO—	3.23
RC(=O)—	1.70	RC(=O)O—	3.13
ROC(=O)—	1.55	R_2N—	1.57

*Dailey, B. P.; Shoolery, J. M. *J. Am. Chem. Soc.* **1955,** *77*, 3977.
†A value of 0.47 was originally suggested (see Dailey and Schoolery) for the CH_3 group. The value for a methyl or alkyl group varies from 0.40 to 0.80, depending on the second substituent.

This occurs because of the inductive effect on the shielding of the protons and is apparent in the methyl halides given at the end of Table 10.2.

3. The inductive (deshielding) effect of a substituent on a proton decreases as the separation between the proton and substituent is increased. This is illustrated by the last four entries in the first column of Table 10.2. Specific examples of this attenuation of the deshielding can be seen in the δ values of the protons in $CH_3CH_2CH_2$—X compounds:

	CH_3—	CH_2—	CH_2—X
X = NO_2	1.03	2.07	4.38
X = OH	0.92	1.57	3.58
X = CHO	0.97	1.67	2.42

The effect of two deshielding substituents on the chemical shifts of methylene protons is cumulative, although not exactly additive. Estimates of the chemical shift in a group X—CH_2—Y can be made by adding the "effective shielding constants" Δδ for X and Y (Table 10.3) to 0.23, which is the chemical shift for the protons in methane. For example, the chemical shift for the benzylic CH_2 group in ethyl phenylacetate is calculated to be: $0.23 + 1.85 + 1.55 = 3.63$ ppm; the observed value (see Fig. 10.2) is 3.48.

Protons on Unsaturated Carbons

Protons on an aromatic ring appear at very low field (e.g., benzene, δ 7.27) because of the aromatic "ring current." Electron-withdrawing groups cause

Table 10.4 Chemical Shifts of Protons on Unsaturated Carbon Atoms*

PROTON	δ(PPM)	PROTON	δ(PPM)
RCH═**CH₂**	4.9–5.2	$C_6\mathbf{H_6}$	7.27
RC**H**═CH₂	5.8–6.0	Ar**H**	6.0–9.0
R₂C═**CH₂**	4.5–5.1	R—C≡C**H**	2.3–2.5†
R**CH**═**CH**R	5.2–5.7	ArC≡C**H**	2.8–3.1
R₂C═**CH**R	5.1–5.5	R′₂NC(═O)**H**	7.9–8.2
ROCH═**CH₂**	4.0–5.0	R′OC(═O)**H**	8.0–8.2
ROC**H**═CH₂	6.0–7.5	RC(═O)**H**	9.4–9.9
RC(═O)CH═**CH**R′	6.8–7.2	ArC(═O)**H**	9.7–10.3
RC(═O)**CH**═CHR′	5.7–6.2		

*Cf. footnote Table 10.2.
†Exception: Propyne 1.80 ppm.

downfield shifts relative to benzene, and electron-releasing groups cause upfield shifts. In addition, substituents with significant (diamagnetic) anisotropy (e.g., —NO_2 and —C═O) deshield the *ortho* protons even more, often as much as 0.5 to 1.0 ppm. Representative values for three compounds are shown below (also see Fig. 10.4, page 155).

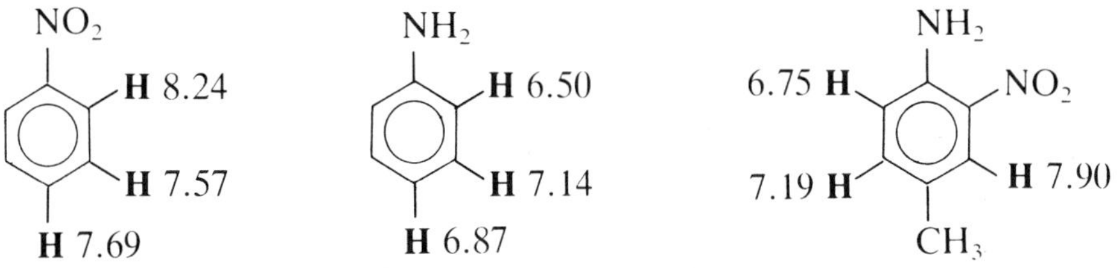

NO_2: **H** 8.24, **H** 7.57, **H** 7.69

NH_2: **H** 6.50, **H** 7.14, **H** 6.87

NH_2, NO_2, CH_3: 6.75 **H**, 7.19 **H**, **H** 7.90

Although the protons on an aromatic ring may not be identical, occasionally near coincidence of chemical shifts can cause the aromatic protons to appear as a singlet. Usually, however, different proton chemical shifts and the splitting that results because of coupling lead to several lines in the aromatic proton region of the spectrum.

Inductive and anisotropic effects on shielding also affect the chemical shifts of protons on other sp^2 hybridized carbons, for example,

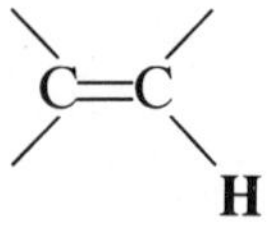

$$\mathrm{>C{=}C<^{H}} \quad \text{and} \quad \mathrm{O{=}C<^{H}}$$

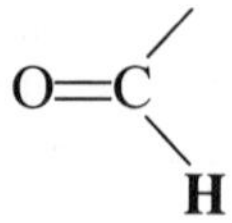

(seen in Table 10.4). This is particularly noticeable in the latter case (formyl protons) where the combined deshielding shifts the resonance to δ 8 to 10 ppm.

Protons Attached to Other Nuclei

Protons attached to atoms other than carbon (i.e., O, N, S) vary widely in chemical shifts, and their position and appearance in the spectrum depend markedly upon the temperature, solvent, concentration, and the presence of acidic or basic impurities. The ranges of chemical shifts shown in Table 10.5 are for 5 to 50% solutions in nonpolar solvents, that is, the concentrations at which they are normally measured. Amide N—**H** protons often appear as broad signals that may be difficult to recognize. Amines and alcohols usually give **OH** and **NH** proton signals that are singlets regardless of the number of adjacent protons. This is caused by the chemical exchange of protons, which occurs at a rate greater than the frequency (observation time) of the NMR instrument. Because of this exchange, protons on N, O, and S can be replaced by deuterium by shaking the sample with D_2O. The result is the disappearance of the signal for these protons from the spectrum and the appearance of a signal for **HOD**.

COUPLING CONSTANTS

The most frequently encountered spin coupling is that between protons on adjacent saturated carbon atoms, as seen with the ethyl protons shown in

Table 10.5 Chemical Shifts of Protons Bound to O, N, S

PROTON	δ(PPM)	PROTON	δ(PPM)
ROH	3–6	RNH_2	0–2.5
ArOH	6–8	$ArNH_2$	3–4.5
R′COOH	10–12	$R'CONH_2$	5.5–7.5
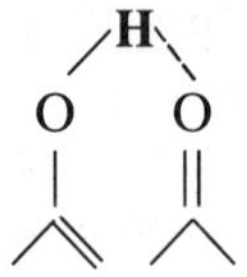 O—H···O=C (hydrogen-bonded enol/chelate)	14–16	RSH	1–2
		ArSH	3–4

Table 10.6 Representative Values of Coupling Constants

Structure	Coupling constant	Structure	Coupling constant
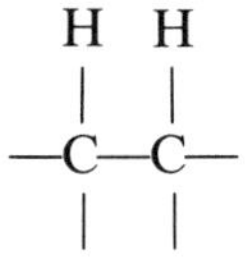 H—C—C—H (vicinal)	J = 5–8 Hz	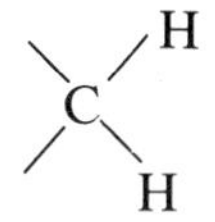 CH_2 (geminal)	J = 12–15 Hz
H—C=C—H (*cis*, *trans*, *gem*)	J_{cis} = 7–11 Hz J_{trans} = 13–17 Hz J_{gem} = 0–3 Hz	Benzene ring (*o*, *m*, *p*)	J_{ortho} = 6–9 Hz J_{meta} = 1–3 Hz J_{para} = 0–1 Hz

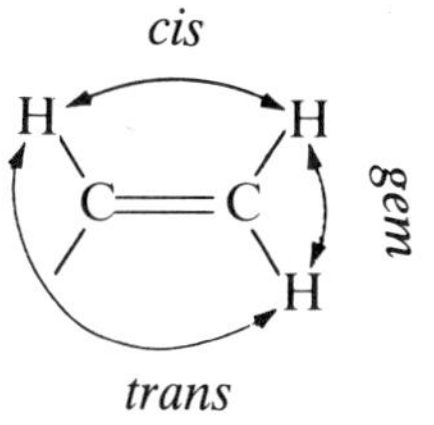

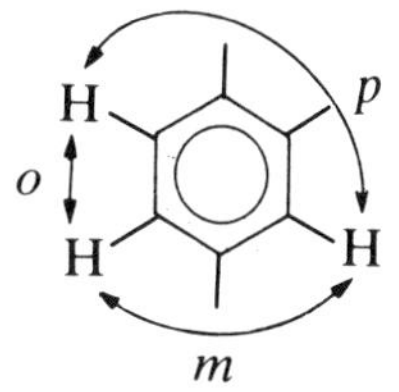

Figure 10.2. In simple systems of this type, the signal of a proton which has *n* identical adjacent protons is split into $n + 1$ peaks (lines). Thus, as seen in Figure 10.2, the OCH_2 protons which are adjacent to three identical methyl protons appear as a quartet (four lines). Similarly the CH_3 proton signal appears as a triplet because of coupling with the two adjacent methylene protons (*not* because it represents three protons). The spacing of the lines (splitting) in the triplet is exactly equal to that in the quartet, and in simple spectra, this spacing is called the **coupling** (or **splitting**) **constant *J***. Typical values of coupling constants between variously related pairs of protons are given in Table 10.6.

The lines in a multiplet are not generally the same height. If two multiplets from coupled protons are very far apart in the spectrum, their relative intensities will correspond to the binomial coefficients: a doublet, 1:1; a triplet, 1:2:1; a quartet, 1:3:3:1; and so on. If the separation (chemical shift difference) between the multiplets is not very large, the lines of each multiplet closest to the other multiplet are more intense (seen in Figs. 10.2 and 10.3). If the chemical shift difference is less than a few times larger than the coupling constant, the simple patterns of doublets, triplets, and so on are no longer observed, and very complex multiplets appear (Fig. 10.3).

Spin coupling can provide a great deal of information about the position of substituents in aromatic compounds. One important example is the symmetrical pattern seen for a *p*-disubstituted benzene, as illustrated in Figure 10.4. The pairs of protons A—A′ and B—B′ *ortho* to the two substituents have the same chemical shifts; each pair gives rise to a doublet because of coupling to an *ortho* proton, with additional small lines due to *meta* and *para* coupling.

A somewhat more complicated situation is seen in the aromatic ring protons of 4-methyl-3-nitroaniline (Fig. 10.5). By comparison of the total

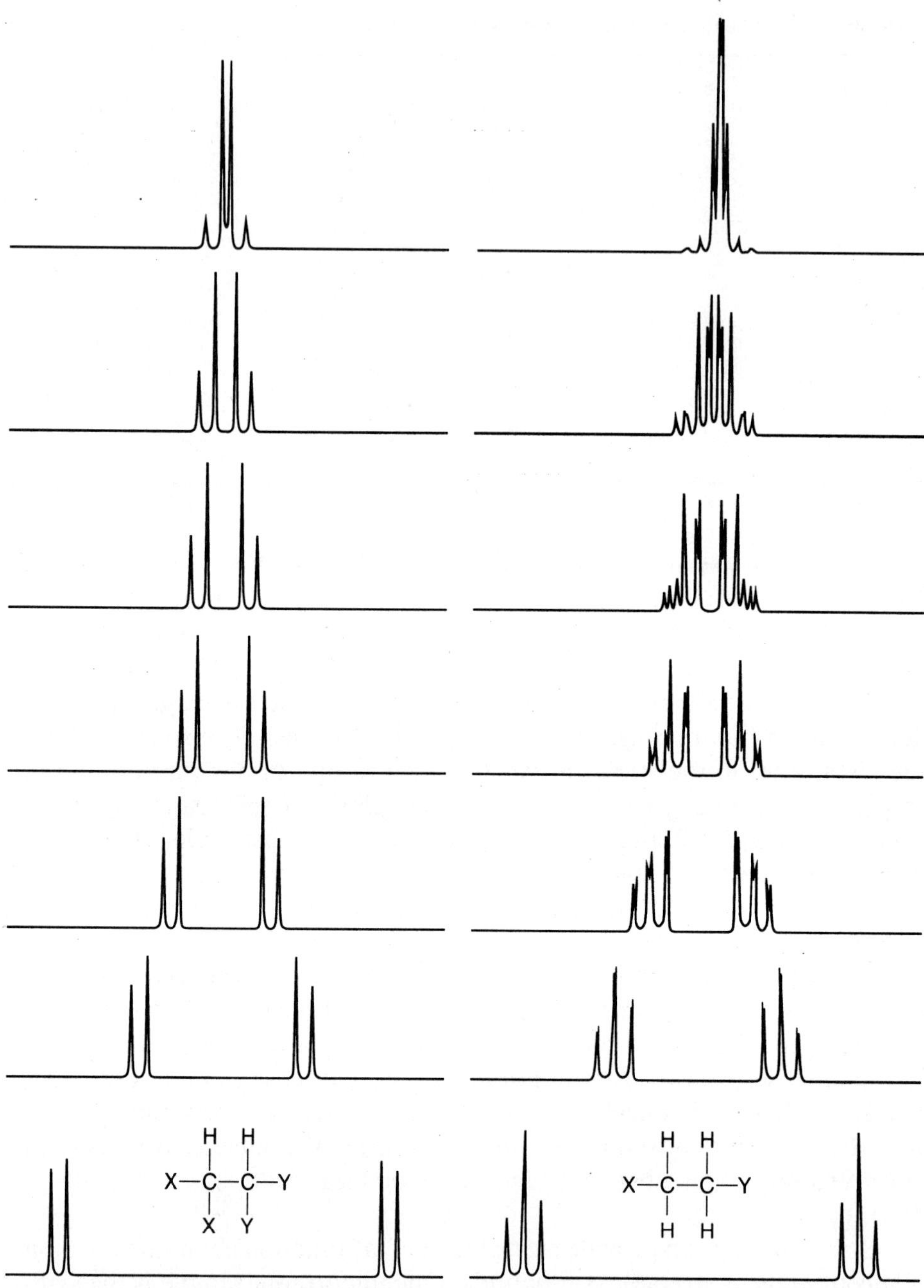

Figure 10.3 Typical splitting patterns with various ratios of chemical shift difference to coupling constant ($\Delta\nu/J$). From top to bottom, $\Delta\nu/J$ = 1, 2, 3, 4, 6, 10, and 20.

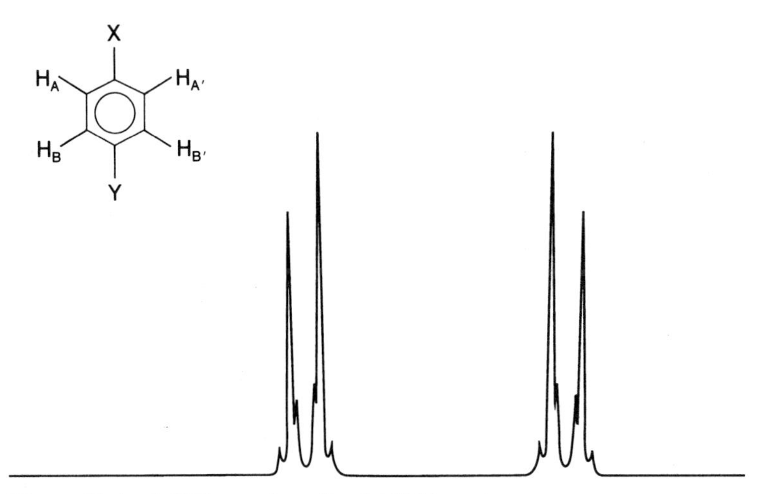

Figure 10.4 Splitting pattern for AA′BB′ protons.

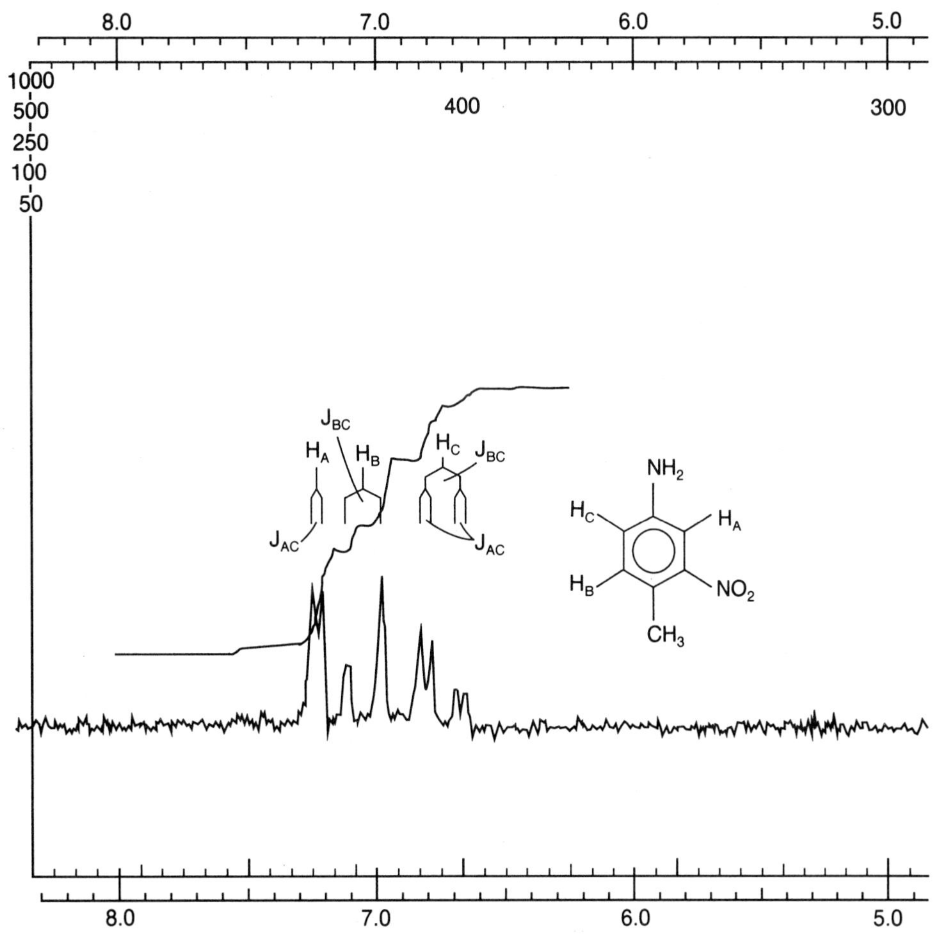

Figure 10.5 Proton NMR spectrum of 4-methyl-3-nitroaniline.

integration shown with that of the methyl group (at δ 2.42, not shown), it is found that there are three protons on the ring. The first proton (A) appears as a doublet (J = 2 Hz) at 7.23 ppm, the second (B) as a doublet (J = 7 Hz) at 7.05 ppm, and the third (C) as a doublet of doublets (J = 2 and 7 Hz) at 6.75 ppm. The magnitudes of the coupling constants (see Table 10.6) indicate that H_A is *meta* to H_C, and H_B is *ortho* to H_C; the coupling constant between the *para* protons, H_B and H_A, is approximately zero. The substituents are therefore in the 1, 2, and 4 positions of the benzene ring. H_A appears at lowest field because it is *ortho* to the deshielding NO_2 group, and H_C at highest field since it is *ortho* to the NH_2 group (see p. 151).

When several nonidentical protons are mutually coupled, very complex

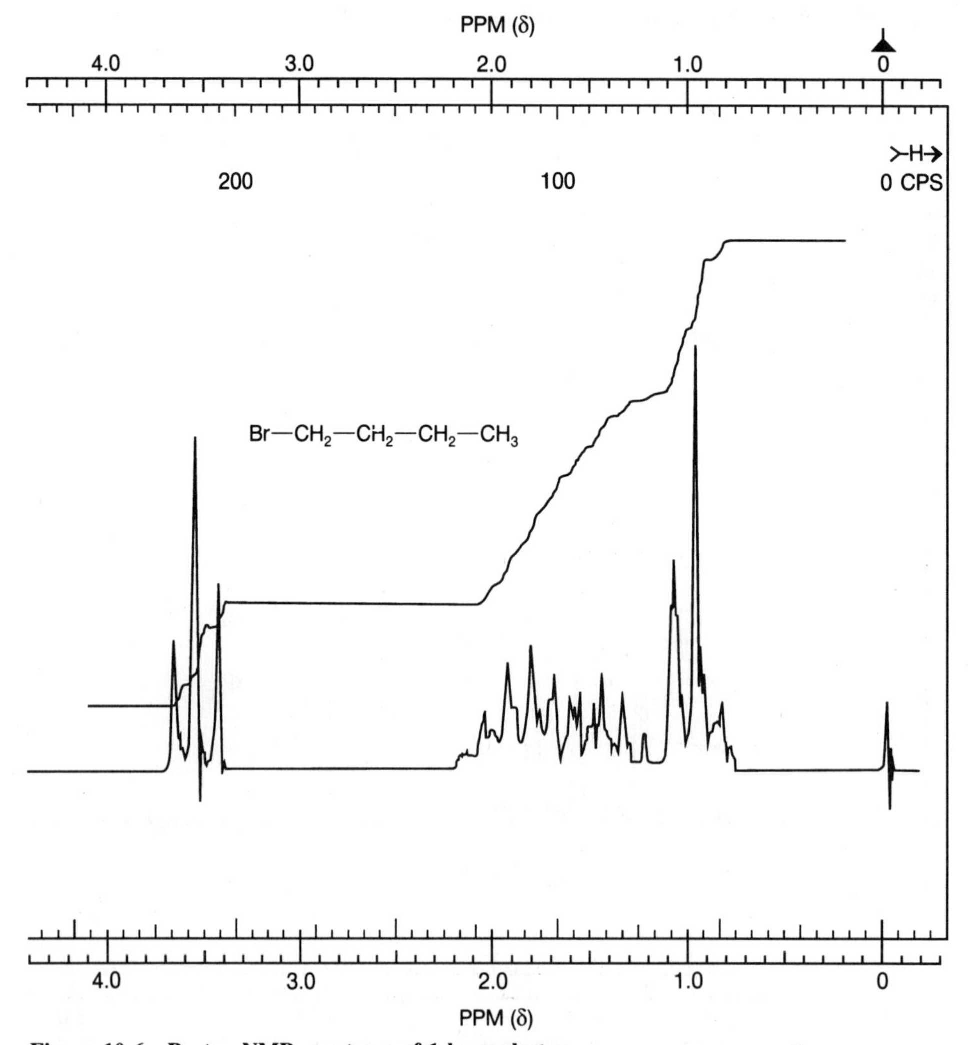

Figure 10.6 Proton NMR spectrum of 1-bromobutane.

multiplets can result, as illustrated by the spectrum of *n*-butyl bromide (Fig. 10.6). However, even in this case the peaks can be seen to fall into three groups at δ 3.7 to 3.4, 2.2 to 1.2, and 1.2 to 0.7 ppm. The integration curve suggests a ratio of 2:4:3 protons in these regions, respectively, and only the relatively deshielded $—CH_2Br$ protons are easily recognizable as a triplet, that is, coupled to an adjacent $—CH_2—$ group. The chemical shift of these protons (δ 3.53) is measured at the center of the triplet. The high field grouping of peaks (three protons) is from the $—CH_3$ protons, and this distorted triplet is typical of longer-chain alkyl groups. The central multiplet of four protons is uninterpretable owing to the extensive coupling and similarity of chemical shifts.

CARBON NMR SPECTROSCOPY (CMR)

In the measurement of carbon NMR spectra, only the minor isotope ^{13}C (1.1% of the total natural carbon) can be detected, and the magnetic moment of ^{13}C is low. For these reasons, ^{13}C resonance is nearly 6000 times more difficult to observe than proton resonance. However, with the more sensitive (and more complex) instrumentation that became available in recent years, it is practical to obtain ^{13}C NMR spectra for routine use in characterizing organic compounds. Some modern spectrometers operate at 76 MHz (7.0 tesla) and 91 MHz (8.5 tesla).

Of the three useful properties of proton NMR spectra (peak areas, chemical shifts, and splitting due to coupling of nuclei), only chemical shifts and splittings are of general utility in carbon NMR spectra. Under the conditions usually used for measuring ^{13}C spectra, the peak areas are not necessarily proportional to the numbers of identical carbons, and therefore they cannot be used to count atoms as can be done with proton spectra.

The splitting of carbon NMR signals into multiplets as a result of spin coupling with nearby protons is observed in Figure 10.7, the ^{13}C spectrum of acetone. As can be seen in the expanded-scale tracings, the carbonyl carbon signal is split into a heptet (outer lines hidden by noise) by the six protons on the methyl groups (J = 6.0 Hz). The methyl carbon signal appears as a quartet of quartets (16 lines), with the large splitting (J = 127.0 Hz) due to the three attached protons and the smaller splitting (J = 1.4 Hz, unresolved) due to the three remote protons. While such splitting can be useful in studying simple molecules, in most compounds it makes the spectrum more difficult to interpret (Fig. 10.8). For this reason, and to increase the signal strength, ^{13}C NMR spectra are normally obtained under conditions in which the protons are "decoupled" from the carbons. The means of accomplishing this will not be discussed here, but the result is a great simplification of the spectrum. Thus, as shown in Figure 10.8, the proton-decoupled spectrum of ethyl phenylacetate shows eight sharp peaks, one for each different carbon atom in the molecule.

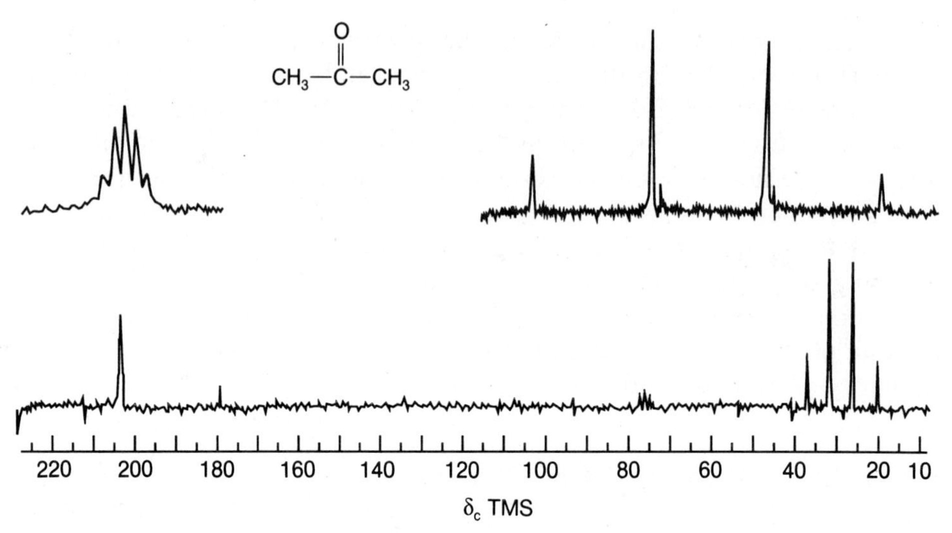

Figure 10.7 **^{13}C NMR spectrum of acetone.**

As can be seen in Figure 10.8, useful information can be gained from a comparison of the decoupled and undecoupled spectra. Since the splitting by protons attached directly to a carbon is much larger (100 to 200 Hz) than the splitting by more remote protons (less than 10 Hz) and stands out clearly in the spectrum, the degree of splitting can be used to determine

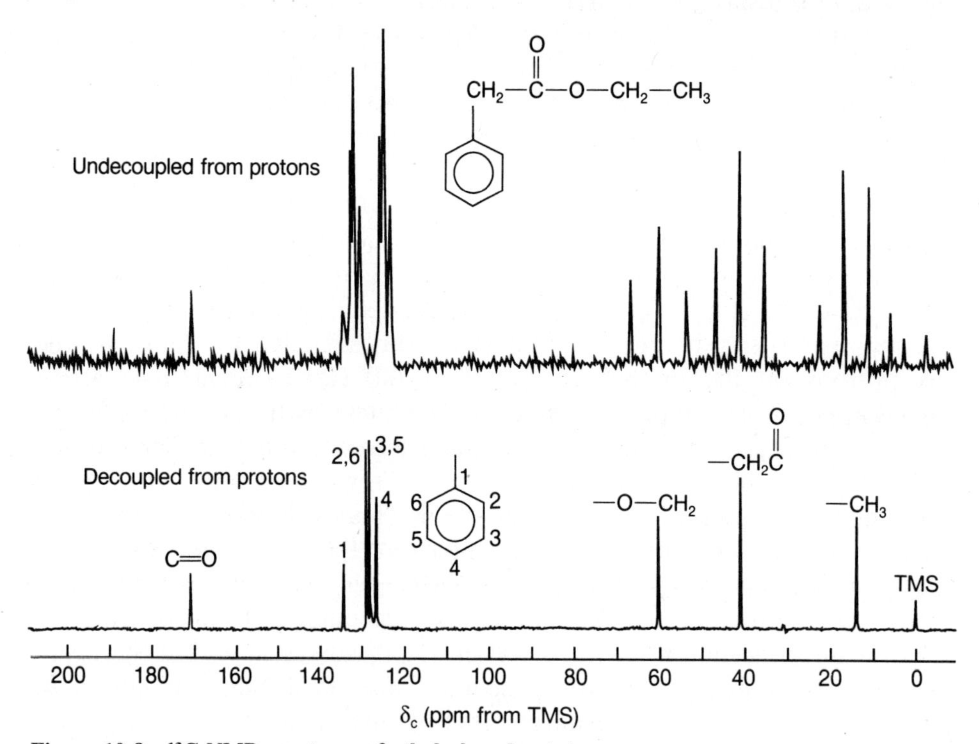

Figure 10.8 **^{13}C NMR spectrum of ethyl phenylacetate.**

Table 10.7 ^{13}C NMR Spectrum of Ethyl Phenylacetate

δ_c (PPM FROM TMS)*	CARBON
171.1 (*s*)	C=O
135.0 (*s*)	Phenyl C-1
129.6 (*d*)	Phenyl C-2,6
128.8 (*d*)	Phenyl C-3,5
127.2 (*d*)	Phenyl C-4
60.6 (*t*)	OCH_2CH_3
41.4 (*t*)	Ph—CH_2—CO
14.2 (*q*)	OCH_2CH_3

**s*, *d*, *t*, or *q* indicates that the carbon appears as a singlet, doublet, triplet, or quartet in the undecoupled spectrum.

the number of protons attached to each carbon: n protons split the signal into $n + 1$ lines. Thus, the methyl carbon on the right is easily identified as a CH_3 group, since it appears as a quartet in the undecoupled spectrum. Obviously such information is useful in assigning peaks to specific carbons in the molecule.

Because of the simplicity of the spectrum and the lack of information from peak areas, it is possible to describe it adequately as a simple list of peak positions (chemical shifts). The number of peaks indicates the number of magnetically different carbons, and the chemical shift of each peak is determined by the electronic environment of the carbon nuclei. If necessary, the splitting due to attached protons can also be measured. The ^{13}C NMR spectra of ethyl phenylacetate given in Figure 10.8 is summarized in Table 10.7.

CARBON CHEMICAL SHIFTS

As in proton NMR, ^{13}C chemical shifts are generally measured and reported in units of parts per million (ppm) downfield from tetramethylsilane (TMS). The range of ^{13}C chemical shifts is, however, much larger than the range of proton shifts (more than 200 ppm vs. less than 20 ppm). With comparable resolution (typically 0.02 ppm), exact coincidence of ^{13}C peaks is extremely unlikely unless the carbons have completely identical environments. Thus, the compound 1-octanol shows a separate peak for each carbon, even though the central carbons in the chain differ only in atoms that are four bonds away (Fig. 10.9).

As in proton NMR, useful deductions can be made about the structural environment of a carbon from the chemical shift. Chemical shift ranges for several types of carbon atoms are listed in Table 10.8. The same data are presented graphically in Figure 10.10, and the major types are discussed

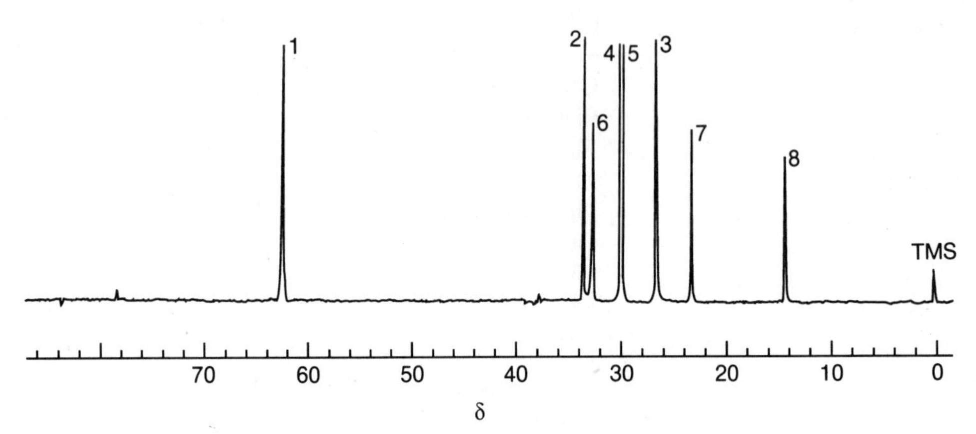

Figure 10.9 **^{13}C NMR spectrum of 1-octanol.**

individually in more detail in the following sections. The chemical shift of each carbon atom is a complex function of its hybridization, the polar and steric effects of its substituents, and of "ring currents," as observed in proton spectra.

SATURATED CARBON ATOMS

Before describing the effect of polar substituents on the ^{13}C chemical shifts of alkanes, some systematization of the effect of skeletal structure must be made. As can be seen in Figure 10.10, there is a general increase in chemical shift in going from primary to secondary to tertiary to quaternary carbon atoms. The shift per replaced hydrogen is in the range of 7 to 10 ppm: CH_4,

Table 10.8 **Chemical Shift Ranges for Various Types of Carbon Atoms***

COMPOUNDS	CARBON	δ_c^{TMS} (PPM)
Alkanes	R_4C	0–55
Alkenes	$R_2C{=}CR_2$	100–165
Alkynes	R—C≡C—R	65–90
Alkylbenzenes	$C_6H_{6-n}R'_n$	125–140
Ketones and aldehydes	$R'_2C{=}O$	180–220
Carboxylic acid derivatives	R′—COX	150–185
Nitriles	R—C≡N	115–125

*R = alkyl or H; R′ = alkyl, aryl, or H; X = OR, NR_2, halogen.

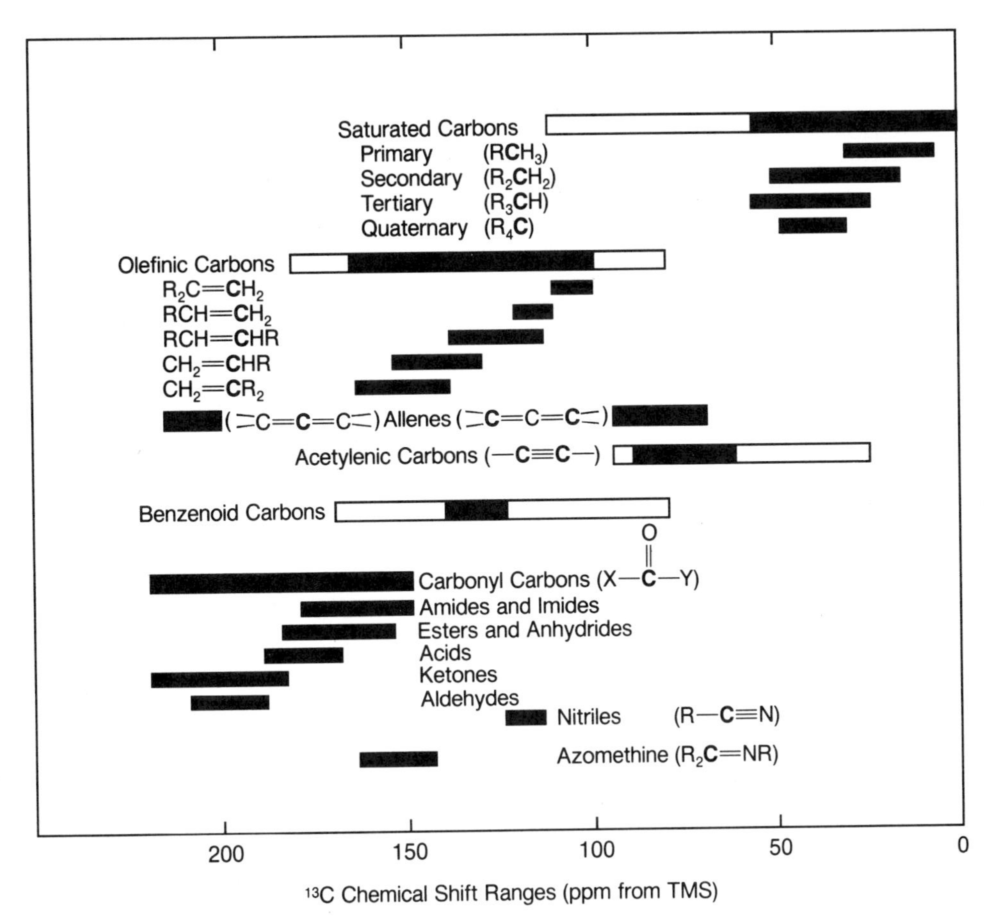

Figure 10.10 ^{13}C **chemical shift ranges (ppm from TMS). Extended ranges when polar substituents are attached to saturated carbon, double or triple bond, or the benzene ring are indicated by lightly shaded areas.**

δ_c^{TMS}–2.3; C_2H_6, 5.7; $CH_2(CH_3)_2$, 15.9; $CH(CH_3)_3$, 25.0; $C(CH_3)_4$, 31.6 ppm. A similar, but smaller, effect was noted in the chemical shifts of protons attached to such carbons (see Table 10.2). The number of carbons two and three bonds away from the atom in question also affects its chemical shift, and the chemical shift of a carbon in an acyclic alkane can be calculated reasonably accurately using the formula

$$\delta_c^{TMS} = 1 + 7n_1 + 8n_2 - 2n_3 \pm 4 \text{ ppm} \qquad \textbf{[10.1]}$$

where n_1, n_2, and n_3 are the number of carbon atoms one, two, and three bonds away from the carbon whose shift is being calculated.

As an example of the application of Equation 10.1, consider the molecule 2-methylpentane. For carbon one, $n_1 = 1$, $n_2 = 2$, and $n_3 = 1$, giving a calculated chemical shift of 22 ppm, compared with 22.3 ppm observed.

Table 10.9 ^{13}C Chemical Shifts of 1-Substituted Pentanes

			δ_c^{TMS}		
1-SUBSTITUENT	C-1	C-2	C-3	C-4	C-5
CH_3	22.7	31.8	31.8	22.7	13.7
Br	33.5	33.2	30.9	22.4	14.2
NH_2	42.7	34.3	29.7	23.1	14.3
Cl	44.8	33.1	29.7	22.6	14.1
OH	62.5	33.2	29.0	23.4	14.4
F	84.3	31.1	28.3	23.2	14.2

Similar calculations for the other carbons give the following results, which are seen to be in quite good agreement with the observed chemical shifts:

$$(CH_3)_2CH\text{—}CH_2\text{—}CH_2\text{—}CH_3$$

Calculated	22	28	39	19	14
Observed	22.3	27.6	41.6	20.5	13.9

The prediction of the effect of functional groups on chemical shifts in alkanes is simplified by the fact that to a good approximation, the polar (inductive) effect of a substituent is restricted to the atom to which it is attached. The effect on more remote carbons is largely steric in nature and is essentially the same ($\pm$ 3 ppm) as that of a methyl group, as seen in Table 10.9.

By comparisons similar to this, the effect of replacing a methyl group in an alkane by a polar substituent has been found to be relatively constant. Listed in Table 10.10 are changes to be added to the chemical shift of R_3C—CH_3 (calculated by Eq. 10.1) to obtain an estimate of the chemical shift of R_3C—X. The general trends in the table can be explained as a decrease in shielding with increasing electronegativity (inductive effect) and decreasing size of the substituent.

As an example of the use of Table 10.10, consider the compound 2-chloroethanol, $ClCH_2CH_2OH$. The analogous (isoskeletal) alkane is butane, for which the chemical shift of carbons two and three can be calculated (Eq. 10.1) to be 23 ppm (observed: 24.8 ppm). Replacing one methyl group by a hydroxyl group should change the chemical shift of the adjacent carbon to 63 ppm (23 + 40). Replacing the other methyl group by a chlorine leads to a predicted chemical shift of 45 ppm (23 + 22). The observed ^{13}C chemical shifts of 2-chloroethanol are 63.4 and 46.7 ppm, respectively.

The cumulative effect of two or more polar substituents on the *same* carbon is generally somewhat less than the sum of the individual effects;

Table 10.10 Changes in ¹³C Shift of Saturated Carbons on Replacing Attached —CH_3 by —X

—X	$\Delta\delta_c$ (PPM)	—X	$\Delta\delta_c$ (PPM)
—I	−16	—NH_3	+17
—CN	−6*	—NH_2††	+19
—C≡CH	−5†	—CHO	+21*
—CH_3	(0)	—$COCH_3$	+22*
—SH	+2	—Cl	+22
—Br	+10	—COCl	+23*
—COOH	+11*	—OH	+40
—CH=CH_2	+12	—OCH_3	+45
—Ph	+13	—OCOR	+45
—$CONH_2$	+13*	—NO_2	+52*
—CO_2^-	+15*	—F	+61

*Also −7 ppm effect on β-carbons
†Also −4 ppm effect on β-carbons
††Applies also to replacement of —CH_2— by —NH—.

for example, CH_2Cl_2: calculated δ_c^{TMS} = 15 + 2(22) = 59; observed, 54.0 ppm.

AROMATIC CARBON ATOMS

As shown in Table 10.8, sp^2-hybridized carbon atoms in alkenes and benzenoid hydrocarbons appear in the same general region of the ^{13}C NMR spectrum (100 to 165 ppm), and for this reason proton NMR is more useful in making a distinction between the two types of unsaturation. However, the ^{13}C chemical shifts of olefinic and aromatic carbons are very useful in determining the identity of polar substituents. Unlike saturated carbon atoms, the polar effects of substituents on unsaturated carbons are longer-ranging, owing to the presence of resonance (mesomeric) effects in addition to inductive effects. For example, a methoxy group inductively deshields the olefinic or aromatic carbon to which it is attached, as is the case with aliphatic carbons. The other carbon of the double bond and the carbons *ortho* and *para* to the methoxy group in an anisole are significantly shielded by the electron-donating ability of the oxygen:

CH_2=CH_2
123.3

$CH_3\ddot{O}$—CH=CH_2
52 153 83

(benzene) 128.5

OCH_3 ← 55
160 ← 114
121 ← 130

Table 10.11 Effect of Substituents on ^{13}C Chemical Shifts of Benzene

	CHANGE IN δ_c^{TMS} (PPM)*		
SUBSTITUENT	C-1	Ortho	Para
—CN	−16	+4	+5
—Br	−5	+3	−2
—H	(0)	(0)	(0)
—CO_2CH_3	+3	+2	+5
—Cl	+6	0	−2
—CHO	+8	+1	+5
—$COCH_3$	+9	+1	+5
—CH_3	+9	+1	−3
—NHCOR	+11	−7	−5
—CH_2CH_3	+16	0	−2
−NH_2	+18	−13	−10
—NO_2	+20	−5	+6
—$CH(CH_3)_2$	+21	−1	−2
—$C(CH_3)_3$	+23	−2	−2
—OH	+27	−13	−7
—OCH_3	+31	−14	−8
—F	+35	−13	−5

* ±2 to 3 ppm, depending on solvent and other substituents.

The effect of substituents on the chemical shifts of the ring carbons of benzene has been studied extensively, and several of these effects on the substituted (C-1), *ortho*, and *para* carbons are listed in Table 10.11. The *meta* carbon is essentially unaffected (± 2 ppm). It has also been found that these effects are nearly additive in disubstituted and trisubstituted benzenes, so that, for example, the chemical shift of C-1 in 1-fluoro-2-nitrobenzene is predicted to be 129 + 35 (attached F atom) − 5 (*ortho*-NO_2 group) = 159 ppm. The observed chemical shift is 155.6 ppm. The observed and similarly calculated chemical shifts of the remaining atoms are given in Table 10.12, and most are seen to agree to within ± 3 ppm.

The effect of a substituent on the chemical shifts of the carbons in

Table 10.12 Observed and Calculated ^{13}C Chemical Shifts (δ_c^{TMS} of 1-Fluoro-2-Nitrobenzene

	C-1	C-2	C-3	C-4	C-5	C-6
Observed	155.6	138.3	126.0	124.8	135.7	118.3
Calculated	159	136	124	124	135	116

Table 10.13 ^{13}C Chemical Shifts of Carbonyl Carbons

CARBONYL	δ_c^{TMS} (PPM)*
Ketone	205–220
Aldehyde	200–210
Carboxylic acid	175–185
Primary amide	170–180
Ester and anhydride	165–175
Secondary amide and imide	160–170

*Ranges given are for aliphatic compounds. For aromatic and α,β-unsaturated carbonyls, subtract 10 ppm from the values given.

ethylene (δ_c^{TMS} 123 ppm) can also be estimated using data from Table 10.11. The effect on the substituted carbon (C-1) is nearly the same (± 5 ppm) as that found for the substituted carbon (C-1) in benzene; the effect on C-2 of the substituted ethylene is usually given by the sum of the *para* plus twice the *ortho* effect in benzene (± 5 ppm).

CARBONYL CARBON ATOMS

The ^{13}C chemical shifts of carbonyl carbons vary from 150 to 220 ppm as the electron-donating or shielding ability of the attached atoms decreases. The data in Table 10.13 illustrate this effect and provide narrower ranges for specific types of carbonyl groups.

Carbon chemical shifts are affected by polar solvent and also by steric effects present in certain specific situations (cyclic compounds, large *cis* or *ortho* substituents, and others), and the generalizations made in this section wil be less accurate under these conditions.

Experiments

SAFETY NOTE ALL SAMPLES SHOULD BE PREPARED IN A WELL-VENTILATED AREA TO AVOID UNNECESSARY BREATHING OF SOLVENT VAPORS. ALTHOUGH THE CAPPED NMR TUBES ARE REASONABLY RESILIENT, CARE MUST BE TAKEN NOT TO CRUSH THE THIN WALLS INADVERTENTLY.

For all of the experiments listed below, the instructor should demonstrate the proper use of the spectrometer, and explain the precautions to be taken in determining a spectrum. A Varian model 360 EM or equivalent spectrometer is appropriate.

A. PROTON SPECTRUM OF ETHYLBENZENE

Procedure

In a 5-mm NMR tube, prepare a solution of 50% ethylbenzene in deuterated chloroform. Add 3 drops of tetramethylsilane (TMS), place the cap on the tube, and mix the contents well by inverting the tube several times. Place the sample tube in the spectrometer and allow it to come to the temperature of the instrument. Determine and record the spectrum and integration curve over the range of 10 ppm. Record the instrument parameters on the spectrum and identify all of the peaks. Secure the spectra in your laboratory notebook using staples or tape.

B. PROTON SPECTRA OF ALCOHOLS

The purpose of this experiment is to illustrate the effect of proton exchange on the splitting patterns of nearest neighbor hydrogens by the hydroxyl hydrogens of alcohols. A suitable deuterated nonhydroxylic solvent such as dimethylsulfoxide (DMSO-d_6), dimethylformamide (DMF-d_7), or chloroform ($CDCl_3$) is necessary. When sufficiently rapid hydrogen exchange involving the hydroxyl hydrogen occurs, the nearest neighbor-splitting patterns due to this hydrogen disappear and the hydroxyl hydrogen itself appears as a broad singlet. Other compounds with exchangeable hydrogens will also demonstrate this effect.

Procedure

Determine the spectra of the following samples:

1. Neat absolute ethanol and 3 drops of TMS.
2. A 25% solution of absolute ethanol in DMSO-d_6.
3. Sample 2 plus 10 drops of water.

Identify each peak and comment on the utility of this effect in the interpretation of NMR spectra. Repeat the experiment using 2-propanol or 2-methyl-2-propanol.

C. CARBON-13 SPECTRUM OF ETHYLBENZENE

This experiment is designed to illustrate the effect of proton decoupling on a carbon-13 spectrum. Decoupled spectra are less complex than "nondecoupled" spectra because the splitting of the carbon signal by hydrogen is absent.

Procedure

1. Determine the ^{13}C spectrum of the ethylbenzene sample used in Part A using the following conditions:
   a. Normal, decoupled power applied.
   b. Decoupling power turned off.

   Overlay the two spectra, line up the TMS peaks and record the identity and splitting patterns for each peak of the sample.
2. Determine the spectrum of an unknown compound supplied by your instructor.

QUESTIONS

1. The partial spectra in Figure 10.11 represent multiplets due to four frequently encountered proton groupings that are among the six listed below. Assign the proper grouping to each spectrum.

CH_3CHXY	$X_2CH—CH_2Y$
$X_2CH—CHY_2$	$XCH{=}CH_2$
CH_3CH_2X	$(CH_3)_2CHX$

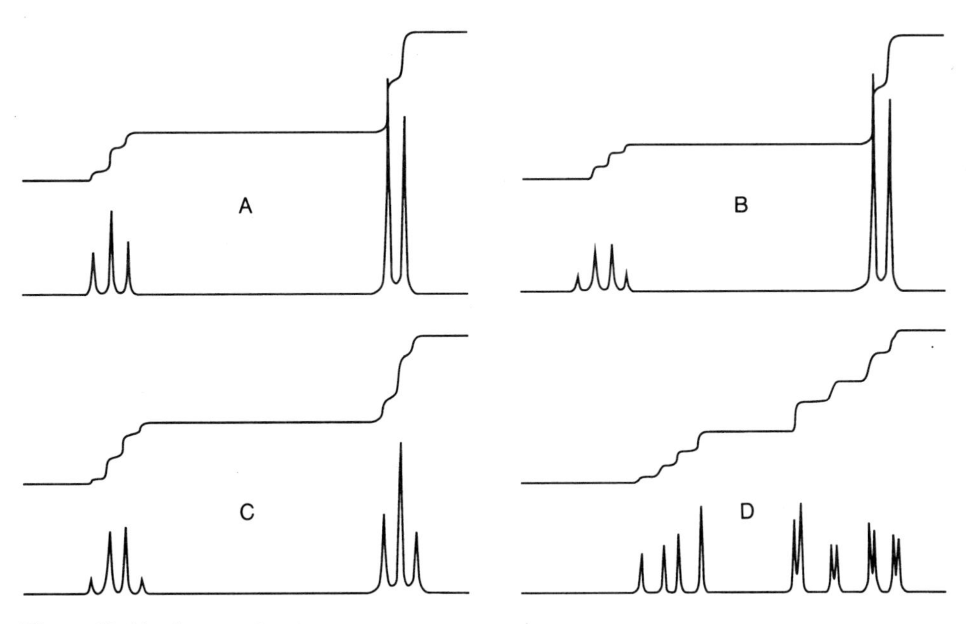

Figure 10.11 Spectra for Question 1.

2. Given the chemical shifts of the ring protons in 4-methyl-2-nitroaniline (p. 151) and assuming J_{ortho} = 7 Hz, J_{meta} = 2 Hz, and J_{para} = 0 Hz, sketch the aromatic region of the proton NMR spectrum. Is this spectrum distinguishable from that of the isomer shown in Figure 10.5?

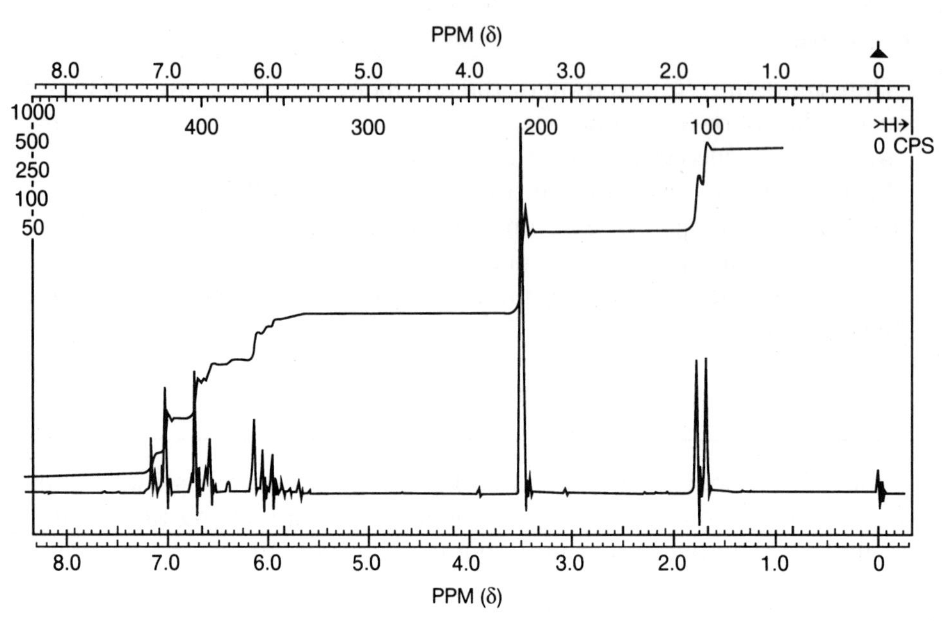

Figure 10.12 Spectrum for compound A.

3. In Figures 10.12 to 10.14 are given the proton NMR spectra of compounds A, B, and C (see Question 6 at end of Chapter 9). Identify the compounds and assign and explain the peaks in the spectra.
4. Based on the ^{13}C NMR spectra given below for two C_6H_{14} hydrocarbons, determine their structures.

Isomer A: δ_c^{TMS} 8.5(q), 28.7(q), 30.3(s), 36.5(t) ppm

Isomer B: δ_c^{TMS} 11.4(q), 18.7(q), 29.4(t), 36.8(d) ppm

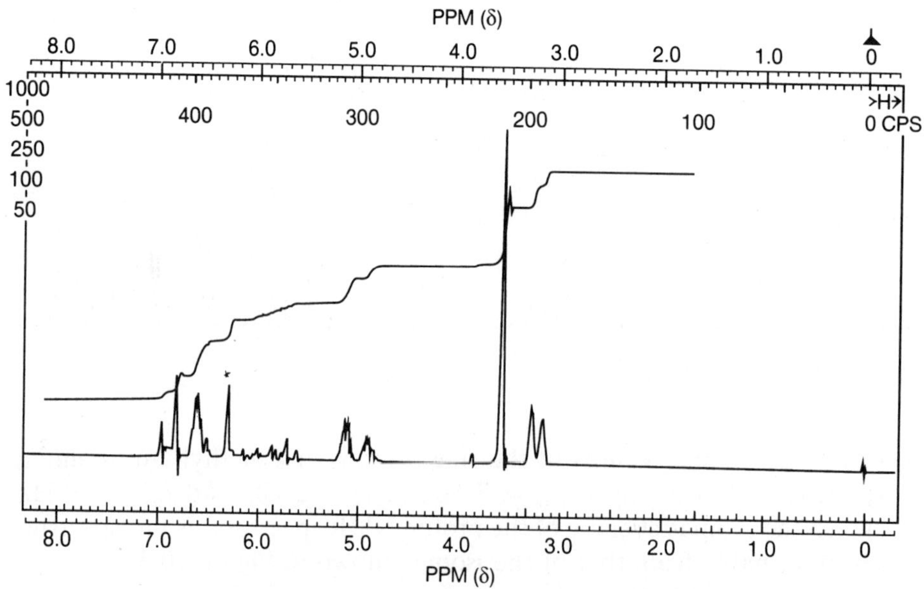

Figure 10.13 Spectrum for compound B. Starred peak exchanges in D_2O.

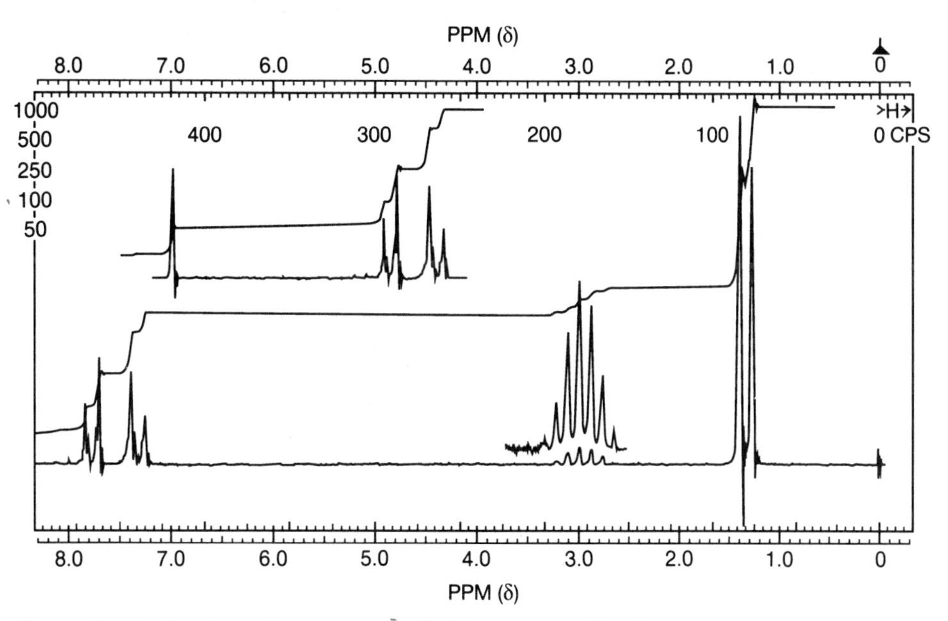

Figure 10.14 Spectrum for compound C. Upper trace offset 3 ppm.

5. Below are given the ^{13}C NMR spectra of six $C_4H_{10}O$ isomers (δ_c^{TMS}, ppm). What are their structures?

A: 14.6, 20.1, 36.0, 62.4 C: 10.4, 23.1, 32.6, 69.4 E: 32.3, 69.4
B: 10.5, 23.3, 58.2, 74.9 D: 22.4, 55.9, 73.6 F: 13.5, 65.4

6. Which of the five compounds given below provided the following NMR spectra?

^1H NMR: δ_H^{TMS} 8.1 (2H, *d*), 7.4 (2H, *d*), 3.8 (3H, *s*), 3.4 (1H, heptet), 1.2 ppm (6H, *d*)

^{13}C NMR: δ_c^{TMS} 19.8(*q*), 35.3(*d*), 55.4(*q*), 113.9(*d*), 129.2(*s*), 130.6(*d*), 163.4(*s*), 202.0(*s*) ppm

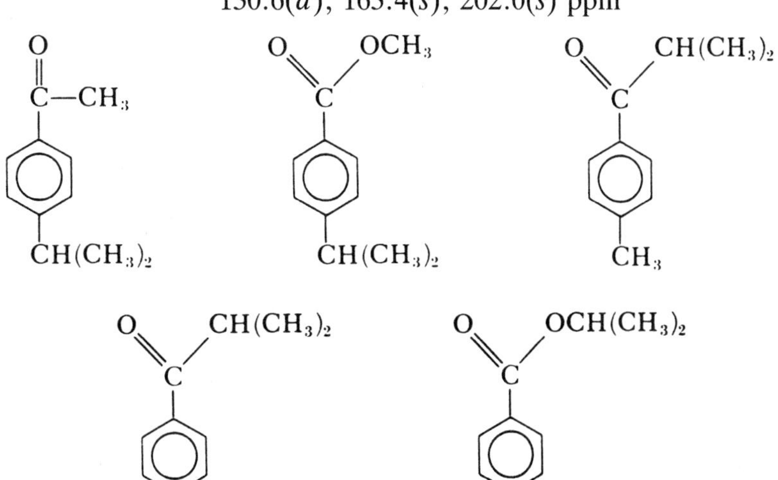

REFERENCES*

Abraham, R. J.; Loftus, P. *Proton and Carbon-13 nmr Spectroscopy*; Heyden and Son: Philadelphia, 1978.

Johnson, L. F.; Jankowski, W. C. *Carbon-13 NMR Spectra*; Wiley-Interscience: New York, 1972.

Levy, G. C.; Lichter, R. L.; Nelson, G. L. *Carbon-13 Nuclear Magnetic Resonance Spectroscopy, 2nd ed.*; Wiley-Interscience: New York, 1980.

Mathieson, D. W., Ed.; *Nuclear Magnetic Resonance for Organic Chemists*; Academic Press: New York, 1967.

Paudler, W. W. *Nuclear Magnetic Resonance*; John Wiley and Sons: New York, 1987.

Pouchert, C. J. *The Aldrich Library of NMR Spectra, 2nd ed.* Aldrich Chemical: Milwaukee, 1985. (A catalog of spectra)

Stothers, J. B. *Carbon-13 NMR Spectroscopy*; Academic Press: New York, 1972.

Varian Associates. *High Resolution NMR Spectra Catalog*. Palo Alto, CA, 1962–1963; 2 vols.

Wehrli, F. W.; Wirthlin, T. *Interpretation of Carbon-13 NMR Spectra*. Heyden: New York, 1976.

*See also general spectroscopy references at end of Chapter Nine.

Name ______________________ Section ____________ Date ________

10 Nuclear Magnetic Resonance Spectroscopy

PRELABORATORY QUESTIONS

1. What is the necessary criterion for an atomic nucleus to exhibit nuclear magnetic resonance?

2. Which of the following nuclei would exhibit nuclear magnetic resonance?

^{2}D ^{14}C ^{32}P ^{32}S ^{36}Cl ^{40}Ca

3. Predict the splitting pattern for ethylbenzene.

4. What is the purpose of adding TMS to a sample?

5. List a common size for NMR sample tubes.

6. What types of compounds exhibit proton exchange?

11

The Literature of Organic Chemistry

The one facet of behavior that distinguishes chemistry from alchemy is that chemists share information (data), whereas alchemists were sublimely secretive; in fact, they went to great lengths to hide their results from others. For a variety of reasons, they were determined to keep experimental details to themselves for fear of persecution or worse. The coming of the research societies in Italy (1657), England (1660), and France (1666), with their attention to the scientific method and their attendant publications, made an immediate impact on science, and chemistry in particular. No longer was it in vogue to hide results; rather, free communication was the proper thing to do. The real beginnings of chemistry lay in the assembling of experimental facts and interpretations along with the publication thereof. The continuous record of experimental fact and interpretation that has been evolving is called the **chemical literature**. One of the major elements in this body of knowledge is the description of the individual compounds that have been encountered and characterized by chemists in the past 150 years.

Students of organic chemistry quickly become aware that a very large number of organic compounds can exist, and they learn to write structures and devise synthetic routes without particular concern as to whether the compound in question has actually been prepared or isolated and described somewhere in the literature. This point becomes of crucial importance, however, if a sample of the compound or information on some property is needed. The problem of finding a specific compound in the literature and locating information on preparative routes, physical constants, or reactions is a practical and exceedingly important one. An introduction to this type of literature search is given in this chapter.

Chemists have been confronted for more than a century with the problem of keeping track of compounds and of coordinating data on new as well as previously known compounds in a systematic way. The organization of literature dealing with organic compounds has become an increasingly complex task, since their number grows exponentially. The sheer bulk of this material is difficult to comprehend; the total number of known compounds is currently estimated to be approaching nine million.

Ultimately, the only possibility for coping with this mass of material in a systematic way is the use of the computer techniques for storing and retrieving information. All new compounds that appear in the current literature are now entered in a computer file by the Chemical Abstracts Service of the American Chemical Society. Each compound is given a unique Registry Number that permits rapid access by computer search to bibliographic data for the compound. This computer registry file includes data on more than eight million compounds, and the number is increasing by about 500,000 per year. Although this computer registry is a powerful tool for exhaustive searches of the literature, information about the properties and preparation of simple compounds is much more readily obtained in other ways described in this chapter.

SEARCHING THE LITERATURE FOR COMPOUNDS

The primary source of chemical information of any kind is the original report of the data in a journal or patent. In order to make this information more accessible, numerous encyclopedias, indexes, and reviews have been and are continuing to be published. Several of the secondary sources that are useful in locating syntheses and properties of organic compounds are described here. An excellent text that provides a broad view of the field of chemical literature is *Chemical Information. A Practical Guide to Utilization*, 2nd ed. by Yecheskel Wolman (John Wiley & Sons: New York, 1988).

Handbooks

For quick reference to physical properties (melting and boiling points, solubility, density) of simple organic compounds, several compilations are available. In most of these, references are given to the source of the data. The following list includes a few of the most common and useful references for specific physical properties:

1. Physical Constants of Organic Compounds. In *Handbook of Chemistry and Physics*, 70th ed.; Weast, R. C., Ed. (CRC Press: Boca Raton, FL, 1989). Contains mp, bp, solubility, and other data for approximately 14,000 organic compounds.
2. Physical Constants of Organic Compounds. In *Lange's Handbook of Chemistry*, 13th ed.; Dean, J. A., Ed. (McGraw-Hill: New York, 1985). Similar to the preceding handbook but less extensive in coverage (approximately 6500 compounds are listed).
3. *Handbook of Tables for Identification of Organic Compounds*, 3rd ed.

(Chemical Rubber Co.: Cleveland, 1967.) Contains mp and bp data for more than 4000 compounds and melting points of derivative compounds for each. Arranged by functional groups.

4. *Merck Index*, 10th ed. Windholz, M., Ed. (Merck: Rahway, NJ, 1983). An encyclopedia with data on nearly 10,000 compounds. Contains information on occurrence, synthesis, and pharmacology of many compounds of importance in biochemistry and medicine. Also provides an extensive cross list of synonyms and trade names.
5. *Dictionary of Organic Compounds*, 5th ed.; I. Heilbron (Oxford University Press: New York, 1982; vols. 1–7 and annual supplements). The Heilbron Dictionary is an alphabetical listing of more than 25,000 compounds with selected reactions, derivatives, and literature references.

Beilstein

The German treatise, *Beilstein Handbuch der Organische Chemie;* (Springer-Verlag: Berlin, 1918 to the present), is the ultimate secondary source on all organic compounds in the earlier literature (before 1950). The fourth edition, in 31 volumes (*Bande*), was published in the period 1918 to 1938, and this *Hauptwerk* (H) covers exhaustively the literature through the year 1909. This edition has been updated periodically by supplements (E)(Table 11.1).

With a rudimentary understanding of German nomenclature, it is possible to locate a compound in Beilstein using the cumulative formula index (*Gesamtregister Formelregister*) for the fourth supplement. This index gives references to entries for the same compound in the H through E-IV series. Alternatively, one can make use of the organization of the volumes to seek out the compound directly. To do the latter, some understanding of the classification system is necessary.

All of the compounds in Beilstein are classified in one or another of three main categories: Aliphatic (*Acyclische*), Carbocyclic (*Isocyclische*, including both alicyclic and aromatic rings), and Heterocyclic (*Heterocyclische*). Heterocyclic compounds are further subdivided according to the type and number of hetero-atoms in the ring(s). Thus *Band* XX (Vol. 20), *Heterocyclische Verbindungen*, *Heteroclasse* 2 O *bis* 9 O, contains heterocyclic compounds with from two to nine oxygen atoms in the ring(s).

Within each of the main series, compounds are arranged according to the functional group present. The major functional group categories are *Kohlenwasserstoffe* (hydrocarbons), *Oxyverbindungen* (alcohols), *Oxoverbindungen* (aldehydes and ketones), *Carbonsauren* (carboxylic acids), *Sulfinsauren* (sulfinic acids), *Sulfonsauren* (sulfonic acids), *Amine*, *Hydroxylamine*, *Hydrazine*, *Azoverbindungen*, and *Phosphine*. The *Stammkerne* sections in the *Heterocyclische Band* contain heterocyclic compounds with none of the foregoing functional groups; that is, they are analogous to the

Table 11.1 The Organization of Beilstein

SERIES	ABBREVIATION	PERIOD OF LITERATURE COMPLETELY COVERED	COLOR OF LABEL ON SPINE
Basic Series	H	Up to 1909	Dark green
Supplementary Series I	E-I	1910–1919	Red
Supplementary Series II	E-II	1920–1929	White
Supplementary Series III	E-III	1930–1949	Blue
Supplementary Series III/IV	E-III/IV*	1930–1959	Blue/black
Supplementary Series IV	E-IV	1950–1959	Black
Supplementary Series V†	E-V	1960–1979	Red on blue binding
Cumulative Subject and Formula Indexes through E-IV			Light green

*Volumes 17–27 of Supplementary Series III and IV, covering the heterocyclic compounds, are combined in a joint issue.
†Supplementary Series V is currently being printed and is in English.

Kohlenwasserstoffe sections of the *Acyclische* and *Isocyclische* categories. In addition to the preceding functional group categories, there are separate sections for compounds containing more than one major functional group. The titles of these are self-explanatory: *polycarbonsauren*, *oxocarbonsauren*, *oxyoxoverbindungen*, *oxoamine*, and so forth.

Compounds containing halogen atoms, nitrogen, nitroso, or azide groups are listed under the parent compound; for example, chloroacetone under acetone. Esters, primary amides, anhydrides, nitriles, and acid halides are listed under the parent carboxylic acids. Secondary and tertiary amides are given under the parent amine. Ethers, peroxides, mercaptans, sulfides, and sulfones are given following the corresponding alcohols.

Within any functional group category, the compounds are ordered according to the number of oxygen, sulfur, or nitrogen atoms contained. These subgroups are further ordered according to the degree of unsaturation (number of rings and double bonds) and, finally, according to the number of carbon atoms present. Isomeric compounds are generally arranged so that the simpler (less branched, fewer rings, and so forth) isomers precede the more complex. The preceding classification is outlined in the Table of Contents (*Inhalt*) at the beginning of each volume.

Finding methyl 5-chloro-2-hydroxybenzoate (methyl 5-chlorosalicylate) can be used as an example of how to use the Beilstein system of organization.

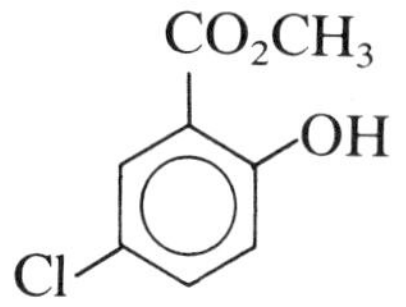

CO_2CH_3
OH
Cl

The compound is first recognized as a carbocyclic (*Isocyclische*) hydroxyacid (*Oxycarbonsauren*) derivative and should be found in *Band X*, which is so titled. The parent compound, salicylic acid, has a molecular formula of $C_7H_6O_3$, that is, $C_nH_{2n-8}O_3$ ($n = 7$). Looking in the Table of Contents of *Band* X, we first locate the subheading *oxycarbonsauren mit 3 sauerstoffatomen* (hydroxyacids with three oxygen atoms). In this section, we then look for compounds with five units of unsaturation (i.e., $C_nH_{2n-8}O_3$) and, finally, the subheading for $n = 7$. At this point we are directed to *Seite* (page) 43.

Turning to page 43, we find salicyclic acid itself; in fact, pages 43 to 59 discuss salicylic acid. Continuing, we find salts of salicyclic acid, ethers of salicylic acid, esters of salicylic acid, and so forth. On page 101 the listing of substituted salicylic acids begins, the first compounds being chlorosalicylic acids. After the 5-chloro isomer is a paragraph on page 103 describing its methyl ester. This is shown in Figure 11.1. Similarly, a supplemental entry for this compound can be found in E-II, *Band* X, Seite 62 (Fig. 11.2). There

5-Chlor-salicylsäure-methylester $C_8H_7O_3Cl = HO \cdot C_6H_3Cl \cdot CO_2 \cdot CH_3$. *B.* Beim Einleiten von Chlorwasserstoff in eine methylalkoholische Lösung der 5-Chlor-salicylsäure (VARNHOLT, *J. pr.* [2] 36, 21). Bei der Einw. von Methyljodid auf das Silbersalz der 5-Chlor-salicylsäure (LASSAR-COHN, SCHULTZE, *B.* 38, 3300).—Nadeln (aus Alkohol). F: 48° (SMITH, *B.* 11, 1227; V.), 50° (L.-C., SCH.). Siedet unter teilweiser Zersetzung bei 249° (V.). Ziemlich leicht löslich in Alkohol (SM.; V.).

Figure 11.1 Beilstein, Hauptwerk, Band X, p. 103.

is no information on this compound in E-I, one entry in E-III, *Band* X, *Seite* 166 (Fig. 11.3), and an entry in Band X of E-IV, *Seite* 209. *Band* X of E-V has not yet been published.

The preceding entries can be located equally well using the *Gesamtregister Formelregister* of E-IV. Nomenclature based on trivial or unsystematic names used in the H, E-I through E-III Series has been replaced by the IUPAC system in E-IV. For further information, consult O. Weissbach, *The Beilstein Guide* (Springer-Verlag: New York, 1976).

Chemical Abstracts

Publication of *Chemical Abstracts* was begun in 1907, and since 1945 the volume and collective indexes have become the "key to the chemical literature." All new compounds appearing in chemical journals (over 10,000 periodicals are now abstracted) and patents are recorded in these indexes, as well as entries on most citations of previously known compounds. The most recent collective index (1982–1986) occupies 92 bound volumes. Despite this huge bulk, a compound mentioned in any abstract published in this five-year period can be found in a few minutes' time after a little practice, and traced back from index to abstract and thence to the original primary journal description. Ten-year collective indexes were published for the first five decades; the collective indexes are now on a five-year basis. Only these collective indexes need be consulted for the period before 1987, but individual volume indexes (now semiannual) must be consulted until the next collective index is published.

Compounds are indexed both by molecular formula and by the chemical name. If a specific compound is wanted, the formula index is generally consulted first, and after the compound in question is located by name out of a list of perhaps 10 or 20 isomeric compounds, an abstract number, the subject index name, and the registry number are obtained. This procedure

5-Chlor-salicylsaure-methylester $C_8H_7O_3Cl = HO \cdot C_6H_3Cl \cdot CO_2 \cdot CH_3$ (H 103). *B.* Bei der Belichtung einer Lösung von Chlorpikrin in Salicylsäuremethylester (PIUTTI, BADOLATO, *R. A. L.* [5] 33 I, 477).–F: 48°.

Figure 11.2 Beilstein, E-II, Band X, p. 62.

5-Chlor-2-hydroxy-benzoesäure-methylester, *5-chlorosalicylic acid methyl ester* $C_8H_7ClO_3$, Formel III (R = H, X = OCH_3) (H 103; E II 62).

B. Aus Salicylsäure-methylester beim Behandeln mit Dichlorcarbamidsäure-methylester in Essigsäure (*Bougault, Chabrier*, C. r. **213** [1941] 400).

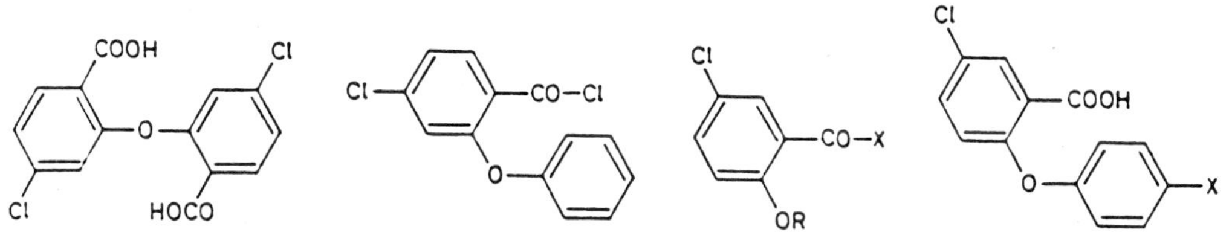

Figure 11.3 Beilstein, E-III, Band X, p. 166.

may immediately turn up references to the desired compound, but it is usually advantageous to go from the formula index to the subject index, using the name obtained in the formula index. The subject index may lead to additional entries, and some indication of the information available in the reference will be given.

Since 1972 (Vol. 76), the subject indexes of *Chemical Abstracts* have been divided into a *Chemical Substance Index* and a *General Subject Index*. The former contains references to specific compounds and the latter to other subjects such as classes of compounds, reactions, concepts, applications, and so on. It should also be mentioned that in the early volumes of *Chemical Abstracts*, the formula indexes did not provide as complete a coverage as the subject index.

The main value of the subject index is its correlative function. The system used for index names is designed to group closely related compounds together; thus, a series of esters of a complex acid will be found under the main entry for the acid; for example, *salicyclic acid, 5-chloro-, methyl ester* is the subject index "inversion name," and other derivatives of the carboxyl group will be listed similarly under this heading. Some experience is required to find a compound in the subject index; a detailed discussion of the nomenclature system used up to 1972 is given in a preface to the index of Volume 56 (1962). Since 1972, *Chemical Abstracts* has used rigidly systematic nomenclature in its indexes, abandoning nearly all trivial names. Thus, toluene is now found as benzene, methyl-; anisole as benzene, methoxy-; and acetone as 2-propanone. An Index Guide, part of the index to Volume 76 and reissued periodically (latest 1989) provides a complete description of the new indexing nomenclature. For this reason, except for simple compounds, the formula index provides a surer way to find a specific compound, but if any of several related compounds would be equally useful, each must be searched for individually in the formula index. Selected *Chemical Ab-*

$C_8H_7ClO_3$ Acetic acid, (chlorophenoxy)-, **16**: 3483^{7}; **21**:1096^{7}; **24**:1344^{7}; **25**:930^{2}; **30**:2963^{5}; **36**:6199^{7}, 6512^{5}; **37**:4426^{3}, 5637^{4}; **38**:2327^{6}; **39**:1199^{4}, 5286^{8}; **40**:1474^{1}, P 2264^{5}, 3846^{8}, $6197^{4,7}$; *Na salt*, **40**:7491^{6}.
Acetophenone, chlorodihydroxy-, **18**: 675^{3}; **23**:2161^{3}; **24**:4012^{9}; **29**:3338^{1}, 5839^{5}; **30**:159^{9}, 3421^{9}; **34**:401^{6}; **37**: 101^{3}, 2358^{2}; **39**:1471^{4}.
Anisic acid, 3-chloro-, **38**:3629^{9}; **40**:2131^{1}.
Benzoic acid, 3-chloro-4-hydroxy-, methyl ester, **20**:3712^{8}.
—, chloromethoxy-, **18**:386^{6}; **20**:1065^{3}; **21**:3189^{8}; **23**:1128^{9}; **24**:1859^{3}; **34**: 7867^{4}; **35**:2125^{6}; **38**:5495^{7}; *and salts*, **33**:$2124^{6,7}$.
Carbonic acid, chloromethyl phenyl ester, **14**:739^{9}.
Cresotic acid, chloro-, **25**: P 4558^{6}; **26**: P 1130^{8}, P 1946^{4}, P 2604^{3}; **27**: P 852^{5}, P 2044^{1}; **33**:6897^{6}.
4-Cyclohexene-1,2-dicarboxylic anhydride, 1-chloro-, **40**: P 3136^{5}.
2-Furoic acid, 2-chloroallyl ester, **32**: P 7925^{5}.
Homoprotocatechuyl chloride, **32**:851^{7}.
Mandelic acid, chloro-, **15**:2632^{2}; **22**: 2746^{4}; **25**:3637^{4}; **32**:4969^{2}; **35**:5637^{2}; **40**:1156^{6}, 3044^{3}.
p-Orsellinaldehyde, 3-chloro-, **28**:6130^{5}; **29**:1078^{7}.
Piperonyl alcohol, 6-chloro-, **33**:1294^{1}.
Salicylic acid, chloro-, Me ester, **31**: 8599^{6}; **33**:2124^{5}; **37**:2010^{3}; **38**:3256^{1}. ←
Sorbic acid, γ-chloroacetyl-δ-hydroxy-, δ-lactone, **35**:7405^{1}.
Toluic acid, chlorohydroxy-, **25**:3325^{5}; **37**:3420^{7}.
Vanillin, chloro-, **19**:2494^{8}; **20**:1980^{6}; **21**:906^{1}; **23**:4456^{5}; **25**:$94^{6,7}$; **40**:3896^{6}.

Figure 11.4 ***Chemical Abstracts*, Collective Formula Index (1920–1946).**

stracts references to methyl 5-chlorosalicylate are shown in Figures 11.4, 11.5, and 11.6.

Another guide to Chemical Abstracts is an audio course, *Chemical Abstracts: An Introduction to Its Effective Use*, by J. T. Dickman, M. O'Hara, and O. B. Ramsay (American Chemical Society: Washington, DC) 1980.

Salicylic acid

droxyphenylazo) - 3 - hydroxy - 4- biphenylylazo] - 5 - hydroxy - 3- methyl - 1 - pyrazolyl}phenylazo}- 5 - hydroxy - 3 - methyl - 1 - pyrazolyl}-, **40**: P 6265^{3}.
——, **chloro-**, effect on dermatophytes, **34**: 5873^{3}.
——, **3-chloro-**, **40**:5423^{4}.
——, **3**(and **5**)**-chloro-**, and derivs., **33**: 2124^{3}.
——, **4**(and **5**)**-chloro-**, **40**:71^{3}.
——, **5-chloro-**, **31**:8117^{7}; **33**:8181^{7}.
4-ethoxybutyl ester, **35**: P 3738^{6}.
methyl ester, **37**:2010^{3}; **38**:3256^{1}. ←
and methyl ester, **31**:8599^{6}.
methyl ester, *N*-(*p*-methoxyphenyl)benzimidate, **32**:1666^{7}.
p-nitrophenyl ester, **33**:1296^{3}.
——, **4-*p*-chlorobenzoyl-**, **35**: P 4218^{5}; **36**: P 2732^{1}.

Figure 11.5 ***Chemical Abstracts*, 4th Decennial Subject Index (1937–1946).**

N,N-**Dichlorocarbamates; chlorination reactions.** J. Bougault and P. Chabrier. *Compt. rend.* **213,** 400–2 (1941); cf. *C. A.* **37,** 87[7].—PhOH is transformed by a small excess of NCl_2CO_2Me (I) in AcOH into 2,4,6-$Cl_3C_6H_2OH$, and *o*-$HOC_6H_4CO_2Me$ into its 5-Cl compd. I and $(ClCH_2CH_2)_2S$ in C_6H_6 give unstable *bis(1,2-dichloroethyl) sulfide*, b_{15}° which readily yields HCl and CHCl:CHSCHClCH$_2$Cl. Carbazole and a small excess of I in AcOH yield *tetrachlorocarbazole,* m. 213°. $BzNH_2$ and I in aq. suspension afford BzNHCl, and $PhCH_2CONH_2$ gives *N-chlorophenylacetamide,* m. 120°. In alk. soln. *2,4-dichloro-3,5-diketo-6-benzyl-*, m. 119°, and *-6-phenylethyl-*, m. 130°, *tetrahydro-1,2,4-triazine* are obtained from the Cl-free parents. *2-Chloro-3,5-diketodibenzyltetrahydro-1,2,4-triazine* m. 153°. I and diphenylhydantoin in alk. soln. afford *1,3-dichloro-5,5-diphenylhydantoin,* m. 166° B. C. P. A.

Figure 11.6 *Chemical Abstracts*. **1943,** *37*, **2010[3].**

SEARCHING FOR PREPARATIVE METHODS

When the preparation of a known compound is required, the first step is to locate the substance in the literature. All sources or preparations of a compound recorded in the time period covered by a volumc of Beilstein will be cited there. The only way to check systematically for more recent syntheses is a search of *Chemical Abstracts* subject indexes. This entails some labor, since a number of entries may be found, and not all of those pertaining to synthesis will necessarily be identified as such. Some entries may refer to uses, analytical methods, and the like and can be ignored; abstracts corresponding to the other entries must then be scanned.

The method or methods described for the compound must then be *evaluated* from the standpoint of **two primary criteria**: (1) simplicity and practicality and (2) economy in terms of yield from readily available starting materials. Under the first heading come such items as need for special apparatus, high pressures or temperatures, and ease of isolating the product. These requirements may not be completely satisfied for any of several reasons. The literature reference may deal with a catalytic process suitable for industrial equipment but impractical for laboratory work. On the other hand, the previous source may not have been a deliberate synthesis; for example, the compound may have been obtained from a more complex molecule or as a minor product in the investigation of a reaction for some other purpose. A practical synthetic method requires that the compound be obtained in useful yield from more readily available or cheaper precursors. Ultimately, of course, the synthesis must go back to commercially available chemicals.

An important consideration, particularly when the literature references are not contemporary, is the possibility that the aforementioned criteria could be better met by a method not previously described for the compound. Improved synthetic procedures and reagents are constantly being developed. For example, the preparation of a primary alcohol may have been carried out before 1948 by reduction of a carboxylic acid ester with sodium in ethanol. The use of lithium aluminum hydride has completely displaced this method for laboratory work.

A number of monographs and compendia deal with preparative reactions, and reference to these sources may be necessary even if a previous preparation is to be repeated, since sufficient detail may not be given in the original paper. In applying a reaction to a compound for which it has not been used previously, general textbook knowledge is seldom sufficient, since reaction conditions and isolation procedures are needed. Some of the more important sources are the following:

1. *Houben-Weyl Methoden der Organische Chemie*, 4th ed.; Miller, E., Ed.; G. Thieme Verlag: Stuttgart, 1952–.

This is an encyclopedic multivolume series, in German, dealing with methods of general laboratory practice and procedures and also with the preparation and reactions of classes of compounds. The fourth edition provides comprehensive coverage for many major functional group classes.

2. *Organic Reactions;* John Wiley & Sons, New York, 1942–.

In this series (currently 36 volumes), more than 150 general reactions of preparative utility are discussed in detail, with typical experimental procedures and extensive tables of examples with references. Cumulative subject indexes in the more recent volumes can be scanned for a desired reaction.

3. Theilheimer, W.; *Synthetic Methods of Organic Chemistry;* S. Karger: Basel; Interscience: New York, 1948–.

The annual volumes in this series contain brief descriptions of useful synthetic transformations taken from the literature of the year covered. The reactions are indexed according to a special system based on the type of bond formed.

4. *Organic Syntheses;* John Wiley & Sons: New York, 1920–.

The annual volumes and six collective volumes of *Organic Syntheses* contain detailed procedures for the preparation of more than 1500 compounds. Apparatus, conditions, and workup procedures are specified, and many of the syntheses illustrate general methods that can be applied to related compounds.

5. *Annual Reports in Organic Synthesis.* Academic Press: New York, 1970–

6. In addition to these compendia, a number of books have been published devoted to synthetic organic chemistry. They vary widely in approach and breadth of coverage, but all of the following can provide references to specific procedures in the literature:

Buehler, C. A.; Pearson, D. E. *Survey of Organic Syntheses;* Wiley-Interscience: New York, 1970, 1977; vols. 1 and 2.

Carey, F. A.; Sundberg, R. J. *Advanced Organic Chemistry;* 2nd ed.; Parts A and B, Plenum: New York, 1984.

Fieser, L. F.; Fieser, M. *Reagents for Organic Synthesis;* John Wiley: New York, 1968–; vols. 1 to 14.

Harrison, I. T.; Harrison, S.; et al. *Compendium of Organic Synthetic Methods;* Wiley-Interscience: New York, 1971–1988; vols. 1–6.

House, H. O. *Modern Synthetic Reactions*, 2nd ed.; W. A. Benjamin: Menlo Park, CA, 1972.

Rodd's Chemistry of Carbon Compounds; Coffey, S., Ed.; Elsevier: New York, 1964–.

Sandler, S. R.; Karo, W. R. *Organic Functional Group Preparations;* 2nd ed.; Academic Press: New York, 1983, 1986, 1989; vols. I, II and III.

To illustrate the process of searching for a preparative method, methods for preparing methyl 5-chlorosalicylate can be examined. Three methods are given in Beilstein (see Figs. 11.1 to 11.3). The first entry cites the preparation by HCl-catalyzed esterification of 5-chlorosalicylic acid in methanol and gives the recrystallization solvent (needles from ethanol) and melting point (48°C).

$$\text{5-Cl-}C_6H_3\text{(OH)}CO_2H \xrightarrow[\text{HCl}]{CH_3OH} \text{5-Cl-}C_6H_3\text{(OH)}CO_2CH_3$$

The original reference, however, to Series 2 of *Journal fur praktische Chemie*, is not easily accessible. In the second Beilstein reference, a preparation is indicated by irradiation of a solution of trichloronitromethane (chloropicrin) in methyl salicylate:

$$C_6H_4\text{(OH)}CO_2CH_3 \xrightarrow[h\nu]{C(Cl_3)NO_2} \text{5-Cl-}C_6H_3\text{(OH)}CO_2CH_3$$

The third Beilstein reference and one of the *Chemical Abstracts* citations (see Fig. 11.6) indicates a preparation by chlorination of the hydroxy-ester with N,N-dichlorocarbamate esters. The other abstracts cited in the subject index (see Fig. 11.5) lead to papers dealing with the bactericidal properties of chlorophenol derivatives.

At this point, we can assess the possibilities. Without consulting the original references (which appear in Italian and French journals), it is apparent that the two chlorination procedures require rather special reagents, neither of which is very commonly used. Further checking on these reagents in contemporary sources reveals that a chlorination step is required for their preparation. On the other hand, the main starting material is a very cheap compound, methyl salicylate (oil of wintergreen).

The alternative is esterification of 5-chlorosalicylic acid. This acid, but not the ester, is listed in chemical supply catalogs. The cost is about eight times that of methyl salicylate. The other reagents, however, are common laboratory chemicals, and the esterification is a standard operation (procedures using sulfuric acid are given in Chapters 23 and 29). For preparation of a small sample of the ester, this method would probably be the best choice.

For specific information on the procedure for carrying out the esterification and isolation of the product, many examples of acid-catalyzed esterifications can be found in the indexes of *Organic Syntheses Collective Volumes*. Another useful step at this point is to consult the section on acid-catalyzed esterification in one of the books on synthetic methods (pp. 182–183). References to esterification of closely similar hydroxybenzoic acids may be given, and from these a procedure in a conveniently accessible journal usually can be found.

KEEPING UP-TO-DATE

A number of journals are designed to allow the reader to scan listings that might contain very recent citations or keywords relating to the area of interest.

Published weekly are the seven editions of *Current Contents*, three of which relate to Chemistry: *Physical Chemical and Earth Sciences*, *Engineering Technology and Applied Science*, and *Life Sciences*. Each edition lists the contents of approximately 1000 journals as well as a subject index and an author index and is also available on diskette. Usually, no more than two or three weeks elapse between the time the original journal article appears and the time of its appearance in *Current Contents*. The articles are not abstracted.

Published biweekly is *Chemical Titles*, containing Keyword in Context Index, Author Index, and Bibliography sections, and this journal covers topics relating to chemistry and chemical engineering. Each issue represents approximately 700 primary journals.

ELECTRONIC METHODS

On-Line Searches

Along with the advent of computers and computer networking has come the ability to conduct literature searches of a large number of data bases using on-line subscription services. Many of these are relatively inexpensive, and the cost is reasonable compared to the expense and expenditure of time necessary to conduct manual searches. Most modern libraries subscribe to one or more automated systems such as DIALOG and CAS ON-LINE, which include the ability to search *Chemical Abstracts*, *Biological Abstracts*, *World Patent Index*, and about 300 other data bases. The cost of these searches depends on the file being searched, and searches that are well planned are well worth the cost. An excellent text that covers on-line data bases, retrieval systems, and other useful information is *Communication, Storage and Retrieval of Chemical Information*, by J. Ash, P. Chubb, S. Ward, S. Welford, and P. Willett (John Wiley & Sons, New York, 1985).

Beilstein is now accessible online. Also available is CASREACT, an online reaction-searching tool from Chemical Abstracts Service that enables one to search for specific reactions, preparations of specific compounds or substructures, and information on reagents, solvents, and catalysts.

Many chemical supply houses have made their inventory of chemicals available for electronic searching, particularly relative to searches that include data on toxicity or disposal methods. These data bases may be online or may be in the form of diskettes for microcomputers. These companies also upgrade the data periodically; thus the information remains current.

A relatively new service with great potential involves the use of CD-ROM, Compact Disk Read Only Memory, in which a large amount of information is stored in a form readable by a laser device similar to a compact disk player. Although a compact disk reader is required, the possibility of having an enormous number of references available at a laboratory work station with minimum additional cost seems to have an advantage over the other methods. Among those now available are *Chemical Abstracts* and *Dissertation Abstracts*.

For detailed information concerning the electronic search capabilities of your library, consult the reference librarian or one of the references cited below.

Dialog Database Catalog 1987; Dialog Information Services, Inc.: 3460 Hillview Avenue, Palo Alto, CA 94304.

Keller, R. J. *The Sigma Library of FT-IR Spectra;* Nicolet Instruments: Madison, WI, 1985. (Computer Search Program.)

Maizell, R. E. *How to Find Chemical Information*, 2nd ed.; John Wiley & Sons: New York, 1987.

Pouchert, C. J. *The Aldrich Library of FT-IR Spectra;* Nicolet Instruments: Madison, WI, 1985. (Computer Search Program.)

LITERATURE PROBLEM

As an exercise in using the literature, you will be asked to find a practical laboratory-scale synthesis for a compound. The compound can be prepared in one or more steps from simple starting materials and without special apparatus. Obtain an assignment of a compound from your instructor, consult the literature, and write up a complete experimental procedure for the synthesis on a scale to provide 50 to 100 mg of the final product. Your report should include all relevant literature references; the experimental descriptions should be sufficiently detailed that the preparation could be carried out without reference to the original literature.

At your instructor's direction, this preparation can then be carried out in the laboratory as a supplemental experiment. The starting materials for the synthesis should be reasonably priced commercial chemicals. Your instructor will determine at what point in the sequence your synthesis should begin; this will depend on the cost of various precursors and the amount of laboratory time available.

Name ______________________ Section ____________ Date ________

11 The Literature of Organic Chemistry

PRELABORATORY QUESTIONS

This exercise is designed to familiarize students with the relevant sections of the library that pertain to chemistry. Students should become familiar with the reference sections and the periodicals locations before attempting the laboratory assignment.

1. Describe the locations of each of the following publications; list the floor, aisle, and call number for each.

a. Beilstein, *Handbuch der Chemie*

b. *Chemical Abstracts*, bound copies

c. *Chemical Abstracts*, loose copies

d. *Heilbron's Dictionary of Organic Compounds*

e. *Lange's Handbook of Chemistry*

f. *The Journal of Organic Chemistry*

g. *The Journal of the American Chemical Society*

2. Which of the above publications can undergraduate students borrow overnight?

3. List five journals abstracted by *Chemical Abstracts*.

12

Properties of Hydrocarbons

Hydrocarbons are those organic compounds composed only of carbon and hydrogen. There are *three main categories of hydrocarbons:* saturated, unsaturated, and aromatic. Saturated hydrocarbons have only carbon–carbon single bonds, whereas unsaturated hydrocarbons have carbon–carbon double or triple bonds. Aromatic hydrocarbons are cyclic compounds whose chemical properties are related to benzene.

Saturated hydrocarbons (alkanes and cycloalkanes) are relatively inert and do not react with common laboratory reagents. Unsaturated hydrocarbons (alkenes and alkynes), however, readily undergo addition reactions and oxidation reactions. Benzene and other aromatic compounds do not undergo addition reactions but are characterized by substitution reactions in which another atom or group of atoms replaces a ring hydrogen.

Although alkanes are relatively inert, they do undergo combustion in the presence of air if ignited. The fact that these reactions are highly exothermic and that huge quantities of alkanes are available as petroleum and natural gas has resulted in their extensive use as fuels. While the chemistry of alkanes is relatively straightforward, their economic impact can hardly be overestimated. Consequences resulting from this economic importance include oil spills, air pollution from automobiles, the greenhouse effect, and the threat of war in certain parts of the world.

Experiments

Performing the following tests will illustrate general properties of hydrocarbons as well as differences in chemical reactivity due to the type of hydrocarbon (saturated, unsaturated, or aromatic) being considered. Perform the first six tests on cyclohexane, cyclohexene, toluene, and on two unknowns. One unknown will be an alkane or cycloalkane and a second will be an alkene or alkyne. Using the information that a lack of reactivity is characteristic of saturated hydrocarbons and addition reactions are common for unsaturated hydrocarbons, decide the identity of each unknown as to type of hydrocarbon.

SAFETY NOTE CONCENTRATED SULFURIC ACID OR BROMINE CAN CAUSE BURNS. IF EITHER IS SPILLED ON THE SKIN, WASH IMMEDIATELY WITH WATER. EXPOSURE TO BROMINE AND METHYLENE CHLORIDE VAPORS SHOULD BE AVOIDED; USE THESE REAGENTS IN A HOOD.

Procedure

1. **Solubility** Place 2 mL of water in a 13 × 100-mm test tube and add 2 or 3 drops of the hydrocarbon to be tested. Shake the mixture to determine whether the hydrocarbon is soluble (a colorless second layer may be hard to see). Record your results and save the mixture for Test 2.
2. **Relative Density** Reexamine the mixtures prepared above and decide in each case whether the hydrocarbon is more dense (sinks) or less dense than water (floats).
3. **Flammability** Test the flammability of each hydrocarbon by placing 2 or 3 drops on an evaporating dish in the hood and igniting it with a match or microburner. Note the nature of the flame: more sooty flames are characteristic of unsaturated compounds.
4. **Addition of Bromine** Dissolve 3 or 4 drops of the hydrocarbon in 1 mL of methylene chloride (dichloromethane) in a 13 × 100-mm test tube. Add dropwise a 2% solution of bromine dissolved in methylene chloride with shaking. The loss of bromine color is an indication of an unsaturated compound (do not confuse diminution of color due to dilution of the Br_2/CH_2Cl_2 solution with an actual loss of color).
5. **Reaction with Potassium Permanganate** Dissolve 3 or 4 drops of the hydrocarbon in 1 mL of reagent-grade acetone and then add dropwise a 1% solution of potassium permanganate with shaking. A loss of the purple color of the permanganate solution indicates that a reaction has taken place and that the hydrocarbon is unsaturated.
6. **Sulfuric Acid** Place 1 mL of concentrated sulfuric acid in a 13 × 100-mm test tube and then add 3 or 4 drops of the hydrocarbon one drop at a time. (Caution!) Reaction is indicated not only by the dissolution of the sample but also by changes in color, production of heat, or the formation of insoluble material.

At this point, information from Tests 4, 5, and 6 should allow you to determine which unknown is an alkane or cycloalkane and which is an alkene or alkyne. Record your conclusions about each unknown.

It should be pointed out that while the reagents in Tests 4, 5, and 6 may be used to distinguish saturated and unsaturated hydrocarbons, they will also react with other organic compounds. Potassium permanganate reacts with easily oxidized compounds (alcohols, aldehydes, amines, etc.), while concentrated sulfuric acid reacts with a variety of Lewis bases (such as alcohols and other oxygen containing compounds and amines). Care should be taken not to use these tests with compounds other than hydrocarbons and then misinterpret the results.

7. **Reactivity Toward Nitration** Perform the following experiment with cyclohexane and with toluene. Do not attempt this reaction with an alkene or alkyne!

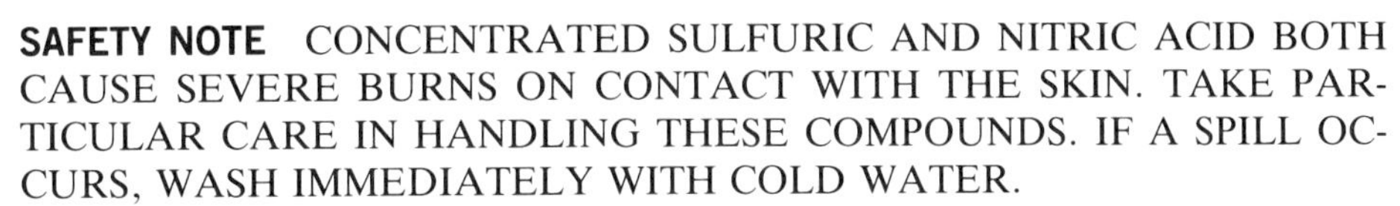

SAFETY NOTE CONCENTRATED SULFURIC AND NITRIC ACID BOTH CAUSE SEVERE BURNS ON CONTACT WITH THE SKIN. TAKE PARTICULAR CARE IN HANDLING THESE COMPOUNDS. IF A SPILL OCCURS, WASH IMMEDIATELY WITH COLD WATER.

Procedure

Mix 1 mL of concentrated sulfuric acid and 0.5 mL (10 drops) of concentrated nitric acid in a small test tube and allow the heat of reaction to subside. Add 4 or 5 drops of the hydrocarbon to be tested and place the test tube in a 55 to 60°C water bath for 10 minutes, shaking it at short intervals. After 10 minutes, pour the contents over approximately 5 grams of crushed ice. Observe the result and decide if the hydrocarbon reacted with the nitric acid.

QUESTIONS

1. Considering your results in the solubility tests, what do you conclude about the solubility of hydrocarbons in water? Predict the solubility of gasoline and motor oil in water.
2. Are hydrocarbons less dense or more dense than water? Does this have any practical importance when oil spills occur?
3. Write equations using 1-butene to illustrate the reactions occurring in Tests 4, 5, and 6 of the experimental procedure (use your text).
4. Write equations using cyclohexene to illustrate the reactions in Tests 4, 5, and 6.
5. Write a balanced equation for the complete combustion of octane. What other product(s) are produced when incomplete combustion of octanes occurs in an automobile engine?
6. Write an equation to illustrate the reaction of toluene with concentrated nitric acid/concentrated sulfuric acid (2,4-dinitrotoluene is produced).
7. How could one distinguish octane from 1-octene by a simple chemical test; 1-octanol from 1-octene; and toluene from 1-octene?

Name ______________________ Section ______________ Date ________

12 Properties of Hydrocarbons

PRELABORATORY QUESTIONS

1 Write the condensed structural formulas for cyclohexane and for cyclohexene.

2. Which would you expect to be more reactive towards chemical reagents, alkanes or alkenes?

3. What is the color of dilute potassium permanganate solution? What is the formula for potassium permanganate?

4. The general formula for an alkane is C_nH_{2n+2}. What is the corresponding general formula for an alkene?

13

Free Radical Chlorination

Although saturated hydrocarbons are inert to most acidic and basic reagents, they can be halogenated in the presence of a free radical initiator. The process is a chain reaction, as shown in Equations 1 through 5 for chlorination with Cl_2, using irradiation to initiate the process.

$$Cl_2 \xrightarrow{h\nu} 2Cl\cdot \qquad \text{Initiation} \qquad [1]$$

$$\left.\begin{array}{l} Cl\cdot + R{-}H \longrightarrow H{-}Cl + R\cdot \quad [2] \\ R\cdot + Cl_2 \longrightarrow R{-}Cl + Cl\cdot \quad [3] \end{array}\right\} \text{Propagation}$$

$$\left.\begin{array}{l} R\cdot + Cl\cdot \longrightarrow R{-}Cl \quad [4] \\ R\cdot + R\cdot \longrightarrow R{-}R \quad [5] \end{array}\right\} \text{Termination}$$

Another method of chlorination involves the use of sulfuryl chloride and benzoyl peroxide ($C_6H_5\overset{O}{\overset{\|}{C}}O O\overset{O}{\overset{\|}{C}}C_6H_5$). At relatively low temperatures, the O—O bond breaks to form two benzoate radicals, and the propagation steps are shown in Equations 6 and 7. (For the initial steps, see Question 2 at the end of this chapter.)

$$Cl\cdot + RH \rightarrow HCl + R\cdot \qquad [6]$$

$$R\cdot + ClSO_2Cl \rightarrow RCl + SO_2 + Cl\cdot \qquad [7]$$

When a molecule contains more than one type of hydrogen atom, as in isopentane, a mixture of alkyl chlorides can result from monochlorination. The observed composition of the mixture is not the one that would be predicted statistically on the basis of random attack of Cl· on the 12 hydrogens (Table 12.1).

$$\underset{①}{H_3C}\!\!\diagdown\!\!\underset{②}{CH}(\underset{①}{CH_3})\text{—}\underset{③}{CH_2}\text{—}\underset{④}{CH_3} \xrightarrow[h\nu]{Cl_2} \text{mixture of chloro-2-methylpentanes}$$

Specifically, the amounts of the primary chlorides (1-chloro or 4-chloro) are smaller than those predicted for random attack, and the amounts of secondary (3-chloro) and tertiary (2-chloro) chlorides are greater than predicted. Analysis of the data indicates that the relative rates of removing a hydrogen atom from positions 1 through 4 are 1:4.0:2.5:1, respectively; thus, the reactivities are 3° > 2° > 1°. This order, which is the reverse of that predicted if the attack were directed by steric effects, is determined by the relative stabilities of the resulting alkyl radicals.

To test the generality of these substitution ratios, 2,2,4-trimethylpentane ("isoctane") is chlorinated in this experiment, and the products are analyzed by gas chromatography.

$$CH_3\text{—}C(CH_3)_2\text{—}CH_2\text{—}CH(CH_3)\text{—}CH_3 \xrightarrow[(PhCO)_2O_2]{SO_2Cl_2}$$

$$Cl\text{—}CH_2\text{—}C(CH_3)_2\text{—}CH_2\text{—}CH(CH_3)\text{—}CH_3$$

$$CH_3\text{—}C(CH_3)_2\text{—}CH(Cl)\text{—}CH(CH_3)\text{—}CH_3$$

$$CH_3\text{—}C(CH_3)_2\text{—}CH_2\text{—}C(CH_3)(Cl)\text{—}CH_3$$

$$CH_3\text{—}C(CH_3)_2\text{—}CH_2\text{—}CH(CH_3)\text{—}CH_2\text{—}Cl$$

Table 12.1 Isopentane Chlorination Products

ISOMER	1-CHLORO	2-CHLORO	3-CHLORO	4-CHLORO
Observed % of mixture	34%	22%	28%	16%
Statistical prediction	50%	8%	17%	25%

(Microscale)

SAFETY NOTE SULFURYL CHLORIDE VAPORS ARE IRRITATING TO THE EYES AND NOSE, AND THE LIQUID CAUSES BURNS IF SPILLED ON THE SKIN. TO REDUCE THE HAZARD SOMEWHAT, THE SULFURYL CHLORIDE WILL BE DISPENSED IN CARBON TETRACHLORIDE SOLUTION. CARBON TETRACHLORIDE IS CLASSIFIED AS A POTENTIAL CARCINOGEN. SHOULD ANY OF THIS SOLUTION GET ON YOUR SKIN, WIPE IT OFF IMMEDIATELY AND WASH THOROUGHLY WITH WATER. BENZOYL PEROXIDE IS UNSTABLE TO HEAT AND FRICTION; TRANSFER BY POURING FROM A PAPER CARTON ONTO A PIECE OF PAPER, AND AVOID CONTACT WITH METAL.

Procedure

Heat a 100-mL beaker of water to 75°C on a steam bath. While the water is heating, obtain 1.0 mL of the sulfuryl chloride/carbon tetrachloride solution (32 g SO_2Cl_2/100 mL solution) in a clean dry 10-mL round-bottom flask. To this solution, add by pipet and bulb 0.5 mL of 2,2,4-trimethylpentane (d = 0.69) and 25 mg of benzoyl peroxide (estimate amount by volume, comparing with a weighed comparison sample). Attach a reflux condenser (see Fig. 1.4) and place a drying tube containing KOH pellets on top of the condenser to exclude moisture and absorb HCl. Clamp the apparatus in the water bath so that about one half of the flask is submerged, and heat the bath at such a rate that the temperature remains at 75 ± 2 degrees for 30 minutes. Remove the flask and condenser from the bath, cool, and add another 25 mg of benzoyl peroxide. Reheat at 75 ± 2 degrees for another 20 minutes. Allow the solution to cool, disassemble the apparatus, and transfer the solution with a pipet and bulb to a 25-mL Erlenmeyer flask. Add 6 mL of water, stopper the flask, and shake vigorously. Add 2 mL of 10% sodium bicarbonate solution, and again shake the mixture well (**CAUTION:** CO_2 evolution).

In a microfunnel prepared from a Pasteur pipet with the tip shortened (see page 34, Fig. 3.4A), gently wedge a very small tuft of cotton at the bottom and add approximately 0.2 g of anhydrous potassium carbonate (enough to make a 1-cm column in the pipet). With a pipet and bulb, remove the lower CCl_4 layer from the flask and filter through the carbonate into a 10 × 75-mL test tube. Remove 1 μL of the filtered solution along with approximately 2 μL of air and inject into a gas chromatograph. Rinse the syringe immediately with ethanol (twice) and then with carbon tetrachloride to prevent corrosion of the needle and plunger.

The order of elution of the compounds from a methyl silicone (OV-101) column is (air), 2,2,4-trimethylpentane and carbon tetrachloride (unresolved), 2-chloro-2,4,4-trimethylpentane, 3-chloro-2,2,4-trimethylpentane, 1-chloro-2,4,4-trimethylpentane, and 1-chloro-2,2,4-trimethylpentane. By triangulation or cutting and weighing (Chapter 6), determine the areas of the chlorotrimethylpentane peaks. (A change in attenuation will be required to keep all the peaks on scale and of measurable size. Consult your instructor for the exact settings.) Assuming that the areas are proportional to the weights of the various compounds, calculate the total percent of chlorination and the percent distribution among the chlorinated products. If some peaks are not completely resolved, the amounts can be estimated with reasonable accuracy by using an overlapping triangulation method.

The product mixture can also be analyzed using high performance liquid chromatography (HPLC). The technique is similar to gas chromatography but is often accompanied by increased resolution. Your instructor will advise you of the procedure for this analytical technique.

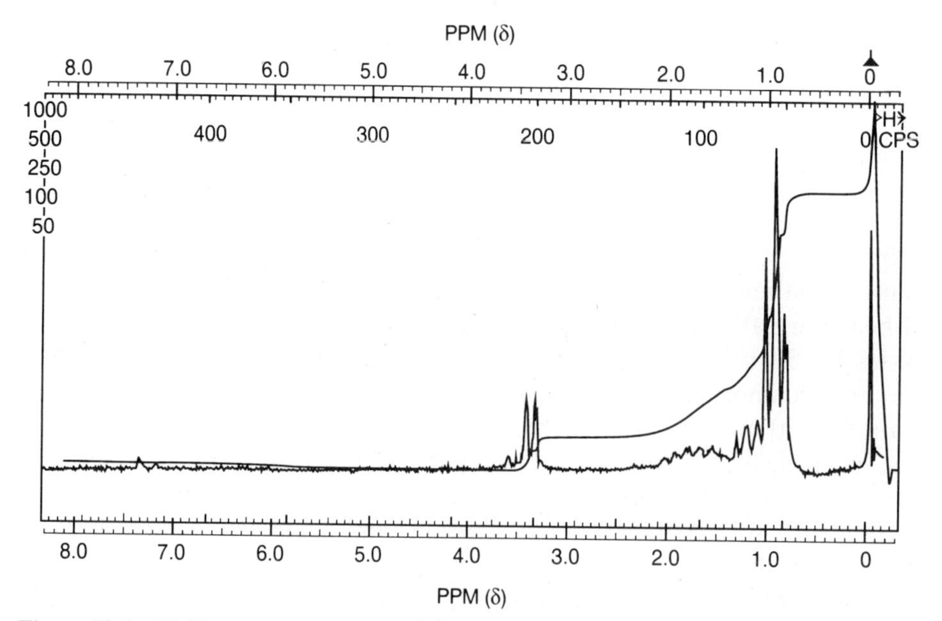

Figure 13.1 NMR spectrum, compound A.

QUESTIONS

1. Using the relative reactivity of 1°:2°:3° hydrogens observed with 2-methylbutane (1.0:2.5:4.0), predict the composition of the monochlorination product mixture from
   a. propane
   b. butane
   c. isobutane
   d. 2,2,4-trimethylpentane
2. The products of chlorination of cyclohexane (RH) with sulfuryl chloride initiated by benzoyl peroxide are cyclohexyl chloride, HCl, SO_2, chlorobenzene, and CO_2. Give equations for the overall reaction (initiation, propagation, and termination steps) that account for these products.
3. How does your observed isomer distribution compare with that which you calculated in Question 1d? Comment on the results. A set of molecular models will be useful here; reference may be made to a paper by Fuller and Hickinbottom and one by Russell and Haffley (see references).
4. In Figures 13.1 to 13.3 are the NMR spectra of the three monochlorodimethylpentanes obtained in a similar chlorination experiment involving 2,4-dimethylpentane. Identify the compounds by their spectra and explain your reasoning.

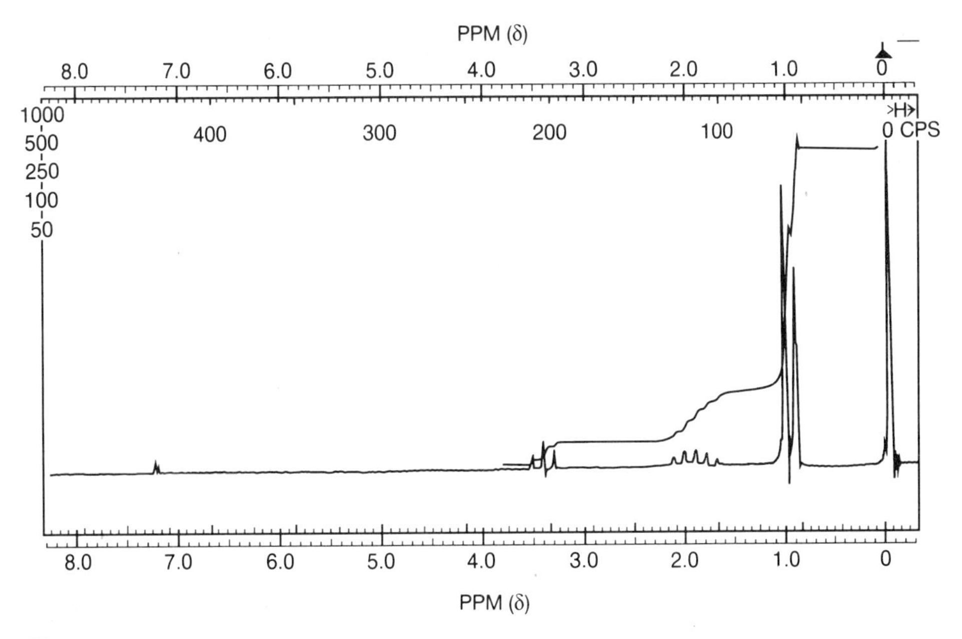

Figure 13.2 NMR spectrum, compound B.

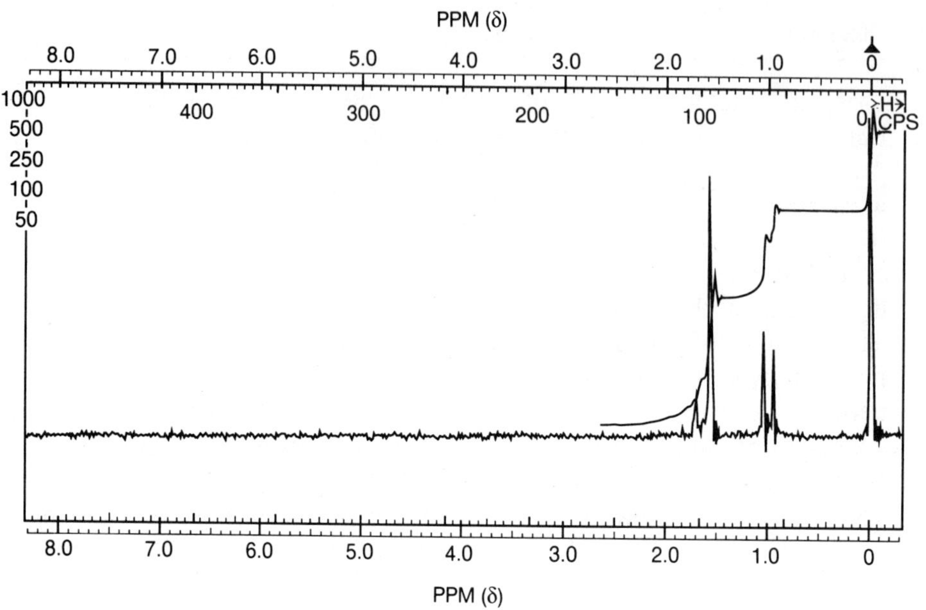

Figure 13.3 NMR spectrum, compound C.

REFERENCES

Fuller, A. E.; Hickinbottom, W. J. *J. Chem. Soc.* **1965**, 3235.
Russell, G. A.; Haffley, P. G. *J. Org. Chem.* **1966**, *31*, 1869.

Name ______________________ Section ____________ Date ________

13 Free Radical Chlorination

PRELABORATORY QUESTIONS

1. Show the initiation, propagation, and termination steps involved in the light-catalyzed monochlorination of ethane using chlorine gas.

2. Draw the structures for all of the possible monochloro isomers of

a. isohexane

b. neopentane

3. In the free-radical chlorination of methane, subsequent chlorination steps (to give higher chlorination products) occur almost as easily as the first. If you wished to utilize this reaction for the commercial production of CH_3Cl, how could the yield of this monochloro product be maximized?

4. Bromine is usually less reactive toward alkanes than is chlorine, but bromine is more selective. For example, chlorination of isobutane yields isobutyl chloride and *tert*-butyl chloride in a ratio of 1.6:1, whereas bromination gives essentially only one monobromo product. Draw this product and explain how you arrived at your answer.

14

Nucleophilic Substitution of Alkyl Halides

Alkyl halides, alkyl sulfates, and alkyl sulfonates serve as reactants or "substrates" in many nucleophilic substitution reactions. These substitution (or displacement) reactions are useful preparative methods, for example, the Williamson synthesis of ethers. For our purpose the term "nucleophilic" will refer to a reaction in which a nucleophile (OH^-, X^-, S^{2-}, HS^-, etc.) attacks a carbon atom and displaces a "leaving group," another nucleophile. Although these reactions proceed in a continuum of slightly different pathways, they can be viewed as occurring in a combination of two limiting mechanisms:

1. Substitution: Nucleophilic, Unimolecular (S_N1)

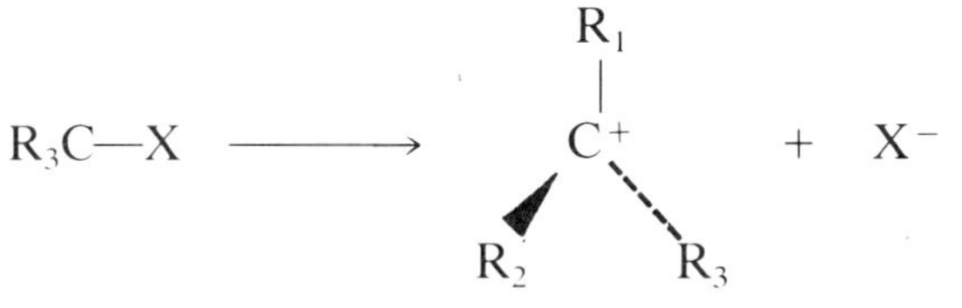

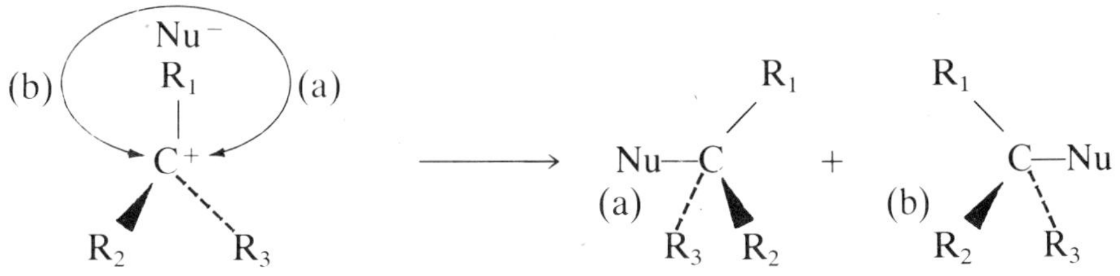

$$R_3C\text{—}X \longrightarrow R_1R_2R_3C^+ + X^-$$

$$Nu^- + R_1R_2R_3C^+ \xrightarrow{(a),\ (b)} Nu\text{—}CR_1R_2R_3\ (a) + R_1R_2R_3C\text{—}Nu\ (b)$$

This mechanism involves more than one step, the first of which is the ionization of the alkyl halide substrate, yielding a carbocation and an anion. The planar carbocation then reacts with a nucleophile, forming the product (in the case of chiral compounds, racemic product generally results). Since the nucleophile is often a solvent molecule, the process is sometimes called **solvolysis**. A typical S_N1 reaction is the solvolysis reaction of tertiary-butyl chloride with water to form *tert*-butyl alcohol.

2. **Substitution: Nucleophilic, Bimolecular (S_N2)**

$$Nu^- + H(R_1)(R_2)C\text{—}X \longrightarrow \overset{\delta-}{Nu}\text{-----}C(H)(R_1)(R_2)\text{-----}\overset{\delta-}{X} \longrightarrow Nu\text{—}C(R_1)(H)(R_2) \text{ (inverted)} + X^-$$

This reaction involves one so-called "concerted" step in which the nucleophile attacks the carbon atom of the substrate on the side opposite the leaving group. An inversion of configuration of chiral compounds results.

Studies of the mechanisms of these reactions were undertaken in the early 1930s in order to explain the vast differences in the rates of seemingly very similar reactions. Since that time, these findings have been refined to encompass the effects of solvent, ionic strength, added electrolytes, the nature of the alkyl group, and other factors. Some of these factors will be illustrated in this experiment. In addition, the complex methodology of determining the rate of a reaction will be undertaken.

Experiments

A. STRUCTURAL EFFECTS ON S_N1 AND S_N2 REACTIVITY

In this experiment the relative reactivities of a series of halides are observed under two sets of conditions, one of which favors the S_N1 mechanism and the other the S_N2. The S_N2 reaction is observed by the displacement of chloride or bromide by an iodide ion in acetone solution. The iodide ion is a good nucleophile for the S_N2 reaction, whereas acetone is a relatively poor ionizing solvent, and S_N1 dissociation is minimized. Sodium iodide is very soluble in acetone, but sodium chloride and sodium bromide have very low solubilities; so, the course of the reactions can be followed by the formation of crystalline NaCl or NaBr (Eq. 14.1).

$$R\text{—}X + NaI \xrightarrow{\text{acetone}} R\text{—}I + NaX \downarrow \qquad \textbf{[14.1]}$$

The S_N1 reaction can be observed by treating the alkyl halide with a solution of silver nitrate in aqueous ethanol. Nitrate ion is a very poor nucleophile so there is little opportunity for S_N2 displacement. Dissociation of the alkyl halide by the S_N1 process is followed by the precipitation of the insoluble silver halide; the carbocation is then captured by alcohol or water (Eq. 14.2).

$$R\text{—}X \rightarrow R^+ + X^- \xrightarrow{AgNO_3} AgX\downarrow + NO_3^- \qquad [14.2]$$

Procedure

SAFETY NOTE ALKYL HALIDES ARE TOXIC; AVOID BREATHING THEM OR SPILLING THEM ON THE SKIN. INSURE THAT THERE IS PROPER VENTILATION.

Label two series of five clean, dry test tubes with the numerals 1 to 5. In each series of tubes, place 0.2 mL of the following halides: (1) *n*-butyl chloride, (2) *n*-butyl bromide, (3) *sec*-butyl chloride, (4) *tert*-butyl chloride, and (5) crotyl chloride [$CH_3CH = CHCH_2Cl$]. Keep the tubes stoppered with corks or parafilm and leave them covered at all times, before and after adding reagents. Obtain 15 mL of 15% NaI-acetone solution and 15 mL of 1% ethanolic $AgNO_3$ solution from the side shelf.

Arrange one series of tubes in order, from 1 to 5. Add 2 mL of the NaI solution to tube number 1 and note the time. (Add the solution from a pipet as rapidly as possible—not dropwise.) After 2 to 3 minutes, add 2 mL of NaI solution to the second tube and again note the time. Continue the addition at 2- to 3-minute intervals with the remaining tubes. After each addition, watch for any rapid reaction and then inspect the other tubes for signs of a precipitate. Note the time as closely as possible when precipitation begins to occur, recording the data in tabular form in your notebook. Cover the tubes and allow them to stand, observing them periodically while the next series is run.

Arrange the second series of tubes, and in the same way add 2-mL portions of the $AgNO_3$ solution to each tube at 2-minute intervals. Again, watch closely for any that change rapidly and then observe the others periodically. If possible, note the time both for the first appreciable turbidity and also for a definite precipitate. If any tubes in the NaI series are still clear at this point, loosen the covers slightly and place the tubes in a water bath at 50°C; note any changes that occur. Record the data in tabular form with column headings: tube number, time turbidity appears, and time of definite precipitate.

When you have completed these test tube reactions, pour the contents of all of the test tubes into the proper liquid waste container.

B. EFFECTS OF SOLVENT ON S_N1 REACTIVITY

In addition to structural features of the halide, reaction conditions can have a large effect on the rate of nucleophile substitution. For example, the solvent plays a major role in both S_N1 and S_N2 reactions, and in this experiment the effect of solvent on the rate of the S_N1 solvolysis of *tert*-butyl chloride will be studied. For this purpose, the most suitable method for comparing solvolysis rates is based on the fact that a strong acid is liberated in the reaction (Eq. 14.3). To determine the extent to which the reaction has proceeded, enough base is added to the reaction mixture to neutralize a small fraction of the acid produced. The solution becomes acidic after that fraction of the *tert*-butyl chloride has reacted, and the change in pH is detected with phenolphthalein indicator. The time required for neutralization is inversely proportional to the rate constant of the reaction, k_1, as shown in Equation 14.4.

$$t\text{-Bu—Cl} + \text{ROH} \rightarrow t\text{-Bu—OR} + \text{HCl} \qquad \textbf{[14.3]}$$

The rate of formation of HCl by S_N1 solvolysis of an alkyl chloride is equal to the rate of disappearance of alkyl chloride and is proportional to the concentration of the alkyl halide (Eq. 14.4).

$$\frac{d[\text{HCl}]}{dt} = k_1[\text{RCl}] = \frac{-d[\text{RCl}]}{dt} \qquad \textbf{[14.4]}$$

Integration of this equation from 0 to 50% reaction (100 to 50% of the RCl) provides Equation 14.5 where $t_{1/2}$ is the time required for 50% of the alkyl halide to react. If 0.50 equivalents of hydroxide ion were present at the start of the reaction, $t_{1/2}$ will be the time at which the solution becomes acidic.

$$\int_{1.0}^{0.5} \frac{d[\text{RCl}]}{[\text{RCl}]} = -\int_0^{t_{1/2}} k_1 dt$$
$$2.3 \log 0.5 = -k_1 t_{1/2} \qquad \textbf{[14.5]}$$
$$k_1 = -2.3 \log 0.5/t_{1/2} = 0.69/t_{1/2}$$

In the procedure to be used, the amounts of alkyl halide and hydroxide are not accurately measured, since only relative rates are desired. If the alkyl halide used were weighed and the NaOH solution were standardized and accurately dispensed, actual rate constants could be obtained.

Suitable solvent systems for this study are mixtures of water with methanol, ethanol, and acetone. The volume percent compositions indicated below will give conveniently measurable reaction rates, with exceptions noted for 50% methanol and 70% acetone. The following solutions will be available.

COMPOSITION (SOLVENT: WATER)

50:50 (not methanol)
55:45 (all solvents)
60:40 (all solvents)
65:35 (all solvents)
70:30 (not acetone)

Procedure

Select three to five solvent mixtures, using either different ratios of one solvent or the same ratio with the three to five solvents, and label each of three to five clean dry 13 × 100-mm test tubes accordingly. To each test tube add 2.0 mL of the solvent system and 3 drops of 0.5 *M* NaOH solution containing phenolphthalein indicator. Cover the tubes with corks or parafilm, and place them in a bath containing a thermometer and water at 30 ± 1°C. A Styrofoam cup placed in an empty beaker for stability makes a convenient insulated container for the bath. After 4 to 5 minutes in the bath to bring the solvent to bath temperature, add 3 drops of *tert*-butyl chloride to each test tube. Note the time of addition, shake or swirl the test tubes to mix the solutions, and replace them in the bath. Add a few mL of hot water as needed to the bath to maintain the temperature at 30 ± 1°C. Record the time required for the pink (basic) color to disappear in each solvent mixture. Tabulate your results and compare them with those of others in the class to obtain a more complete picture of the solvent effects.

C. EFFECT OF TEMPERATURE ON REACTION RATES

The rate of any reaction increases with increasing temperature because of the greater kinetic energy of the reacting molecules. The effect of temperature on the rate constant of a reaction is generally expressed in terms of the Arrhenius Equation (Eq. 14.6) where E_a is the activation energy, T is the temperature in K, A is a term related to the probability of a reaction occurring if the molecules have sufficient energy, and R is the gas constant (1.99 cal/mole deg). The activation energy can be obtained if the rate constant is known at two or more temperatures. Using the logarithmic form of Equation 14.6 (Eq. 14.7), it is seen that a plot of log k versus $1/T$ gives a straight line with slope equaling $-E_a/2.3R$.

$$k = A \exp\left(-\frac{E_a}{RT}\right) \qquad [14.6]$$

$$\log k = \log A - \frac{E_a}{2.3R}\left(\frac{1}{T}\right) \qquad [14.7]$$

The effect of temperature on the rate of solvolysis of *tert*-butyl chloride can be studied using the procedure and solvent systems of Part B at different temperatures. To obtain conveniently measurable rates, a solvent should be used that gives an end point at about 10 minutes in Part B. Although numerical values of the rate constants are not obtained by the comparative procedure in Part B, the neutralization times measured are inversely proportional to the rate constants (Eq. 14.5). Thus, a plot of log t values versus $1/T$ gives a straight line with slope $+ E_a/2.3R$ from which the value of E_a can be obtained.

Procedure

From your results in Part B, select the solvent system for which the neutralization time at 30°C is closest to 10 minutes. Obtain two 2-mL mixtures of this solvent composition and add 3 drops of phenolphthalein indicator solution as in Part B. (In this part of the experiment, duplicate samples should be run.)

Adjust the temperature of the water bath to 20 ± 1°C, insert both tubes, and allow 4 to 5 minutes for temperature equilibration. Note the time, add 3 drops of *tert*-butyl chloride to each tube, mix, and measure the time required for the color to disappear. Repeat the procedure with two other samples after adjusting the bath temperature to 40 ± 1°C.

Using the times required for the samples at 20°C and 40°C and the time at 30°C from Part B, make a plot of log t versus $1/T$ (°K), and from the slope, calculate the activation energy for the solvolysis in the solvent used.

D. REACTION OF ETHYL BROMIDE WITH HYDROXIDE ION, AN S_N2 REACTION

This is a variation on the classic experiments of Grant and Hinshelwood,

$$EtBr + OH^- \rightarrow EtOH + Br^- + \text{other products}$$

(*Note:* Students should work in pairs.) For each pair of students, prepare the following solutions:

a. 0.5 L of 0.10 *M* HCl
b. 0.1 L of 0.40 *M* ethyl bromide
c. 0.1 L of 0.40 *M* potassium hydroxide (CO_2 free)

Procedure

Place the ethyl bromide and KOH solutions into a water bath at 40°C and allow them to come to temperature equilibrium.

In a 250-mL volumetric flask, mix exactly 100 mL of the ethyl bromide solution and exactly 100 ml of the KOH solution and bring the volume to the mark with absolute ethanol. Mix the contents of the flask thoroughly, and place it securely in the water bath. Immediately take a 10-mL aliquot and deliver it to a 250-mL Erlenmeyer flask containing 50 mL of ice water. This will quench the reaction so that a titration can be performed. Take as the starting time when the pipet is half empty. Titrate rapidly with 0.10 *M* HCl to the phenolphthalein end point. Record the number of mL of acid necessary to allow a faint pink color to remain in the solution and record this as the initial reading. Repeat, taking aliquots and titrating them at 20-minute intervals up to 2 hours. Record the data in tabular form with these column headings: mL acid used, OH^- concentration, and $1/OH^-$.

Plot the value of $1/OH^-$ versus time. The slope of the line is the rate constant. Compare the values obtained with those of others in the class.

QUESTIONS

1. Account for the relative reactivities observed in Part A for the alkyl halides with NaI–acetone in terms of the structure of the transition state.
2. Similarly, account for the relative reactivities of the alkyl halides with $AgNO_3$–ethanol.
3. Based on the data obtained by your class in Part B, arrange the following solvents in order of increasing S_N1 solvolytic power: water, acetone, methanol, and ethanol. Explain in terms of the S_N1 reaction mechanism how the properties of the solvents can account for the observed trend.
4. With the procedure used in Part B, would the color change occur sooner, later, or at the same time if (a) twice as much *tert*-butyl chloride were used (6 drops); (b) twice as much NaOH–phenolphthalein solution were used?
5. Draw a sketch of the reaction coordinate diagram for the solvolysis of *tert*-butyl chloride, labeling the maxima and minima with appropriate structures, and indicating the activation energy measured in Part C.

REFERENCES

Grant, G. H.; Hinshelwood, C. N. *J. Chem. Soc.* **1933**, 258.

March, J. *Advanced Organic Chemistry*, 3rd ed.; Wiley-Interscience: New York, 1985.

Moore, J. W.; Pearson, R. G. *Kinetics and Mechanism;* John Wiley & Sons: New York, 1981.

Name ______________________ Section ____________ Date ________

14 Nucleophilic Substitution of Alkyl Halides

PRELABORATORY QUESTIONS

1. Give the structures of the products of the following reactions:

a. *tert*-butyl iodide and ethanol, heated

b. ethyl bromide and KOH(aq), room temperature

c. butyl chloride and KOH(aq), room temperature

2. Explain how the rate of a reaction such as that in Question 1a can be determined.

3. How many mL of a 0.10 *M* solution of HCl are necessary to neutralize 10.00 mL of 0.16 *M* NaOH?

4. What is the fate of the silver when $AgNO_3$ reacts with ethyl bromide?

15

Properties of Alcohols

Alcohols are an interesting and important group of organic compounds. They are used in beverages, in medicines, as solvents, and as synthetic intermediates. The characteristic functional group of alcohols is the hydroxyl group (OH) attached to a saturated carbon.

Alcohols are classified into three categories: primary (1°), secondary (2°), and tertiary (3°); the classification is based on the type of carbon to which the hydroxyl group is attached. If the carbon is attached to only one other carbon, it is a primary carbon and the alcohol is a primary alcohol. If the carbon is attached to two other carbons, it is a secondary carbon and the alcohol is a secondary alcohol and so on.

Alcohols have much higher boiling points than ethers or alkanes of similar molecular weight. This is due to the association of alcohol molecules through hydrogen bonding. For example, the boiling point of butyl alcohol is 118°C whereas the boiling point of the isomeric diethyl ether is 36°C.

A number of alcohols are common materials of commerce. Ethyl alcohol (grain alcohol) has been consumed in wines and other alcoholic beverages for thousands of years. Most methyl alcohol was once produced by the destructive distillation of wood, hence the name "wood alcohol." It is the major component of a number of commercial preparations used to avoid gasoline line freeze-up in automobiles. Ethylene glycol has a high boiling point, low freezing point, and is miscible with water; therefore, it finds wide use as an antifreeze in automobile radiators. Glycerol is frequently used in cosmetics and medicinal preparations.

Experiments

Some important reactions or properties of alcohols are demonstrated in the following experiments. The alcohols to be used in these experiments are ethanol (ethyl alcohol), 2-propanol (isopropyl alcohol), 1-butanol (*n*-butyl alcohol), 2-butanol (*sec*-butyl alcohol), 2-methyl-2-propanol (*tert*-butyl alcohol), and cyclohexanol.

A. REACTION WITH SODIUM (DEMONSTRATION)

SAFETY NOTE SODIUM IS VERY REACTIVE AND MUST BE TREATED WITH CAUTION. HANDLE IT WITH FORCEPS AND USE ONLY SMALL PIECES. SODIUM REACTS VIOLENTLY WITH WATER. CARRY OUT ALL REACTIONS IN A HOOD. IT IS NORMALLY STORED UNDER MINERAL OIL.

Procedure

1. Drop a small piece (2- to 3-mm cube) of sodium metal into a beaker of water (use a hood). After the reaction is complete, add 1 or 2 drops of phenolphthalein indicator to the remaining solution. Write an equation for the reaction of sodium with water.
2. Place 2 mL of ethanol in a large, dry test tube and add a small piece of sodium. Test the resulting solution with phenolphthalein. Write an equation for the reaction of sodium with ethanol. Is this reaction more or less vigorous than with water?
3. Place 2 mL of 1-butanol in a large, dry test tube and add a small piece of sodium. Note the rate of this reaction compared with the previous ones. Test the resulting solution with phenolphthalein. Write an equation for the reaction of sodium with 1-butanol.

Note that many organic liquids give a reaction with sodium metal even if the compound does not have a reactive hydrogen such as alcohols do. This is normally due to small amounts of water in the sample.

B. SOLUBILITY IN WATER

To 2 mL of water in a 13 $\times$ 100-mm test tube, add ethanol, drop by drop, and shake the tube to mix the two materials. Note how many drops must be added until the alcohol is no longer soluble in water, but do *not* use more than 10 drops.

Repeat this experiment using in turn 2-propanol, 1-butanol, and cyclohexanol. Record your results as very soluble (6 to 10 drops), soluble (2 to 5 drops), or insoluble (1 drop).

C. LUCAS TEST

The OH group of an alcohol is a poor leaving group and is not readily displaced by nucleophiles. However, in strongly acidic solutions, protonation of the OH group can occur to give $—\overset{+}{O}H_2$ and water may be displaced. The **Lucas reagent**, prepared by dissolving 16 g of anhydrous zinc chloride in 10 mL of concentrated hydrochloric acid with cooling, will cause conversion of alcohols to alkyl halides if the alcohol is sufficiently soluble in the reagent. With this reagent, tertiary alcohols react immediately and secondary alcohols in 3 to 10 minutes, while primary alcohols may take an hour or longer.

SAFETY NOTE IN THE FOLLOWING EXPERIMENTS CONCENTRATED HYDROCHLORIC ACID AND SULFURIC ACID, GLACIAL ACETIC ACID, AND ACETIC ANHYDRIDE ARE USED. THESE ARE CORROSIVE MATERIALS AND SHOULD BE USED WITH CARE EVEN THOUGH SMALL AMOUNTS ARE REQUIRED. CHROMIC ACID AND ITS SALTS ARE CLASSIFIED AS SUSPECTED CARCINOGENS. IF ANY OF THE ABOVE MATERIALS ARE SPILLED ON THE SKIN, RINSE THOROUGHLY WITH WATER THEN WASH WITH SOAP AND WATER.

Procedure

Place 2 mL of the Lucas reagent in a 13 × 100-mm test tube, add 4 drops of alcohol, shake the tube to mix the materials, and note the time required for the mixture to become cloudy or to separate into two layers. Try the test on 1-butanol, 2-butanol, 2-methyl-2-propanol, and your unknown. (*Note:* The alcohol must dissolve initially.)

D. OXIDATION USING CHROMIC ACID

Primary and secondary alcohols are oxidized rapidly to acids and ketones, respectively, by **chromic acid reagent**. This reagent, also called the **Jones reagent**, is prepared by dissolving 13.4 g of CrO_3 in 10 mL of water, adding 12 mL of concentrated H_2SO_4, and diluting the solution to 50 mL with water to give 2.68 *M* CrO_3. Tertiary alcohols are not oxidized by this reagent; however, they dehydrate in the acid environment and then the dehydration products oxidize after 2 or 3 minutes.

Procedure

Place 1 mL of acetone in a 13 × 100-mm test tube (acetone that is free of traces of alcohol must be used); add 1 drop of the alcohol and then 2 drops of the chromic acid reagent. Note whether the mixture stays clear and orange [CrO_3 color] or turns turbid blue-green [formation of Cr(III)]. Make the decision about the test within 5 seconds after the reagent is added. Try this test on 1-butanol, 2-butanol, 2-methyl-2-propanol, and your unknown.

Based on the result of the Lucas test and the chromic acid test, what can you conclude about the nature of your unknown alcohol (1°, 2°, or 3°)?

E. ESTER FORMATION

Alcohols react with carboxylic acids to yield esters in a condensation reaction in which water is eliminated. This reaction is catalyzed by mineral acids. Many low molecular weight, aliphatic esters have fragrant odors and are the materials primarily responsible for the odor of flowers and ripe fruits.

Procedure

1. **Acid and alcohol**. Mix 5 drops of ethanol and 5 drops of glacial acetic acid in a 13 × 100-mm test tube; add 1 drop of concentrated sulfuric acid, warm the mixture gently (3 to 4 minutes in a hot water bath), and then add 2 mL of cold water. Note the odor (if you have trouble detecting the odor or can only smell traces of acetic acid, pour the mixture into a small beaker and make it slightly basic with dilute NaOH). Write an equation for the reaction of acetic acid and ethanol.
2. **Acid Anhydride and Alcohol**. Place 5 drops of 1-butanol and 5 drops of acetic anhydride in a 13 × 100-mm test tube, stir well, and warm briefly. Add 2 mL of cold water, make the mixture slightly basic with dilute NaOH, and note the odor. Write an equation for the reaction of acetic anhydride and 1-butanol.

QUESTIONS

1. How does the size of the alkyl group in an alcohol affect the vigor of its reaction with sodium metal?
2. Why does the length of the carbon chain in an alcohol effect its solubility in water?
3. Write an equation for the reaction of 2-propanol with the Lucas reagent (ignore $ZnCl_2$).
4. Do primary or tertiary alcohols react more rapidly with the Lucas reagent? Why?
5. Which of the following alcohols would not be oxidized by chromic acid reagent within 5 seconds: 1-pentanol, cyclohexanol, 2-methyl-1-propanol, 2-methyl-2-butanol?
6. Using the tests described in this experiment, how could one distinguish between the following pairs of compounds? Tell what reagent or test you would use and what you would see.
   a. 1-butanol and 2-butanol
   b. isobutyl alcohol and *tert*-butyl alcohol
   c. 1-propanol and 1-heptanol
   d. cyclohexanol and cyclohexane

Name ______________________ Section ____________ Date ________

15 Properties of Alcohols

PRELABORATORY QUESTIONS

1. Write an equation for the reaction of sodium with water. Write a similar equation for the reaction of sodium with ethanol.

2. What is the color of chromium(VI) in solution? What is the color change when it is reduced to Cr(III) in the test for primary and secondary alcohols?

3. What precautions should be taken in using acetic anhydride? Why?

4. Write the equation for the reaction of acetic anhydride and ethanol.

5. Tertiary alcohols will give a reaction with chromic acid reagent after a few minutes. What initial reaction occurs in this acid environment?

16

Dehydration of Alcohols

Alcohols may be dehydrated under acid-catalyzed conditions to form alkenes. In the presence of Bronsted acids, such as sulfuric or phosphoric acid, dehydrations involve the initial protonation of the hydroxyl group (Eq. 16.1), converting it to an excellent leaving group. Loss of water then normally takes place by a unimolecular process, resulting in a highly reactive carbocation that can stabilize itself by the elimination of a proton from an adjacent carbon to give the alkene.

$$\underset{\text{alcohol}}{\overset{\mathrm{H}\quad\mathrm{OH}}{-\mathrm{C}-\mathrm{C}-}} \xrightarrow{\mathrm{H}^+} \overset{\mathrm{H}\quad\overset{+}{\mathrm{O}}\mathrm{H}_2}{-\mathrm{C}-\mathrm{C}-} \xrightarrow{-\mathrm{H_2O}} \underset{\text{carbocation}}{\overset{\mathrm{H}\quad +}{-\mathrm{C}-\mathrm{C}-}} \xrightarrow{-\mathrm{H}^+} \mathrm{>C{=}C<} \tag{16.1}$$

Carbocations can also rearrange, by alkyl and/or hydride shifts, to form more stable cationic species that then yield other alkenes. A typical example of such a multiple-product reaction is shown in Equation 16.2.

$$\mathrm{CH_3{-}C(CH_3)_2{-}CH(OH){-}CH_3} \underset{-\mathrm{H_2O}}{\xrightarrow{\mathrm{H}^+}} \underset{0.4\%}{\mathrm{CH_3{-}C(CH_3)_2{-}CH{=}CH_2}} + \underset{80.0\%}{\mathrm{CH_3{-}C(CH_3){=}C(CH_3){-}CH_3}} + \underset{19.6\%}{\mathrm{CH_3{-}CH(CH_3){-}C(CH_3){=}CH_2}} \tag{16.2}$$

The ratios and types of alkenes formed depend on the catalyst used and the reaction conditions. Many types of catalysts other than Bronsted acids have been used in alcohol dehydrations. These include iodine, alumina, molecular sieves, and ion exchange resins. Dehydrating agents such as $POCl_3$ and acetic anhydride have also been used in which case the elimination usually proceeds via the corresponding ester intermediate.

With secondary and tertiary alcohols, direct Bronsted acid–catalyzed dehydration is often a simple and convenient procedure. If the elimination product is sensitive to further reaction with acid, as may be the case if the alcohol contains other functional groups, procedures that avoid strong acids must be used.

Dehydration of alcohols using Bronsted acids is illustrated in the following experiments. The procedures described use phosphoric acid; sulfuric acid can cause charring and can give off sulfur dioxide fumes.

Experiments

SAFETY NOTE THE DEHYDRATION PRODUCTS FORMED IN THESE REACTIONS ARE HIGHLY FLAMMABLE. IF A BURNER IS USED IN THE DISTILLATION, TAKE CARE THAT VAPORS OF THE PRODUCT ARE NOT EXPOSED TO THE FLAME. HANDLE PHOSPHORIC ACID WITH CARE.

A. PREPARATION OF CYCLOHEXENE (Macroscale)

OH

H_3PO_4 / heat →

cyclohexanol
bp 160–161°C

cyclohexene
bp 83°C

Procedure

Weigh 10.0 grams of cyclohexanol into a 25-mL round-bottom flask. Carefully add 5 mL of 85% phosphoric acid, then a boiling stone, and mount the flask for fractional distillation (see Fig. 5.5; either use no column packing or a low hold-up packing). Slowly heat the mixture until it comes to a gentle boil. After 10 minutes of gentle boiling, increase the heat sufficiently to cause distillation (the temperature of the distilling vapor should not exceed 100°C) and collect the distillate in a cooled receiver.

To the distillate, add 2 mL of 10% sodium carbonate solution to neutralize any traces of acid which may have been carried over. Transfer the liquids to a separatory funnel, add 10 mL of cold water, swirl the mixture gently, and drain off the lower aqueous layer. Pour the cyclohexene into a small, dry Erlenmeyer flask and dry it over anhydrous calcium chloride for 10 to 15 minutes.

Decant the dried cyclohexene into a small dry distilling flask, add a boiling stone, attach the flask to a simple distillation assembly, and distill carefully. Collect the material distilling at 80 to 85°C. Determine the weight of the product and calculate the percentage yield.

Test the reactions of the product with bromine and with potassium permanganate (see Chapter 12, Tests 4 and 5). Also examine the purity of product by use of vapor phase chromatography, IR, or NMR as designated by your instructor.

B. DEHYDRATION OF 2-METHYLCYCLOHEXANOL (Macroscale)

CH_3, OH — $\xrightarrow[\text{heat}]{H_3PO_4}$ — CH_3 and/or CH_3

2-methylcyclohexanol bp 165–168°C

1-methylcyclohexene bp 110°C

3-methylcyclohexene bp 104°C

The dehydration of 2-methylcyclohexanol can occur in two directions, giving rise to 1-methylcyclohexene or 3-methylcyclohexene. In this experiment, gas chromatography (GC) will be used to determine the product composition. A disadvantage of using sulfuric acid in this experiment is that it is a sufficiently strong acid to reprotonate the double bond of the initial product and cause isomerization to other alkenes; therefore, phosphoric acid is used. In addition to the reactants, it is advantageous to use decalin or bromobenzene as a "chaser" in this experiment. These relatively high-boiling substances are inert under the reaction conditions and allow a more complete distillation of the product.

Procedure

To a 25-mL round-bottom flask, add 6 mL of 2-methylcyclohexanol, 2 mL of 85% phosphoric acid, 3 mL of bromobenzene, and a boiling stone. Attach the flask to a fractional distillation assembly (see Fig. 5.5; either use no column packing or a low hold-up packing). Slowly heat the contents of the flask to boiling and distill out the product. The distillation temperature should be kept below 96°C by regulating the rate of heating. Continue the distillation until 4 to 6 mL of liquid have been collected (a test tube or distillation receiver [centrifuge tube] is a convenient collection vessel).

Separate the product from the water layer by using a Pasteur pipet. Transfer the product to a clean, dry test tube and let it dry over $CaCl_2$ for 10 minutes.

Inject 0.5 μL of the dried liquid onto a nonpolar gas chromatograph column. Before doing so, be sure to fill and empty the syringe several times, discarding the sample each time, to clean out the previous contents. Draw approximately 1 μL of air into the syringe after the liquid when you are ready to analyze the sample. Insert the needle as far as possible into the injection port before injecting the sample to insure placing the sample directly onto the column.

Mark the point of injection on the recorder chart paper and measure accurately the distance (or time) from this point to the air peak and the olefin peak(s). By comparing the latter with the retention times of 1-methylcyclohexene and 3-methylcyclohexene, determine the identity of the dehydration product(s). If both are present, determine the approximate composition by triangulation of peak areas (see Fig. 6.5). Explain your results. If additional peaks are present, attempt to identify these as well.

If GC analysis is unavailable, measure the refractive index of the product and assume a linear relationship between the molar concentration and the refractive index. Refractive index values for 1-methylcyclohexene and 3-methylcyclohexene are 1.4503 and 1.4414, respectively.

C. DEHYDRATION OF 2-METHYLCYCLOHEXANOL (Microscale)

See the previous discussion in Part B. In the following procedure, phosphoric acid is used as a catalyst; however, the quantity used also allows it to function as a "chaser." Because of the relatively high ratio of phosphoric acid, a separate water phase will not be seen in the distillate.

Procedure

To a 5-mL round-bottom flask, add 1 mL of 2-methylcyclohexanol, 2 mL of 85% phosphoric acid, and a boiling stone. Mount the flask for simple distillation (see Fig. 5.4), slowly heat the contents of the flask to boiling, and distill out the product keeping the distillation temperature below 96°C. Continue the distillation until 0.5 to 0.7 mL of product is collected (a small test tube or centrifuge tube is a convenient receiver). Dry the product for 10 minutes over $CaCl_2$. Examine the dried material by gas chromatography as described in paragraphs 3 and 4 of the Procedure in Part B above.

QUESTIONS

1. What alkene(s) will be produced when each of the following alcohols is dehydrated:
   a. cyclohexanol
   b. 2-methylcyclopentanol
   c. 3-methylcyclopentanol
   d. 2,3-dimethyl-2-butanol
2. Illustrate the mechanism for the acid catalyzed dehydration of cyclohexanol.
3. Illustrate the mechanism for the acid-catalyzed dehydration of 2-methylcyclohexanol. How many products are possible?
4. Give equations to show how sulfuric acid can convert 4-methylcyclohexene into a mixture of four isomeric C_7H_{12} compounds (including starting material). Predict the relative amounts of each isomer.
5. Explain how the methylcyclohexene(s) can be distilled from the reaction mixture if the distillation temperature is kept well below 100°C.
6. The 2-methylcyclohexanol used in Parts B and C is actually a mixture of two isomers. What are they? Is it likely that this makes a difference in the product composition? Explain your answer.
7. Write equations to illustrate the reaction of cyclohexene with $KMnO_4$ and with bromine.
8. Infrared spectra of the two methylcyclohexene isomers obtained in this experiment are given in Figures 16.1 and 16.2. Assign vibrational modes for as many peaks as possible in both spectra.

   As an alternative to GC analysis, the relatively intensities of peaks in the infrared spectrum of a mixture may be used to determine its composition. Which peaks in Figures 16.1 and 16.2 could be used to analyze a mixture of these isomeric methylcyclohexenes?
9. The two methylcyclohexenes formed in Part B or C give the ^{13}C NMR spectra below. Decide which isomer gives which spectrum and assign as many signals as possible to specific carbon atoms. The chemical shifts of carbons 1,3, and 4 of cyclohexene are 127.4, 25.7, and 23.3 ppm from TMS, respectively.

   Isomer A: $\delta_c(CDCl_3,TMS)$ 133.6(*d*), 126.3(*d*), 31.5(*t*), 30.3(*d*), 25.3(*t*), 21.8(*q*), 21.6(*t*).

   Isomer B: $\delta_c(CDCl_3,TMS)$ 134.0(*s*), 121.2(*d*), 30.2(*t*), 25.4(*t*), 24.0(*q*), 23.2(*t*), 22.5(*t*).

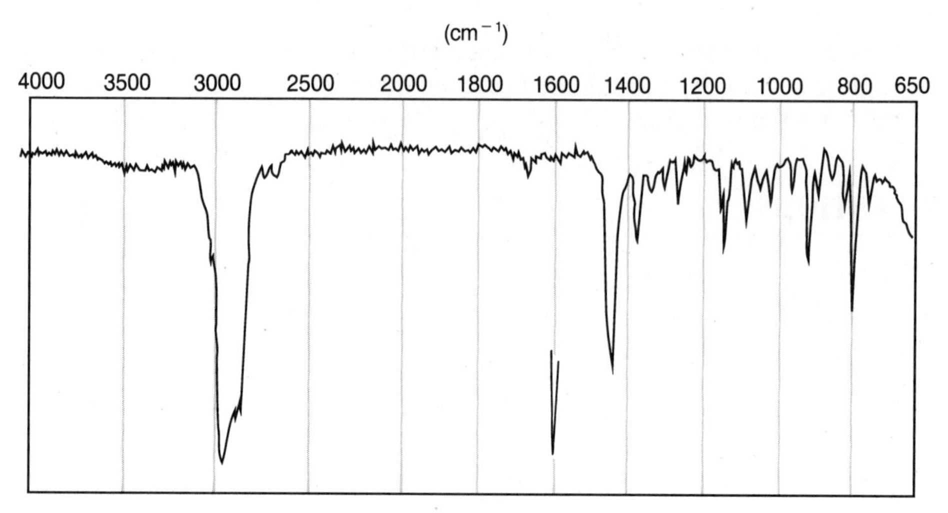

Figure 16.1 Infrared spectrum of 1-methylcyclohexene.

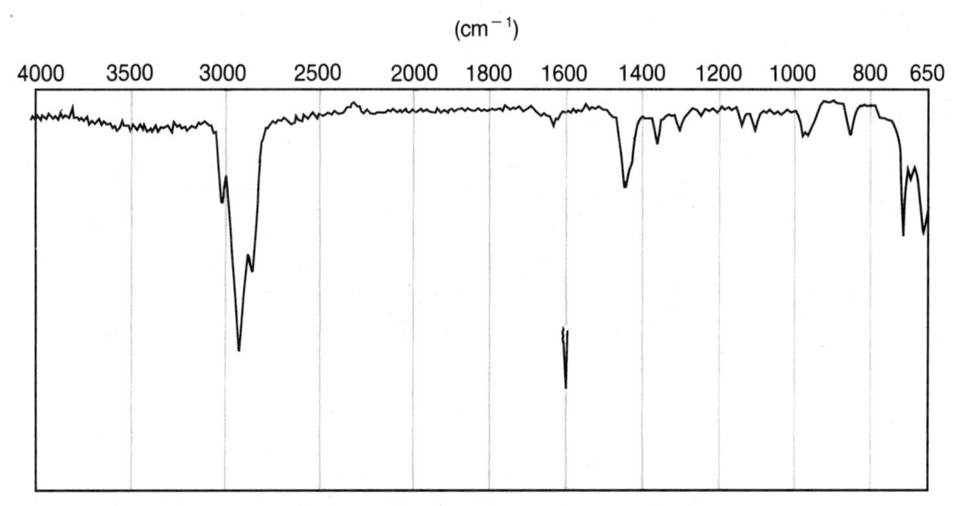

Figure 16.2 Infrared spectrum of 3-methylcyclohexene.

REFERENCES

Banthrope, D.V. *Elimination Reactions;* Elsevier: New York, 1963.
Taber, R.L.; Champion, W.C. *J. Chem. Ed.* **1967**, *44*, 620.
Taber, R.L.; Grantham, G.D.; Champion, W.C. *J. Chem. Ed.* **1969**, *46*, 849.

Name ______________________ Section ____________ Date ________

16 Dehydration of Alcohols

PRELABORATORY QUESTIONS

1. What alkene(s) will be produced when each of the following alcohols is dehydrated?

   a. propanol

   b. 2-butanol

2. In the preparation of cyclohexene, where are traces of sulfuric acid removed from the product?

3. Why is sulfuric acid or phosphoric acid used as a catalyst in these dehydration reactions rather than hydrochloric acid?

4. What safety measures should be taken if sulfuric or phosphoric acid is spilled or splashed on the skin?

17

The Grignard Reaction

Grignard reagents play a commanding role in organic synthesis. These compounds can be adapted to the preparation of a large variety of functional systems, and the formation and reaction of organomagnesium derivatives are one of the major uses of alkyl halides in organic synthesis.

The reaction of a halide and magnesium occurs on the metallic surface and is formally an oxidation of the metal. The reaction is usually carried out in ether solution, the ether solvent functioning as a Lewis base by solvating the Grignard reagent and permitting it to diffuse away from the metal.

$$RX + Mg \xrightarrow{Et_2O} \underset{\underset{OEt_2}{\vdots}}{\overset{\overset{OEt_2}{\vdots}}{RMgX}}$$

The formation of the organometallic reagent requires an active metal surface, and some difficulty may be encountered in getting the reaction started. Grinding the magnesium, either dry or in the presence of halide, usually is effective in providing a clean surface. Other tricks include the addition of a small crystal of iodine or the addition of a small amount of 1,2-dibromoethane, which reacts rapidly with the magnesium to form ethylene.

$$BrCH_2CH_2Br + Mg \rightarrow CH_2{=}CH_2 + MgBr_2$$

A major pitfall in preparing a Grignard compound is the presence of water, alcohol, or any other "active hydrogen" compound that can react rapidly as an acid with the organomagnesium compound.

$$R'OH + RMgX \rightarrow RH + MgXOR'$$

This reaction consumes and wastes the Grignard compound by the formation of hydrocarbon, and it may seriously complicate the preparation, since the product MgXOR′ can coat the surface of the metal and prevent attack of the halide. For this reason, the apparatus and all reagents and solvents must be scrupulously dry.

Grignard reagents are most frequently prepared from alkyl bromides, which are usually more costly than the chlorides but also somewhat more reactive with magnesium. The experiments in this chapter involve two syntheses using phenylmagnesium bromide.

Experiments

A. TRIPHENYLCARBINOL AND BENZOIC ACID

Br (bromobenzene, d 1.5) + Mg $\xrightarrow{\text{ether}}$ MgBr (phenylmagnesium bromide)

MgBr $\xrightarrow[\text{2) } H_3O^+]{\text{1) } C_6H_5COOCH_3}$ $(C_6H_5)_3C{-}OH$ (triphenylcarbinol)

MgBr $\xrightarrow[\text{2) } H_3O^+]{\text{1) } CO_2}$ COOH (benzoic acid)

SAFETY NOTE THE MAJOR POTENTIAL HAZARD IN THIS EXPERIMENT IS FIRE. ETHER IS HIGHLY FLAMMABLE, AND THERE SHOULD BE **NO FLAMES** IN THE LABORATORY AT ANY TIME DURING THIS EXPERIMENT.

(Macroscale)

Procedure

Clamp a 100-mL two-or three-neck round-bottom flask (or a single-neck flask fitted with a Claisen adapter) by the center neck to a ring stand, leaving room to slide an ice bath under the flask. (Do not support the flask on a ring.) Place a reflux condenser in one of the side necks, a dropping funnel in the center neck, and a stopper in the third neck. *All of the apparatus must be thoroughly dry* and you may want to place it in a drying oven before assembly. All ground glass joints should be carefully greased. Fit a drying tube containing calcium chloride to the top of the condenser.

Obtain 10.0 g of bromobenzene and 1.50 g of magnesium turnings. Grind the magnesium in a mortar to provide a fresh surface, place it in the flask, and add 20 mL of *anhydrous* ether. With a pipet, add approximately 2 mL of the bromobenzene through the side neck of the flask below the surface of the ether, to obtain a high concentration of halide close to the metal. Replace the stopper or condenser, and warm the flask gently with warm water or the palm of your hand.

If no visible sign of reaction (bubbling, turbidity) is seen after 5 minutes, add a small crystal of iodine and carefully crush it against the magnesium with a stirring rod. Continue the warming, and if the reaction again does not begin after 5 minutes, a "starter" solution should be added. To prepare a **starter solution**, place 0.2 to 0.3 mL of the bromobenzene, 1 mL of ether, and two to three chips of magnesium in a small test tube. With a glass rod, gently grind the metal against the glass to expose fresh surface. If this mixture does not begin to react, add 1 drop of 1,2-dibromoethane. (This compound is classified as a suspected carcinogen.) When a vigorous reaction occurs (solution begins to bubble and becomes warm), quickly pour the contents of the test tube into the reaction flask.

When the reaction is actively in progress (heat evolved, turbulence), begin the flow of condenser water. Mix the remaining bromobenzene with 23 mL of anhydrous ether and place this solution in the dropping funnel. Add the bromobenzene solution dropwise, with occasional swirling of the flask, at a rate to maintain a steady reflux. After the halide has been added, warm the mixture on the steam bath as needed to maintain reflux for another 20 minutes. If the solution is black at this point, it is due to finely dispersed impurities from the magnesium; a few bits of metal may also remain. Proceed from this point with Part B or C.

⦿ (Microscale)

Procedure

On a ring stand, set up the apparatus shown in Figure 17.1 consisting of a 5-mL round-bottom flask fitted with a Claisen adapter. Cap the shorter arm of the adapter with a septum, and place a condenser with a drying tube containing calcium chloride in the other. The apparatus must be absolutely dry to assure a successful reaction, and you may want to dry the pieces in an oven before assembly.

Weigh out about 50 mg of magnesium turnings which had been previously ground in a mortar to expose a fresh surface. Because of the size of the individual turnings, it may be difficult to obtain exactly 50 mg; however, this is not important. Get an accurate weight, calculate the mil-

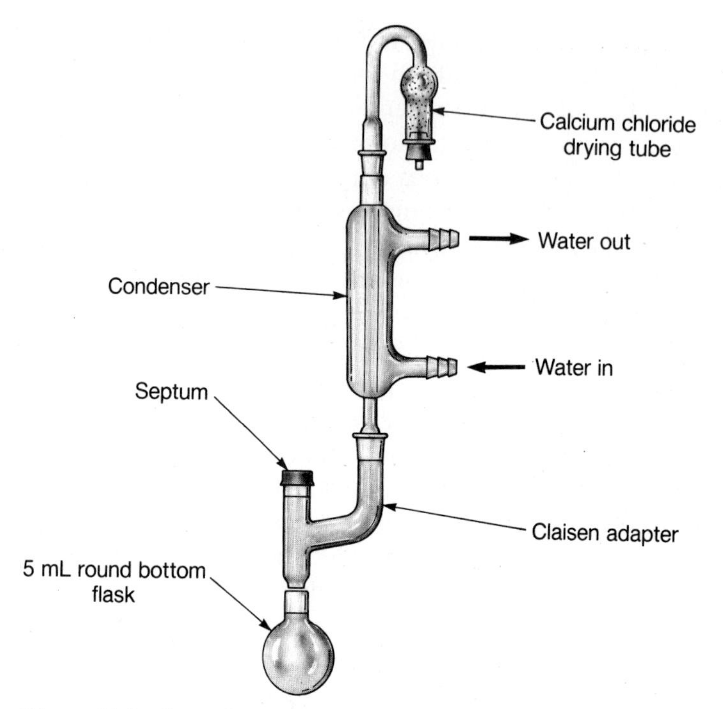

Figure 17.1 Setup for the microscale Grignard reaction.

limoles, and use an equal millimolar amount of bromobenzene where it is called for in the procedure below. Place the magnesium in the 5-mL flask, and assemble the apparatus as shown in Figure 17.1. All ground glass joints (except the one capped by the septum) must be properly greased to avoid their fusion in the presence of the strongly basic Grignard reagent. With a syringe, add 0.5 mL of *anhydrous* ether through the septum. In a separate container mix 0.5 mL of *anhydrous* ether and the equimolar amount of bromobenzene needed to react completely with the magnesium. Add approximately 10 drops of this mixture through the septum to the magnesium and gently shake the apparatus to mix the contents. If the reaction does not start in several minutes, remove the septum and gently rub the magnesium against the bottom of the flask with a dry stirring rod. If the reaction does not begin after several additional minutes, one of the methods mentioned in the Macroscale procedure should be tried. When the reaction is vigorously under way, turn on the condenser water and add the remainder of the bromobenzene–ether mixture through the septum with the syringe at such a rate that the reaction remains at a moderate reflux. If the reaction becomes too vigorous, it may be moderated by immersing the flask in a beaker of cold water. (Should reaction cease completely, it may be necessary to warm the flask to start it again.) When the reaction begins to slow, place it in a beaker of warm water (40°C) and continue the gentle reflux until the magnesium is essentially consumed. Replace any anhydrous ether that may have been lost due to evaporation, and proceed with Part B or C.

B. PREPARATION OF TRIPHENYL CARBINOL

In synthetic chemistry, the Grignard reaction is one of the most useful and versatile, particularly when it involves carbonyl systems. For example, reaction with formaldehyde gives primary alcohols, with other aldehydes it yields secondary alcohols, and with ketones it produces tertiary alcohols. The reaction with esters also yields tertiary alcohols, but with two groups the same, both arising from the Grignard reagent (Table 17.1).

(Macroscale)

Procedure

While the Grignard solution is being refluxed, weigh out methyl benzoate equivalent in moles to one-half the amount of bromobenzene used. (Remember that the reaction of 1 mole of ester requires 2 moles of Grignard reagent.) Mix the ester with three volumes of anhydrous ether and place it in the dropping funnel. When the formation of the Grignard reagent is complete, cool the flask and its contents briefly in an ice bath. Add the ester solution dropwise with swirling of the apparatus to insure mixing. The addition should be at a rate to prevent excessive refluxing, and if necessary, an ice bath can be used to control the reaction. When the addition is complete, heat the mixture at reflux for 30 minutes or let stand overnight.

Pour the reaction mixture into a 125-mL Erlenmeyer flask containing 25 mL of 10% sulfuric acid and approximately 15 g of ice. Rinse the flask with a small amount of ordinary ether, followed by a small amount of 10% sulfuric acid, and add these rinses to the flask containing the reaction mixture. Swirl the mixture well to promote hydrolysis and then transfer the contents with a pipet to a separatory funnel, leaving the magnesium and remaining ice behind. Separate the layers, wash the ether layer with 10% sulfuric acid, followed by saturated sodium chloride solution, and let it dry over magnesium sulfate. Filter the mixture to remove the drying agent, add 15 mL of high-boiling petroleum ether, and evaporate the solution slowly on a steam bath until crystallization of the triphenyl carbinol begins. Cool in an ice bath to complete the crystallization and collect, dry, and weigh the product.

(Microscale)

Procedure

In a test tube mix 1.0 mL of anhydrous ether and an amount of methyl benzoate equivalent in moles to one-half the amount of bromobenzene used (recall that two moles of Grignard reagent react with each mole of ester). When the formation of the Grignard reagent is complete, let the mixture cool and add the ester solution dropwise through the septum with a syringe.

Table 17.1 The Reaction of a Grignard Reagent with Various Carbonyl Systems

PREPARATION OF A	GRIGNARD REAGENT UTILIZED		CARBONYL COMPOUND REQUIRED				PRODUCT OBTAINED
Primary alcohol	RMgX	+	HCHO	$\longrightarrow$	RCH_2OMgX	$\xrightarrow{H_3O^+}$	RCH_2OH
Secondary alcohol	RMgX	+	R′CHO	$\longrightarrow$	$R\text{-}CH(R')\text{-}OMgX$	$\xrightarrow{H_3O^+}$	$RCH(R')\text{-}OH$
Tertiary alcohol	RMgX	+	$R'\text{-}C(=O)\text{-}R''$	$\longrightarrow$	$R\text{-}C(R'')(R')\text{-}OMgX$	$\xrightarrow{H_3O^+}$	$R\text{-}C(R'')(R')\text{-}OH$
	RMgX	+	$R'\text{-}C(=O)\text{-}OR''$	$\longrightarrow$	$R\text{-}C(R)(R')\text{-}OMgX$	$\xrightarrow{H_3O^+}$	$R\text{-}C(R)(R')\text{-}OH$

Because of the low vapor pressure of ether, the ester solution may tend to dribble or even squirt out of the syringe—hold the plunger tightly and work carefully so as not to lose any of the material. The addition rate should be such that the reaction always remains under control. If the mixture begins to reflux excessively, the reaction can be moderated by cooling in a beaker of water or even briefly in ice. When the addition is complete, heat the mixture at gentle reflux on a steam bath for 15 minutes or let it stand overnight.

Transfer the mixture to a centrifuge tube that contains 1.0 mL of 10% sulfuric acid and is approximately half filled with ice. Rinse the reaction flask with approximately 0.5 mL of ordinary ether and add this to the tube containing the reaction mixture. Stopper the tube and shake gently to hydrolyze the alkoxide salt. Unstopper the tube and let it stand. When the ice has melted, remove the lower (aqueous) layer with a pipet and bulb and discard. Wash the ether layer with water, followed by saturated sodium chloride solution, and dry it over magnesium sulfate. Add 0.5 mL of high-boiling petroleum ether and a boiling stone, and evaporate some of the solvent on a steam bath until the triphenylcarbinol crystallizes. Complete the crystallization by cooling the tube in ice. Filter the mixture using a Hirsch funnel, dry the product, and weigh.

C. PREPARATION OF BENZOIC ACID

Carbonation of a Grignard reagent with CO_2 is a general method for preparing carboxylic acids. The initial product is the halomagnesium salt of the acid, which undergoes further addition of Grignard reagent at a very much slower rate than does CO_2. Carbon dioxide can be bubbled into the Grignard solution from a cylinder of liquid CO_2 under pressure, or more conveniently for small reactions, the Grignard solution can simply be poured onto dry ice.

$$C_6H_5\text{MgBr} + O{=}C{=}O \longrightarrow C_6H_5CO_2\text{MgBr} \xrightarrow{H_3O^+} C_6H_5\text{COOH}$$

benzoic acid
mp 121°C

(Macroscale)

Procedure

While the Grignard solution is being refluxed, obtain approximately 30 g of crushed dry ice in a towel. Cool the Grignard solution and remove the condenser and dropping funnel. Blot the dry ice to remove condensed ice crystals, place it in a dry 250-mL Erlenmeyer flask, and pour the Grignard solution into the flask. Stopper the flask *loosely* with a cork and allow it to stand until the excess dry ice has sublimed. (If necessary, the reaction mixture can be stored at this point until the next laboratory period.)

Add 20 mL of ordinary ether to the residue and then add 20% hydrochloric acid in small portions, with shaking, until all solid has dissolved. Separate the layers, and wash the ether solution with water. To isolate the acid, extract the ether solution thoroughly with two 15-mL portions of 10% NaOH solution and then with 10 mL of water. Combine the aqueous extracts, back-extract once with ether, and then acidify with hydrochloric acid. Extract the mixture with ether, dry the extract with Na_2SO_4, and evaporate the ether in a tared round-bottom flask in a hood. Remove the last traces of ether with aspirator vacuum, and weigh the crude acid. Cool in ice to crystallize if the product is not solid. The crude benzoic acid contains a small amount of biphenyl ($C_6H_5—C_6H_5$, mp 69 to 72°C), which can be removed by recrystallizing the product from aqueous acidic solution.

(Microscale)

Procedure

Wipe the frost from several small pieces of dry ice, crush them in a cloth towel, and add enough to a centrifuge tube to fill it about one-fourth full. With a syringe, squirt the Grignard reagent directly onto the dry ice. Rinse the reaction flask with approximately 1 mL of anhydrous ether, add this to the dry ice, and allow the excess dry ice to sublime. (At this point, the reaction mixture can be stored until the next period, if required.)

Add 1 mL of ordinary ether to the residue, and then add dropwise, with shaking, 10% hydrochloric acid until all of the solid has dissolved. With a syringe or pipet, remove and discard the aqueous layer, and wash the ether solution with water. The benzoic acid is purified by extracting the ether solution with two 1-mL portions of 10% aqueous NaOH and then with 1 mL of water. Combine the aqueous solutions, back-extract with 1 mL of ordinary ether, and then acidify with 10% hydrochloric acid. Extract the acidified mixture with three 1-mL portions of ordinary ether, and dry the combined ether layers over Na_2SO_4. With a syringe or pipet, carefully remove the ether solution from the drying agent, and place it in a tared 10-mL round-bottom flask. Evaporate the ether on a steam bath, preferably in a hood, and remove the last traces of solvent under aspirator vacuum. Weigh the crude acid, and place the flask in ice to effect crystallization if the product is not solid. The crude acid can be recrystallized from a minimum amount of slightly acidified water if required.

QUESTIONS

1. In a reaction of phenylmagnesium bromide with 2-butanone, the Grignard reagent was prepared from 25 g of bromobenzene and was then treated with a solution of 10 g of the butanone in 50 mL of ether that contained 1% by weight of water. Calculate the amount of benzene (in g) that would be produced and the maximum theoretical yield of tertiary alcohol, corrected for loss due to benzene formation.
2. Give a detailed procedure, with amounts of reagents and so forth, for the conversion of 15 g of 1-bromoheptane to 3-decanol using a Grignard reaction.
3. A side reaction in the preparation of phenylmagnesium bromide (favored if the bromobenzene is added too quickly to the magnesium) leads to biphenyl (C_6H_5—C_6H_5). Give a mechanism that explains the formation of this product.
4. Triphenylcarbinol can also be prepared by the reaction of phenylmagnesium bromide with benzophenone (C_6H_5—$\overset{\overset{\displaystyle O}{\|}}{C}$—$C_6H_5$) and with diethyl carbonate (C_2H_5O—$\overset{\overset{\displaystyle O}{\|}}{C}$—$OC_2H_5$). Write stepwise mechanisms for these reactions.

REFERENCE

March, J. *Advanced Organic Chemistry*; 3rd ed.; Wiley: New York, 1985, pp. 825–826.

Name ______________________ Section ______________ Date ________

17 The Grignard Reaction

PRELABORATORY QUESTIONS

For questions 1 through 4, circle the most correct answer.

1. Comparing alkyl bromides and alkyl chlorides in the preparation of Grignard reagents, alkyl bromides are **a.** more reactive, **b.** less reactive, **c.** both have about the same reactivity, **d.** cannot predict the reactivity because it depends on the structure of the alkyl group, or **e.** are totally unreactive.

2. The common solvent for Grignard reactions is **a.** methylene chloride, **b.** benzene, **c.** hexane, **d.** acetone, or **e.** ether.

3. 1,2-Dibromoethane **a.** is sometimes used in the preparation of Grignard reagents as a complexing agent for the reagent after it has formed, **b.** can activate the magnesium to help start the reaction, **c.** is a useful cosolvent to help dissolve the Grignard reagent, **d.** increases the boiling point of the reaction mixture, thereby enhancing the formation of the Grignard reagent, or **e.** is not normally useful in the formation of a Grignard reagent since it forms a Grignard reagent itself, being an alkyl halide.

4. The presence of a small amount of ethanol **a.** should not affect the formation of the Grignard reagent, **b.** should actually enhance the formation of the Grignard reagent, **c.** enhances the activity of the magnesium, **d.** would react with the Grignard reagent, or **e.** would stabilize a Grignard reagent by complexing with it.

5. Write equations for the reaction of phenylmagnesium bromide with the following compounds: acetone, methyl formate, water, benzaldehyde, acetic acid.

18

Preparation and Stereochemistry of Bicyclic Alcohols

Reactions and interconversions of compounds in the monoterpene series have been of great importance in studying the mechanism of carbocation rearrangements and the stereoselectivity of various reagents and synthetic reactions. Typical monoterpenes are the ketone camphor and the corresponding epimeric alcohols borneol and isoborneol. A related series of compounds, not occurring in nature, also has been studied extensively; these are the norbornane derivatives **1** through **4**.*

The norborneols **2** and **3** can be prepared by two of the major synthetic reactions for alcohols: hydration of alkenes and reduction of carbonyl groups (see Eq. 18.1). Since separation of the epimeric norborneols is difficult, it is important that preparative methods for each isomer proceed with high stereoselectivity. The two reactions in this experiment fulfill this requirement and are complementary, each giving a different isomer as the major product. Since only two isomers are possible, the outcome of either preparation will establish the steric course of both reactions. The isomer obtained in the hydration can be oxidized to norcamphor (Chapter 19), supplying the starting material for the other alcohol.

Hydration of alkenes in the presence of acid proceeds by way of carbocations, and mixtures of products are often encountered (see Chapter 16, Question 4). Although carbocation rearrangements occur frequently in

*The prefix "nor-" in the names of these compounds indicates the presence of three hydrogen atoms in place of the methyl groups in the natural terpenes. (The usual usage for this prefix is to denote the absence of a single CH_2 group from the parent compound.) Systematic names for these bridged bicyclic compounds (and the terpenes as well) are based on the parent hydrocarbon bicyclo[2.2.1]heptane; the numbers in brackets designate the number of atoms in each of the three "bridges" of the bicyclic skeleton.

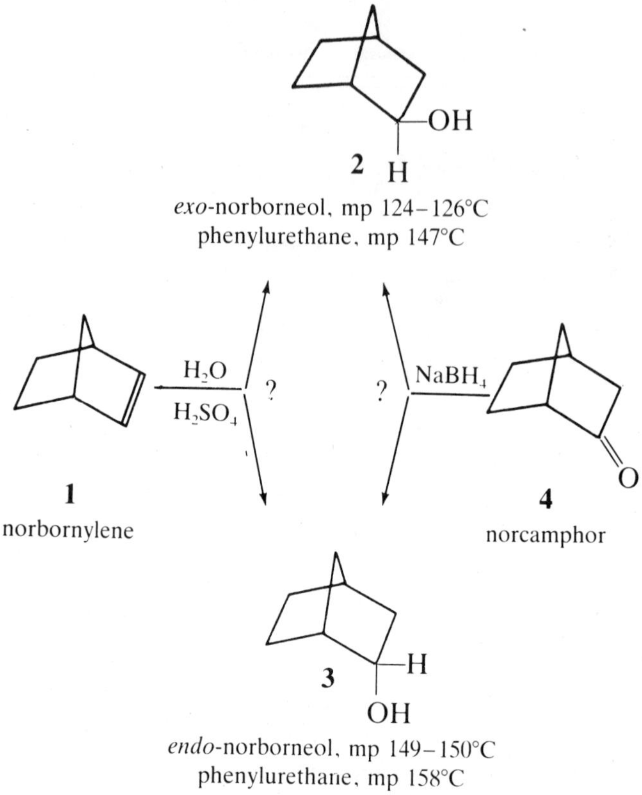

OH
2 H
exo-norborneol, mp 124–126°C
phenylurethane, mp 147°C

H_2O / H_2SO_4 ? ? $NaBH_4$

1
norbornylene

O
4
norcamphor

[18.1]

3 H
OH
endo-norborneol, mp 149–150°C
phenylurethane, mp 158°C

reactions of terpenes, the hydration of norbornylene (**1**) proceeds cleanly and in one steric direction, which is to be determined in this experiment. The mechanism of the reaction and the structure of carbocations in this bridged bicyclic system have been subjects of some controversy.

The second norborneol isomer is obtained by the hydride reduction of norcamphor. The complex hydrides $NaBH_4$ and $LiAlH_4$ are among the most useful reagents available for the conversion of carbonyl compounds to alcohols. Sodium borohydride is the less reactive of the two; for example, esters and acids are not affected. Sodium borohydride is very convenient to use, since reactions can be carried out in aqueous or alcoholic solutions. The reduction of bicyclic ketones such as camphor and norcamphor with these hydrides is quite stereoselective, one of the two diastereomeric alcohols being formed in over nine times the amount of the other. In these rigid cyclic compounds, the stereochemistry of the reduction is controlled by the shielding of one side of the carbonyl group from attack by the reagent. Thus, in camphor, the methyl groups on the one-carbon bridge screen the approach of hydride from the "top" or *exo* side of the two-carbon bridge, and the hydrogen atom is added to the *endo* side, giving the *exo*-alcohol isoborneol (see Eq. 18.2).

Since the separation of norborneol mixtures is impractical, accurate product ratios can be established only by refined analytical methods. When one isomer greatly predominates, however, as in these experiments, it can easily be isolated in sufficiently pure form to permit identification on the basis of the melting point of the alcohol and, if necessary, a derivative.

$$\text{camphor} \xrightarrow{MH_4} \text{isoborneol} \qquad [18.2]$$

Experiments

SAFETY NOTE CONCENTRATED SULFURIC ACID WILL CAUSE BURNS ON CONTACT WITH THE SKIN; WASH IMMEDIATELY WITH COLD WATER IF A SPILL OCCURS.

A. ACID-CATALYZED HYDRATION OF NORBORNYLENE (Macroscale)

Carefully add 4 mL of concentrated H_2SO_4 to 2 mL of water in a 50-mL Erlenmeyer flask, cool in an ice bath to 15 to 20°C, and add 2.0 g of norbornylene. (*Note:* This hydrocarbon is highly volatile, and it must be added to the acid as soon as it is weighed to avoid loss.) Swirl the mixture until all of the hydrocarbon has dissolved, cooling briefly in cold water if the mixture becomes perceptibly warm; do not cool below room temperature. Allow the solution to stand, and prepare a solution of 3 g of KOH in 15 mL of water. Cool both solutions in an ice bath, and slowly add the KOH solution to the acid-reaction mixture. Transfer the mixture to a separatory funnel and extract with ether (40- and 10-mL portions). Three liquid phases may be present in the separatory funnel at this point; if this occurs, the addition of a few mL of water will cause the lower two layers to coalesce. Wash the ether solution with 3 mL of water, followed by 10 mL of $NaHCO_3$ solution, dry over anhydrous $MgSO_4$, filter into a heavy-walled 125- or 250-mL side-arm filter flask, and concentrate to an oil on the steam bath (see Fig. 4.4). Cool the flask to room temperature, stopper, and evacuate (using an aspirator) to remove residual solvent. Purify the product by sublimation as described in the following section.

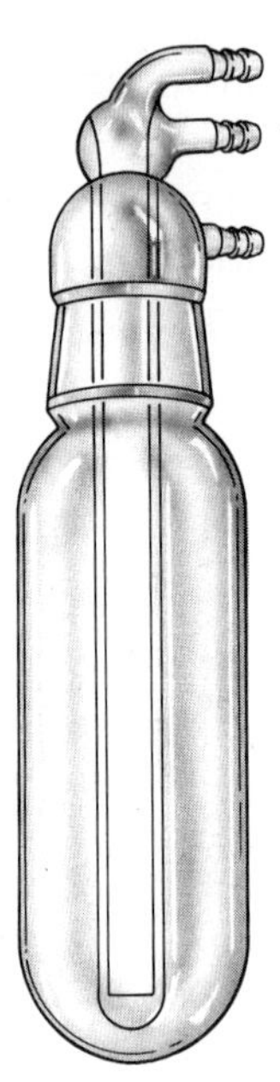

Figure 18.1 All-glass sublimer.

Sublimation is very effective for the purification of norborneols. In this technique, a solid is placed in a vessel that can be evacuated and heated, and in which a cold finger is positioned a short distance above the material to be sublimed. A typical all-glass sublimer is shown in Figure 18.1. For this experiment, a very simple sublimer can be set up using ordinary glassware, as shown in Figure 18.2. A 125- or 250-mL filter flask serves as the subliming vessel and an 18 $\times$ 150-mm test tube as the cold finger. The test tube is fitted with a rubber stopper having a 16-mm hole. Lubricate the rubber sleeve with glycerin, slip it over the test tube, and adjust the position so that the cold finger is approximately 2 cm above the bottom of the flask when it is securely seated. Wash off the excess glycerin with water, wipe the test tube clean, and dry thoroughly.

Insert the cold finger, wedge it securely in the neck, and again evacuate the flask with the aspirator; warm the flask to ensure that all the ether is removed. Half fill the cold finger with chipped ice, and gently heat the flask on the steam bath until the alcohol has completely sublimed. Keep the cold finger well filled with ice during the sublimation process. It may be necessary to draw off the water from the melted ice with a Pasteur pipet.* Cool the flask, disconnect it from the aspirator, and pour the water out of the cold finger. Loosen the cold finger from the neck and pull it out gently, holding it in a horizontal or slightly inverted position. The crystalline deposit is soft,

*During these manipulations it is vitally important that no water enter the flask, either by condensation or leakage, otherwise an inferior product will result.

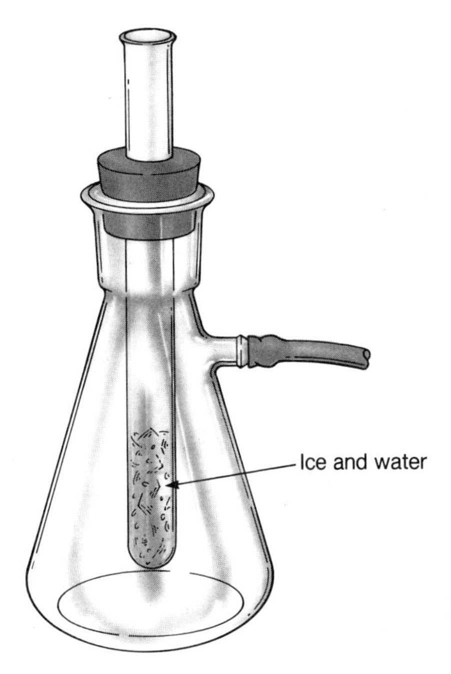

Figure 18.2 Filter flask–test tube sublimer.

and, if need be, it can be compressed somewhat without crumbling. Scrape the crystals from the test tube, weigh, and determine the melting point in a sealed capillary. (Fill the capillary in the usual way, and then heat it in a microburner flame approximately 3/4 inch from the open end until soft. Pull it out, and fuse it in the flame at the constriction.)

Identify the isomer that was obtained by comparing the melting point with the known values. To confirm the identification, a sample of the alcohol can be converted to the phenylurethane (see Chapter 46).

(Microscale)

Carefully add 2.0 mL of sulfuric acid to 1.0 mL of water in a 25-mL Erlenmeyer flask and cool the mixture in an ice bath. Transfer 0.75 mL of this mixture to a distillation receiver or centrifuge tube with a Pasteur pipet. Weigh 200 mg of norbornylene and add this immediately to the distillation receiver, stirring the contents with a microspatula until the alkene is completely dissolved. Return the mixture to the ice bath.

Dissolve 1.5 g of potassium hydroxide in 7.5 mL of water in another 25-mL Erlenmeyer flask. Cool this mixture in the ice bath, and carefully add 2 mL of this aqueous base to the contents in the distillation tube. (**CAUTION:** Recall that the addition of a strong base to a strong acid produces considerable heat!) Cool the mixture and extract it right in the distillation receiver with two portions of ether (5 mL and 1 mL), each time transferring the ether layer with a Pasteur pipet to a 25-mL centrifuge tube. Wash the ether extracts with 1 mL of water, followed by 1 mL of $NaHCO_3$

(when removing the lower layer in such washes, squeeze out a small stream of air bubbles from the pipet while it is being lowered through the upper layer to avoid removing any of the upper layer). Remove the ether layer with a clean Pasteur pipet and filter into a 50-mL filter flask through a plug of cotton containing a small layer of anhydrous $MgSO_4$. Wash the $MgSO_4$ with 1 to 2 mL of fresh ether, and remove the ether solvent by heating on a steam bath. Insert a 13 × 100-mm test tube as a cold finger (see Fig. 18.2), and evacuate the system to remove residual solvent. Add ice to the cold finger, and while continuing to evacuate, place the entire apparatus into the steam bath. The oily residue should sublime as white crystals on the cold finger. Periodically draw off the water that accumulates in the cold finger with a Pasteur pipet and replace with more ice as needed.* Continue the evacuation and heating until all of the product has sublimed, remove the apparatus from the steam bath, pour out the water and ice remaining in the cold finger, and carefully remove the cold finger from the flask. Scrape the product from the cold finger, and record the percent yield and melting point. Follow the procedure given in the Macroscale section for determining a melting point in a sealed capillary. Identify the isomer obtained by comparing the melting point with the reported values. If required, the identification can be confirmed by preparing the phenylurethane derivative (see Chapter 46).

B. SODIUM BOROHYDRIDE REDUCTION (Macroscale)

In a 50-mL Erlenmeyer flask, dissolve 1.1 g of norcamphor in 5 mL of ethanol and cool the solution in ice. In a small test tube, dissolve 0.15 g of $NaBH_4$ in 2 mL of water. Add the borohydride solution to the camphor, and allow the solution to warm and stand at room temperature for 10 minutes. Add 6 mL of water, and heat the solution for 5 minutes on the steam bath. Cool, extract with 20 mL of pentane, wash with water, dry over $MgSO_4$, filter into a 125- or 250-mL filter flask, and concentrate to an oil on the steam bath. Cool the flask to room temperature, stopper, and evacuate (using an aspirator) to remove residual solvent. Purify the norborneol by sublimation as described in the Macroscale acid-catalyzed hydration procedure.

(Microscale)

In a centrifuge tube or distillation receiver, dissolve 200 mg of norcamphor in 1 mL of ethanol and cool the mixture in an ice bath. In a 10 × 75-mm test tube dissolve 30 mg of $NaBH_4$ in 0.5 mL of water, and add this mixture

*During these manipulations it is vitally important that no water enter the flask, either by condensation or leakage, otherwise an inferior product will result.

to the norcamphor solution. Warm to room temperature, and let stand for 10 minutes. Add 1.5 mL of water, and heat the mixture on a steam bath for 5 minutes. Cool the solution to room temperature, and extract with 4 mL of pentane using the Pasteur pipet procedure described in the Microscale procedure for Part A. Wash with 1 mL of water, and filter the pentane mixture through anhydrous $MgSO_4$, again as described in Part A. Remove the pentane solvent by evaporation (finally under reduced pressure), and sublime the remaining oil. Record the percent yield and melting point using a sealed capillary as described in the Macroscale section. Identify the product by comparing the melting point with the known values.

QUESTIONS

1. The major source of norbornane derivatives is the Diels–Alder reaction (Chapter 30). Formulate the preparation of norbornylene, and endo-norborneol, using ethylene and vinyl acetate ($CH_2{=}CHOC(=O)CH_3$), respectively, with a suitable diene.
2. Compare the steric course of hydride reductions of camphor and norcamphor. Suggest a reason for the stereochemistry of the norcamphor reaction.
3. Another method for hydrating olefins utilizes hydroboration–oxidation, which brings about *cis*-addition of water via an organoborane:

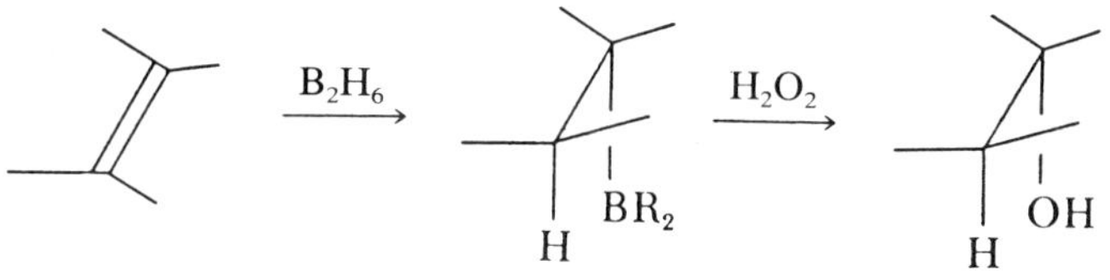

$$\xrightarrow{B_2H_6} \quad (H,\ BR_2) \quad \xrightarrow{H_2O_2} \quad (H,\ OH)$$

This procedure is reported to give exclusively one alcohol with norbornylene (see Brown, H.C.; Zweifel, G. *J. Am. Chem. Soc.* **1961**, *83*, 2544). From the outcome of the borohydride reduction of norcamphor, which norborneol isomer would you predict from the B_2H_6/H_2O_2 reaction with norbornylene?

Name ______________________ Section ______________ Date ________

18 Preparation and Stereochemistry of Bicyclic Alcohols

PRELABORATORY QUESTIONS

1. Write the stepwise mechanism for the acid-catalyzed hydration of norbornylene to give either the *exo*- or *endo*-norborneol.

2. Give the systematic names for each of the following bridged bicyclic compounds.

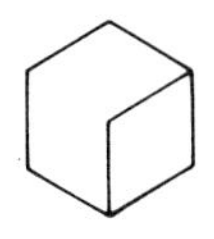 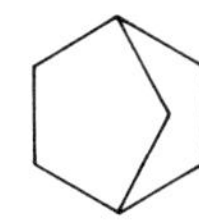 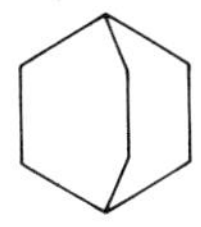 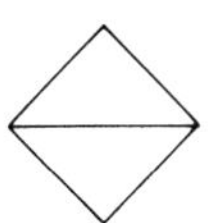

3. What is the limiting factor (i.e., that component in the lowest molar ratio) in both the macroscale and microscale procedures for Parts A and B? Support your answers with the appropriate calculations.

4. Phenylurethanes are excellent derivatives of alcohols and can serve to confirm their identities. Using structural formulas, write the reaction (including necessary reagents) for the preparation of the phenylurethane of *exo*-norborneol. (*Hint:* See Chapter 46 for the preparation of phenylurethanes.)

19

Oxidation

Oxidation is a common and widely used reaction in organic chemistry. For example, primary alcohols may be oxidized to aldehydes or carboxylic acids, secondary alcohols to ketones, aldehydes to carboxylic acids, alkenes to glycols, aromatic side chains to carboxylic acids, and so forth. Alkenes, alkynes, and 1,2-glycols may be cleaved by oxidizing agents. These and numerous other reactions make oxidation of organic compounds an important category.

Two of the most important and general oxidizing agents are chromic acid and potassium permanganate. Both are strong oxidizing agents and can be used for a variety of purposes. Other oxidizing agents include hypochlorous acid, nitric acid, Tollens reagent, Benedict's reagent, and various reagents containing chromium trioxide such as the Jones reagent (CrO_3 in aqueous H_2SO_4), Collins reagent (CrO_3 in pyridine), and the Corey reagent (pyridinium chlorochromate).

Alcohol oxidations are usually carried out with chromic acid, H_2CrO_4, generated either from the reaction of $Na_2Cr_2O_7$ or CrO_3 (chromic anhydride) with sulfuric or acetic acid. Permanganate in alkaline solution is more often the preferred reagent for side-chain oxidations or cleavage of a carbon chain at a double bond or carbonyl group (the latter via the enol). In both cases the oxidations proceed by the formation of intermediate esters with the reagent followed by breakdown to the product and a reduced inorganic species.

In the chromic acid oxidation of an alcohol, the intermediate chromate ester gives initially an unstable Cr(IV) species that reacts with Cr(VI) to give two equivalents of a Cr(V) species, $HCrO_3$. The $HCrO_3$ oxidizes additional alcohol and is finally reduced to Cr(III). The overall electron change for chromium is thus VI $\rightarrow$ III, or a gain of three electrons. The color change in the oxidation provides a sensitive test to detect a primary or secondary alcohol; aliphatic aldehydes also give a positive test. Under controlled conditions, other functional groups, including tertiary hydroxyl groups, amines, aromatic aldehydes, and double bonds, do not react. The reaction is the basis for the determination of ethanol in exhaled breath in devices used by law-enforcement officials.

$$R_2C(OH)H + H_2CrO_4 \xrightarrow{-H_2O} R_2C(H)\text{—}O\text{—}CrO_3H \longrightarrow R_2C\text{=}O + H_2CrO_3$$

(unstable) Cr(IV)

$$H_2CrO_3 \downarrow HCrO_3 \xrightarrow{R_2CHOH} Cr(III) + R_2C\text{=}O$$

Permanganate oxidation of an alkene or enol involves a cyclic permanganate ester. Under mild conditions, a glycol or ketol can be obtained; more vigorous oxidation gives carboxylic acids. The stable reduction product of permanganate in alkaline solution is MnO_2, which forms an insoluble brown precipitate. The electron change for this process is thus a gain of three; however, the actual stoichiometry may be more complex, since molecular oxygen is often formed in a side reaction during permanganate oxidations:

$$RCH\text{=}CHR' + MnO_4^- \longrightarrow RCH\text{—}CHR' \text{ (cyclic ester, O—Mn(=O)(O}^-\text{)—O)} \longrightarrow RCH(OH)\text{—}CH(OH)R'$$

$$\xrightarrow{MnO_4^-} RCO_2H + R'CO_2H + MnO_2$$

An important consideration in carrying out an oxidation reaction is the amount of oxidant required. For this, of course, a balanced equation for the oxidation is needed. To balance such a reaction, keep in mind that for any C—H or C—C bond that is broken (and replaced by a C—O bond), the molecule loses two electrons. For example, in the chromic acid oxidation of 2-propanol to acetone, one C—H bond is broken, with loss of two electrons. Since chromic acid gains three electrons, one then requires a ratio of three moles of alcohol (3 × 2) to two of chromic acid (2 × 3). The equation is completed by adding acid to balance charges and water to balance oxygen atoms:

$$3(CH_3)_2CHOH + 2H_2CrO_4 + 6H^+ \rightarrow 3(CH_3)_2C\text{=}O + 2Cr^{3+} + 8H_2O$$

In the vigorous oxidation of ethylbenzene with $KMnO_4$, assuming complete oxidation of the CH_3 group to CO_2, there are five C—H bonds and one C—C bond broken for an overall electron change of 12 electrons, requiring four moles of permanganate.

$$C_6H_5CH_2CH_3 + 4MnO_4^- \rightarrow C_6H_5CO_2^- + 4MnO_2 + 2H_2O + OH^- + CO_3^{2-}$$

In this chapter, procedures for a medium-scale oxidation and several smaller scale oxidations are described. It should be noted that many oxidation reactions are exothermic and caution must be exercised in keeping the reaction under control. This is particularly important in larger-scale reactions.

In the oxidation of cyclohexanol (Part A), hypochlorous acid is used. Chromic acid has been widely used for this and similar reactions; however, chromic acid and its salts are suspected carcinogens so their use in other than small-scale reactions is not advisable. The procedure employs household bleach, a relatively safe material, as a source of the hypochlorite ion.

In the second reaction (Part B), cyclohexanone is oxidized further by permanganate with ring opening to give a dicarboxylic acid. The reaction is carried out on a microscale. Since the reaction is exothermic, careful addition of one of the reactants would be necessary to keep the reaction under control if it were being carried out on a large scale.

The technique of small-scale oxidation with chromic oxide in aqueous H_2SO_4 (Jones method) is illustrated in the third experiment (Part C). This procedure is particularly convenient for small-scale oxidations when an accurately measured amount of Cr(VI) is required.

In the fourth experiment (Part D), the usefulness of the Jones reagent in a qualitative test for primary and secondary alcohols is described.

In the last experiment (Part E), the haloform reaction of acetophenone is illustrated. This interesting reaction results in the oxidation of the side chain to a carboxylic acid.

Experiments

SAFETY NOTE ALKALINE SOLUTIONS, STRONGLY ACID SOLUTIONS, AND CHROMIC ACID SOLUTIONS (SUSPECTED CARCINOGEN) ARE USED IN THE FOLLOWING EXPERIMENTS. TAKE CARE TO AVOID CONTACT WITH THESE REAGENTS; IF SOME IS SPILLED ON THE HANDS, RINSE THOROUGHLY WITH WATER AND WASH WITH SOAP AND WATER.

A. PREPARATION OF CYCLOHEXANONE (Macroscale)

cyclohexanol (OH) $\xrightarrow{HOCl}$ cyclohexanone (O)

cyclohexanol
bp 161°C

cyclohexanone
bp 155°C

Procedure

In a 250-mL Erlenmeyer flask, place 10.0 g of cyclohexanol (10.4 mL, 0.1 mole) and then 25 mL of glacial acetic acid. Place the flask in an ice bath, and using a separatory funnel, add 75 mL of 5% NaOCl solution (Clorox) dropwise, with occasional swirling, while keeping the temperature at 30 to 35°C. Swirl the mixture well, return it to the ice bath, and add another 75 mL of 5% NaOCl, dropwise. At this point the mixture should give a positive reaction with potassium iodide–starch paper. If it does not, add more NaOCl solution (1 to 5 mL) until a positive reaction is obtained. Let the mixture stand 15 minutes at room temperature. Add saturated sodium bisulfite solution until the mixture gives a negative potassium iodide–starch test.

Transfer the mixture to a 250-mL distilling flask and distill until the distillate appears free of droplets (approximately 75 mL). It is convenient to collect the distillate in a 250-mL Erlenmeyer flask.

To free the distillate of any acetic acid, add anhydrous sodium carbonate carefully with stirring until foaming ceases. Then add 8 g of sodium chloride, and stir approximately 10 minutes to further salt out the cyclohexanone. Decant the mixture into a separatory funnel, and separate the layers; place the cyclohexanone in a 25-mL Erlenmeyer flask. Return the aqueous layer to the separatory funnel and extract it with 5 mL of ether. Separate the two layers, and add the ether extract to the cyclohexanone. Dry the product over anhydrous magnesium sulfate, decant through a cotton plug into a 25-mL distilling flask, and after distilling off the ether (CAUTION: fire hazard), collect the fraction boiling at 150 to 155°C in a tared receiver. Record the weight of product obtained, and calculate the percent yield. Examine the product by gas chromatography and infrared spectroscopy.

B. PREPARATION OF ADIPIC ACID (Microscale)

cyclohexanone (O) $\xrightarrow{KMnO_4}$ COOH / COOH

d 0.95
bp 155°C

adipic acid
mp 153–154°C

Procedure

Place 245 mg (0.0025 mole) of cyclohexanone in a 50-mL Erlenmeyer flask, add a mixture of 790 mg of $KMnO_4$ (0.0050 mole) in 15 mL of water, and then add 0.1 mL (2 to 3 drops) of 10% NaOH. Let the mixture stand for 10 minutes with occasional swirling, and then place it in a boiling water bath for 20 minutes. Test for residual potassium permanganate by placing a drop of the reaction mixture from the tip of a stirring rod on a piece of filter paper. Unreacted potassium permanganate will appear as a purple ring around the brown manganese dioxide. If unreacted permanganate remains, decompose it by adding a small portion of solid sodium bisulfite.

Vacuum filter the mixture through a small Hirsch funnel to remove most of the manganese dioxide, and wash the manganese dioxide with approximately 2 mL of hot water. Place the filtrate in a 25-mL beaker, add a boiling stone, and boil it gently on a hot plate until the volume is reduced to approximately 5 mL. If the solution is colored at this stage, add decolorizing charcoal (after letting the solution cool), boil briefly, and remove the charcoal by filtration through a filter pipet. Add concentrated HCl to the clarified solution until it is acid to litmus, and then add an additional 10 drops of HCl. Allow the solution to cool to room temperature, and collect the product by vacuum filtration. Dry the product, and determine its weight and melting point.

C. PREPARATION OF NORCAMPHOR (Macroscale)

$$\text{OH (norborneol)} \xrightarrow[H_2SO_4]{CrO_3} \text{O (norcamphor)}$$

mp 88–90°C

The following procedure is based on a 0.01-mole scale; if a smaller amount of alcohol is used, reduce the amount of CrO_3 solution and solvents proportionately. A semimicroscale preparation such as this, of a low molecular weight, relatively volatile compound, places a premium on thoughtful experimentation and good technique. All glassware should be clean and dry. Extraction must be thorough, but if an excessive volume of solvent is used, losses will be incurred during evaporation. Volumes of wash water should be kept to a minimum, and separations should be sharp.

Procedure

In a 25 × 100-mm test tube, dissolve 1.12 g (0.01 mole) of norborneol in 8 mL of reagent-grade acetone and place the solution in an ice bath. Measure (using a graduated pipet) 2.50 mL of 2.68 *M* CrO_3 in sulfuric acid into a 10 × 75-mm test tube. Add the CrO_3 solution dropwise (using a transfer pipet and bulb) to the norborneol, shaking and swirling it in an ice bath. As soon as addition is complete, dilute the reaction mixture with 8 mL of pentane, stopper the test tube with a rubber stopper, shake, and carefully pour off the upper layer into a small separatory funnel. Extract the aqueous residue in the same way with three additional 4-mL portions of pentane. (Extracting the solution in the test tube and decanting will help avoid losses due to transferring the mixture in and out of the separatory funnel.) Wash the combined pentane solutions with three or four 1- to 2-mL portions of water, adding each wash to the original aqueous phase.

Extract the combined aqueous phases once more in the test tube with 6 mL of pentane, decant the pentane as completely as possible into the funnel containing the initial pentane solutions, separate any droplets of the lower phase, wash the pentane with 3 mL of $NaHCO_3$ solution, and transfer the pentane solution to an Erlenmeyer flask, rinsing the funnel with a minimum volume of fresh pentane. Dry with 1 to 2 g of $MgSO_4$ (only a small amount of drying agent is needed, since pentane dissolves little water), and filter the dried solution through a small plug of cotton into a dry 125-mL side-arm flask. Rinse the drying agent with a few milliliters of pentane. Evaporate the solution on the steam bath (see Fig. 4.4) to a thin oil, then remove, tip the warm flask slightly, and rotate it to spread the oil over the bottom and lower part of the sides as the residual pentane vapor flows out. Cool the flask to room temperature, stopper, connect to the aspirator, and evacuate *briefly* until the residue forms a solid crust. Release the vacuum immediately, and sublime the product (see Chapter 18). Record the weight and melting point (sealed capillary).

D. CHROMIC ACID TEST FOR ALCOHOLS

The Jones reagent, CrO_3 dissolved in sulfuric acid/water, may be used to distinguish primary and secondary alcohols from tertiary alcohols, as the former are readily oxidized by this reagent. The procedure is described in Chapter 15, Part D.

E. HALOFORM REACTION (Microscale)

$$C_6H_5\text{—}\overset{\overset{\displaystyle O}{\|}}{C}\text{—}CH_3 \xrightarrow[\text{2) HCl}]{\text{1) NaOCl}} C_6H_5\text{—}\overset{\overset{\displaystyle O}{\|}}{C}\text{—}OH + CHCl_3$$

benzoic acid
mp 122°C

Procedure

Place 100 mg of acetophenone in a 15 × 125-mm test tube, add 4.0 mL of 5% NaOCl solution (Clorox or other household bleach), and then add 0.30 mL of 10% NaOH. Heat the mixture in a 80 to 90°C water bath for 15 to

25 minutes with frequent shaking; note when all the acetophenone appears to have dissolved. Add 4 drops of acetone to destroy excess hypochlorite, and cool to room temperature. Add concentrated HCl dropwise until the mixture is acid to litmus, and then add 2 drops in excess. Chill the acidified solution in an ice bath for 10 to 15 minutes to fully precipitate the benzoic acid. Vacuum filter the product, allow it to dry, and then determine the weight and melting point. Calculate the percent yield. If necessary, the benzoic acid may be recrystallized from water.

QUESTIONS

1. Considering that the Cr in CrO_3 and $Cr_2O_7^{2-}$ is in the 6+ oxidation state and that each step in the sequence $CH_4 \rightarrow CH_3OH \rightarrow CH_2{=}O \rightarrow HCOOH \rightarrow CO_2$ corresponds to a two-electron change, complete and balance the following reactions:
   a. $CH_3CH_2CHOHCH_3 + Na_2Cr_2O_7 + H_2SO_4 \rightarrow CH_3CH_2COCH_3 + Cr^{3+}$
   b. $C_6H_5CH_2OH + CrO_3 \rightarrow C_6H_5COOH + Cr^{3+}$
   c. $CH_3CH_2{-}C(CH_3){=}CH_2 + H_2Cr_2O_7 \rightarrow CH_3CH_2COCH_3 + CO_2 + Cr^{3+}$
2. Write a balanced equation for the oxidation of cyclohexanone to adipic acid, and calculate the theoretical amount of permanganate needed for this procedure.
3. In an oxidation of cyclohexanol in which the temperature was allowed to rise above 50°C, the yield of cyclohexanone was low, and an acid by-product, with a melting point of 153 to 154°C, was isolated. Suggest the structure of this product.
4. In basic solution, sodium bisulfite reduces permanganate to manganese dioxide, whereas in acidic solution the product is manganous ion (Mn^{+2}). Write balanced equations for these reactions.
5. What organic product would you expect from the permanganate oxidation of the following ketones: **a.** acetophenone, **b.** 2-pentanone, **c.** 1-phenyl-2-propanone, and **d.** camphor?

REFERENCES

Chinn, L.J. *Selection of Oxidants in Synthesis;* Marcel Dekker: New York, 1971.
Perkins, R. *J. Chem. Ed.* **1984**, *61*, 551.
Perkins, R.A.; Chau, F. *J. Chem. Ed.* **1982**, *59*, 981.
Trahanovesky, W.S., Ed. *Oxidation in Organic Chemistry*, Parts B, C, D; Academic Press: New York, 1973, 1978, 1982.
Wiberg, K.B., Ed. *Oxidation in Organic Chemistry*, Part A; Academic Press: New York, 1965.
Zuczek, N.M.; Furth, P.S. *J. Chem. Ed.* **1981**, *58*, 824.

Name ______________________ Section ______________ Date ________

19 Oxidation

PRELABORATORY QUESTIONS

1. a. What is the reduction product of CrO_4^{2-} in acidic solution? How many electrons are gained in this reaction?

b. What is the reduction product of MnO_4^- in basic solution? How many electrons are gained in this reaction?

2. Complete and balance the following reaction:

$$MnO_4^- + Fe^{2+} + H^+ \rightarrow Mn^{2+} + Fe^{3+} + H_2O$$

3. a. In the oxidation of a secondary alcohol to a ketone, how many electrons are lost?

b. Write a balanced equation for the oxidation of cyclohexanol with hypochlorous acid, HClO. (*Note:* The reduction product of hypochlorite ion is chloride ion.)

4. Give two reasons why particular care should be taken in working with chromic acid reagent (CrO_3 in aqueous H_2SO_4).

[illegible] Oxidation

PREPARATORY QUESTIONS

1. [illegible]

2. [illegible]

3. [illegible]

4. [illegible]

5. [illegible]

6. [illegible]

7. [illegible]

20

Aldehydes and Ketones

Aldehydes and ketones are two of several types of compounds that contain the carbonyl group. Reactions that occur because of the presence of the carbonyl group include nucleophilic addition reactions and base-catalyzed condensations. Aldehydes are also easily oxidized, which provides a convenient means to distinguish them from ketones.

The carbonyl group in aldehydes and ketones is highly polarized; the carbonyl carbon bears a substantial partial positive charge and is susceptible to nucleophilic attack. Further, since it is sp^2 hybridized it is relatively open to attack. Because the carbonyl carbon contains no good leaving group, nucleophilic addition occurs rather than nucleophilic substitution. Hydrogens attached to a carbon adjacent to the carbonyl group (α-hydrogens) are unusually acidic because of the polarization of the carbonyl group and the resonance stabilization of the resulting anion. The acidity of these α-hydrogens leads to a variety of base-catalyzed condensation reactions.

Because of the carbonyl group, aldehydes and ketones are polar compounds; however, the pure compounds do not undergo hydrogen bonding as alcohols do. The boiling points of aldehydes and ketones are therefore lower than alcohols of comparable molecular weight but higher than alkanes or ethers. The carbonyl oxygen will form hydrogen bonds with water so that low molecular weight aldehydes and ketones are appreciably soluble in water.

Aldehydes and ketones are compounds of considerable commercial and biological interest. Formaldehyde, acetaldehyde, and acetone are important industrial chemicals and the annual production of each is in the range of 1- to 3-million tons. Naturally occurring compounds include vanillin, camphor, cinnamaldehyde, and carvone (spearmint oil).

Each of the tests in the Experiments section will be performed on known compounds to familiarize you with the test procedure and the interpretation of results. You will also perform these tests on an unknown, to determine whether the compound is an aldehyde or ketone and, if it is a ketone, whether it is a methyl ketone. You may be instructed to identify the specific compound, in which case the data in Table 19.1 will be helpful.

Table 20.1 Derivatives of Aldehydes and Ketones

COMPOUND	2,4-DNP (°C)	SEMICARBAZONE (°C)
Acetone	126	187
Propanal	154	154
Butanal	123	106
2-Butanone	117	146
2-Pentanone	144	112
3-Pentanone	156	139
Cyclopentanone	146	203
Cyclohexanone	162	167
Heptanal	108	109
Benzaldehyde	237	222
Acetophenone	250	198
p-Tolualdehyde	234	215

Experiments

SAFETY NOTE ALTHOUGH ONLY SMALL AMOUNTS OF REAGENTS ARE USED IN THIS EXPERIMENT, SOME ARE STRONGLY ACIDIC OR BASIC AND APPROPRIATE CARE SHOULD BE TAKEN IN HANDLING THEM.

A. REACTIONS WITH NITROGEN COMPOUNDS

Aldehydes and ketones react with a number of nitrogen containing compounds through nucleophilic addition and subsequent loss of water to give products that have a carbon-nitrogen double bond. These reactions are useful in distinguishing aldehydes and ketones from other functional groups and in the identification of specific aldehydes and ketones. Two reagents that frequently yield crystalline derivatives are 2,4-dinitrophenylhydrazine and semicarbazide.

1. **2,4-Dinitrophenylhydrazones.** Prepare a solution of 4 drops of sample, or a comparable amount of solid, in 1 mL of ethanol. Add 1 mL of 2,4-dinitrophenylhydrazine reagent (see p. 123), mix thoroughly, and allow to stand. Many aldehydes and ketones give a solid derivative immediately. If no precipitate appears within 5 minutes, heat the solution in a hot water bath for 5 minutes and then allow it to cool to room temperature. The solid 2,4-dinitrophenylhydrazone (2,4-DNP) may be isolated by vacuum filtration, rinsed with a little cold water, and recrystallized

from ethanol. Perform this test on benzaldehyde, methyl ethyl ketone, and your unknown.

$$R{-}\overset{\overset{\displaystyle H(R')}{|}}{C}{=}O + H_2N{-}NH{-}C_6H_3(NO_2)_2 \longrightarrow R{-}\overset{\overset{\displaystyle H(R')}{|}}{C}{=}N{-}NH{-}C_6H_3(NO_2)_2$$

2. **Semicarbazones** Add 4 drops of sample, or a comparable amount of solid, to 1 mL of semicarbazide reagent (see Chapter 46) and mix thoroughly. Allow the mixture to stand until the product has crystallized well; it may be necessary to chill it in an ice bath. Add 10 drops of water to dissolve any NaCl, chill the solution, and vacuum filter the solid semicarbazone. Prepare this derivative for benzaldehyde, methyl ethyl ketone and your unknown.

$$R{-}\overset{\overset{\displaystyle H(R')}{|}}{C}{=}O + H_2N{-}NH{-}\overset{\overset{\displaystyle O}{\|}}{C}{-}NH_2 \longrightarrow R{-}\overset{\overset{\displaystyle H(R')}{|}}{C}{=}N{-}NH{-}\overset{\overset{\displaystyle O}{\|}}{C}{-}NH_2$$

B. OXIDATION REACTIONS

Aldehydes are very easily oxidized to yield carboxylic acids or their salts if the oxidation is done in basic solution. Since ketones are not readily oxidized, this difference provides a convenient means to distinguish aldehydes from ketones. Strong oxidizing agents such as chromic acid are frequently employed; however, much less vigorous oxidizing agents such as Tollens reagent have been used.

1. **Chromic Acid Reagent** Dissolve 1 drop of sample in 1 mL of reagent-grade acetone, and to this solution, add 1 drop of chromic acid reagent (Chapter 15, Part D). Rapid formation of a green to blue-green precipitate constitutes a positive test for an oxidizable compound, that is, in this case an aldehyde. Aliphatic aldehydes react usually within 15 seconds, whereas aromatic aldehydes require approximately 1 minute. Ketones react only after several minutes. Note that other relatively easily oxidized compounds, such as primary and secondary alcohols, also react readily. Perform this test on propionaldehyde, benzaldehyde, methyl ethyl ketone, and your unknown.

$$3R{-}\overset{\overset{\displaystyle O}{\|}}{C}{-}H + 2CrO_3 + 6H^+ \longrightarrow 3R{-}\overset{\overset{\displaystyle O}{\|}}{C}{-}OH + 2Cr^{3+} + 3H_2O$$

2. **Tollens Test** To prepare the reagent, add dilute (1:10) ammonium hydroxide with stirring to 3 ml of 5% silver nitrate solution until the initial precipitate just dissolves. Add 2 drops of sample to 1 ml of the reagent in a thoroughly clean test tube, and let stand for 10 minutes. (For water-insoluble aldehydes, dissolve 2 drops of sample in 0.5 mL of 95% ethanol, then add 1 mL of the reagent.) If the test is negative, place the test tube in a boiling water bath for 1 minute. The formation of a silver mirror on the wall of the tube through reduction of silver ion constitutes a positive test for aldehydes. Discard all unused reagent and test solutions *immediately* after use, as these solutions can become explosive on standing. Excess Tollens reagent can be destroyed by the addition of dilute nitric acid. Perform this test on propionaldehyde, benzaldehyde, methyl ethyl ketone, and your unknown.

$$R{-}\overset{\overset{\displaystyle O}{\|}}{C}{-}H + 2Ag(NH_3)_2OH \longrightarrow R{-}\overset{\overset{\displaystyle O}{\|}}{C}{-}O^-NH_4^+$$

$$+ 2\ Ag + H_2O + 3NH_3$$

C. IODOFORM TEST

Methyl ketones can be distinguished from other ketones by their reaction with iodine in basic solution to yield iodoform. Alcohols, such as 2-propanol and 2-butanol, that can be oxidized to methyl ketones also give a positive reaction.

In a small test tube, dissolve 3 drops of liquid unknown, or about 100 mg of solid, in 2 mL of water (if insoluble in water, add 1 to 2 mL of 1-propanol and warm) and then add 1 mL of 10% NaOH. Add dropwise, with shaking, a solution of iodine in aqueous potassium iodide until a dark iodine color persists. Note the formation of any yellow precipitate (CHI_3). If no precipitate appears within 3 to 5 minutes, warm the solution slightly, and if the color fades, add more I_2/KI until the color remains for 1 to 2 minutes. Then add a few drops of NaOH to remove the excess iodine, and dilute the mixture with cold water. Iodoform, if present, will precipitate as a yellow solid, mp 119 to 121°. Perform this test on methyl ethyl ketone and your unknown.

$$R{-}\overset{\overset{\displaystyle O}{\|}}{C}{-}CH_3 + 3I_2 + 4NaOH \longrightarrow R{-}\overset{\overset{\displaystyle O}{\|}}{C}{-}O^-Na^+$$

$$+ CHI_3 + 3NaI + 3H_2O$$

D. INFRARED SPECTRA

Procedure

1. Record the infrared spectrum of benzaldehyde (neat). Note the absorption bands due to the carbonyl stretch, the aromatic C—H stretch, and the aldehyde C—H stretch. Also examine the spectrum to determine whether benzoic acid appears to be present as a result of air oxidation (a common occurrence with aldehydes).
2. Record the infrared spectrum of your unknown. Identify the absorption band present that indicates it is a carbonyl compound. Is the unknown aliphatic or aromatic? Is there evidence of an aldehyde versus a ketone?

E. ALDOL CONDENSATION (Microscale)

$$2\ C_6H_5\text{—CHO} + CH_3\text{—}\overset{O}{\overset{\|}{C}}\text{—}CH_3 \xrightarrow{OH^-}$$

$$C_6H_5\text{—CH=CH—}\overset{O}{\overset{\|}{C}}\text{—CH=CH—}C_6H_5$$

dibenzalacetone
mp 110–111°C

In a centrifuge tube or distillation receiver, combine 636 mg of benzaldehyde (0.006 mole), 174 mg of acetone (0.003 mole), and 3 mL of 95% ethanol. Shake to dissolve the components, and then add 1 mL of 10% sodium hydroxide solution. Shake the mixture until a precipitate begins to form, and then let it stand with occasional shaking for 20 minutes.

Cool the mixture in an ice bath for 5 to 10 minutes, and then remove the liquid by using a Pasteur pipet with some cotton in the tip pressed against the bottom of the test tube or centrifuge the mixture and remove the supernatant liquid using a Pasteur pipet. Add 2 mL of ice water to the centrifuge tube, stir to wash the crystals thoroughly, and remove the water using a Pasteur pipet. Repeat with another 2 mL of ice-water wash. After removing as much of the water as possible, recrystallize the solid in the centrifuge tube using 95% ethanol.

Collect the product by vacuum filtration, and allow it to dry. Determine the weight, melting point, and percent yield. Submit the product to your instructor in a properly labeled container.

QUESTIONS

1. Using your text or a chemistry handbook, list the following compounds in order of increasing boiling point: butanal, 2-butanone, butyl alcohol, and diethyl ether. Explain this order.
2. Write a balanced equation for the oxidation of benzaldehyde with CrO_3.
3. Write a stepwise mechanism to illustrate the aldol condensation of benzaldehyde with acetophenone.
4. Distinguish between the following pairs by simple, chemical tests. Tell what you would do and see.
   a. pentanal and 2-pentanone
   b. pentanal and 1-pentanol
   c. 2-pentanone and 3-pentanone
5. How would you distinguish between the following pairs by use of infrared spectroscopy? (See Chapter 9.) Tell what absorption would be present or absent in each case.
   a. 2-pentanone and 1-pentanol
   b. acetophenone and *p*-tolualdehyde
   c. 3-pentanone and benzophenone

REFERENCE

Hathaway, B.A. *J. Chem. Ed.* **1987**, *64*, 367.

Name ______________________ Section ______________ Date ________

20 Aldehydes and Ketones

PRELABORATORY QUESTIONS

1. Write equations to illustrate the reaction of 2,4-dinitrophenylhydrazine with
   a. acetone

   b. benzaldehyde

2. Write an equation for the reaction of benzaldehyde with semicarbazide to form a semicarbazone. What is the melting point of this derivative?

3. In the Tollens test or "silver mirror test," silver ion, in the form of $Ag(NH_3)_2^+$, is reduced to silver metal while an aldehyde is oxidized to the ammonium salt of a carboxylic acid. Write an equation to illustrate this reaction with propionaldehyde.

4. Write an equation to illustrate the iodoform reaction involving acetophenone.

21

Equilibrium Constants for Esterification

Direct acid-catalyzed esterification is a classical method for preparing esters, but it is a reversible reaction and is governed by an equilibrium. For preparative purposes, water is removed, usually by codistillation with an immiscible solvent (such as benzene), driving the reaction to the right:

$$\underset{\text{acid}}{\mathrm{RC(=O)OH}} + \underset{\text{alcohol}}{\mathrm{R'OH}} \underset{}{\overset{H^+}{\rightleftharpoons}} \underset{\text{ester}}{\mathrm{RC(=O)OR'}} + \underset{\text{water}}{H_2O}$$

The equilibrium constant is expressed by the equation

$$K = \frac{[\text{ester}]\,[H_2O]}{[\text{acid}]\,[\text{alcohol}]} \qquad \textbf{[21.1]}$$

If equimolar amounts of acid and alcohol are used, and one assumes no volume change (permitting use of molar amounts directly instead of concentrations in Eq. 21.1), the equilibrium constant K can be determined simply by titrating the amount of acid present at the beginning and at equilibrium. The amount of acid can be obtained directly from the volume of base consumed by an aliquot; the base used in the titration is not accurately standardized since the exact normality is not required (Question 2).

The following relationships will be used.

1. n = moles of acid = moles of alcohol present initially = $V_0 \times M$ of base, where V_0 is the volume of the base used for the initial aliquot.
2. x = moles of acid = moles of alcohol at equilibrium = $V_E \times M$ of base, where V_E is the volume of the base used for the final aliquot.
3. $(n - x)$ = moles of ester = moles of water at equilibrium.
4. $K = (n - x)\,(n - x)/(x)\,(x) = (n - x)^2/x^2$.

In this experiment, the equilibrium constant for the reaction of 1-butanol and acetic acid will be determined, and the ester will then be isolated.

$$CH_3CO_2H + CH_3CH_2CH_2CH_2OH \overset{H^+}{\rightleftharpoons} CH_3CO_2CH_2CH_2CH_2CH_3 + H_2O$$

Experiments

A. EQUILIBRIUM CONSTANT (Macroscale)

Procedure

In a 100-mL boiling flask, place 25.9 g (0.35 mole) of 1-butanol and 21 g (0.35 mole) of glacial acetic acid. After mixing thoroughly (swirling), transfer a 1.00-mL sample of the solution to a 125-mL Erlenmeyer flask containing 20 mL of water and a few drops of phenolphthalein solution, and titrate to a pink end point with 0.5 *M* sodium hydroxide to obtain V_0.

Add four drops of concentrated sulfuric acid (this is approximately 100 mg) to the alcohol–acid mixture, add a boiling stone, and reflux for 40 minutes. A very small upper layer of ester may separate during this reflux period. Cool the solution, and remove another 1.00-mL sample; be sure to take this sample from the lower layer (insert pipet below surface, expel a few bubbles of air, and then fill pipet). Titrate this sample as before, with the same NaOH solution.

Reflux the solution (with a fresh boiling stone) for another 20 minutes, cool, remove a 1.00-mL aliquot from the lower layer, and titrate. If the volume of base in this titration deviates more than 0.3 mL from that in the previous one, repeat the reflux for another 30 minutes and again titrate a 1.00-mL sample. The final titration gives V_E, representing acid present at equilibrium.

From these data, calculate the equilibrium constant for the esterification, neglecting the small error that may be introduced by the separation of two phases. (Assume volume remains constant throughout.)

B. ISOLATION OF *n*-BUTYL ACETATE (Macroscale)

SAFETY NOTE ETHER IS HIGHLY FLAMMABLE; THERE SHOULD BE NO FLAMES IN THE LABORATORY WHEN ETHER IS IN USE.

Procedure

Place the ester solution under a fractionating column with condenser, adapter, and thermometer, and distill the mixture (boiling stone) until approximately 20 mL of distillate has been collected. Separate the two layers in the distillate, record the volume of aqueous (lower) phase, and return the upper phase to the distilling flask. This upper layer contains mainly ester and unreacted alcohol.

Repeat the distillation, collecting 15 to 20 mL, and again return the upper phase to the flask and record the volume of the lower phase. If an appreciable volume of water is obtained in this second distillation, repeat the distillation again.

Combine the organic material from the flask and the distillate. Rinse the column and condenser with 20 mL of ether, and add the rinse to the other organic mixture. Transfer the ether solution to a separatory funnel, and wash with 10-mL portions of aqueous bicarbonate until the wash is no longer acidic. (**CAUTION:** Pressure will build up in the separatory funnel during the washing due to CO_2 evolution. Vent the funnel frequently.) Dry the ether solution over $MgSO_4$, filter through a cotton plug into a 100-mL round-bottom flask, and assemble a setup for a simple distillation, using a 50-mL round-bottom flask as a receiver (see Fig. 5.4). A fractionating column is not needed because of the large difference (90°C) in the boiling points of ether and butyl acetate. The flask used to collect the ether fraction should be immersed in an ice bath to minimize the escape of ether vapor. The distillation should be carried out slowly until the temperature reaches 120°C; at this point change receivers. Collect the ester, record the boiling range, and determine the yield.

QUESTIONS

1. The apparatus shown in Figure 21.1 is a Dean–Stark trap, designed for removing water from a reaction mixture containing a water-immiscible solvent. Sketch an equipment setup for the preparation of 500 g of

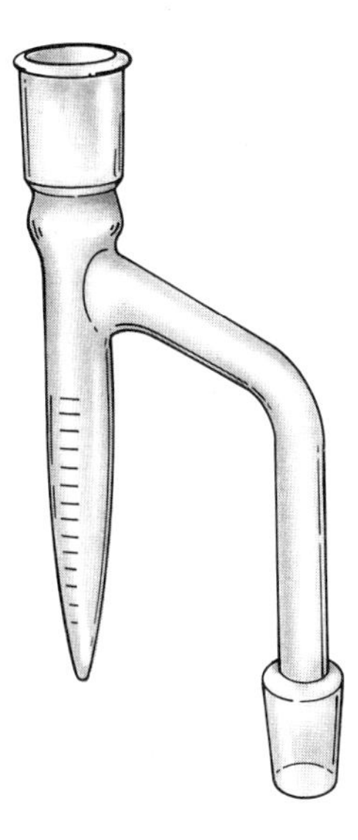

Figure 21.1 Dean–Stark trap.

n-butyl acetate, incorporating this type of trap into the apparatus, and show how it would function to permit removal of water during the reaction.

2. Why is the molarity of the aqueous base not required in the titration of the reaction aliquots?

3. Assume that the amount of H_2SO_4 added as a catalyst is exactly 100 mg. Calculate the error that was introduced in the titration by neglecting to correct for this added acid.

A meticulous student decided to avoid this error by titrating the solution to obtain V_0 both before and after adding the H_2SO_4. The student added the sulfuric acid, swirled the solution for a minute, and then removed a 1-mL aliquot and titrated it. This titration required *less* NaOH than did the aliquot removed before adding the H_2SO_4. What explanation could you offer?

4. Using the equilibrium constant you determined, what would be the equilibrium composition of the reaction mixture if 2.6 g of 1-butanol and 21 g of acetic acid were heated (with a drop of sulfuric acid catalyst)?

Name ______________________ Section ______________ Date __________

21 Equilibrium Constants for Esterification

PRELABORATORY QUESTIONS

1. Write an equilibrium expression, *K*, for the following reaction:

$$\text{acetic acid} + \text{ethanol} \rightleftharpoons \text{ethyl acetate} + \text{water}$$

2. In the reaction above, a small amount of sulfuric acid is added. Why?

3. An esterification reaction, such as that shown in Question 1, has an equilibrium constant of 2.4. At equilibrium, is the concentration of the carboxylic acid or the ester higher? Explain.

4. List three ways to increase the amount of ethyl acetate produced in the reaction in Question 1.

5. Write equations for the reaction of sodium bicarbonate with all of the acidic components in an esterification reaction.

6. Sketch a reflux apparatus.

22

Determination of pK_a and Equivalent Weights—Effects of Substituents on Acidity

The acidity or acid strength of carboxylic acids, not to be confused with acid concentration, is a property often used to examine the relationship between structure and chemical properties. In a series of acids, RCO_2H, with different R groups, measurement of the acidity provides a direct means of determining the influence of the group R. In this experiment, the effect of substituents on acidity will be determined for an unknown aliphatic acid and then for a series of benzoic acids. Moreover, the method for determining the equivalent weight of an unknown acid will be illustrated.

The *acid strength* of a carboxylic is expressed quantitatively as the **equilibrium constant, K_a**, for the dissociation of the acid in water. (Another way of looking at this is to describe, in Bronsted–Lowry terms, the ability of the acid to donate a proton to the base, water.)

$$RCO_2H + H_2O \rightleftharpoons RCO_2^- + H_3O^+ \qquad \textbf{[22.1]}$$

$$K_{eq} = \frac{[RCO_2^-]\,[H_3O^+]}{[RCO_2H]\,[H_2O]}$$

$$K_{eq}[H_2O] = \frac{[RCO_2^-]\,[H_3O^+]}{RCO_2H} = K_a$$

or

$$K_a = \frac{[H_3O^+]\,[RCO_2^-]}{[RCO_2H]} \qquad \textbf{[22.2]}$$

A logarithmic form of Equation 22.2 is usually used, just as hydrogen ion concentration (actually hydrogen ion *activity*) is expressed in pH units. Recall that **pH** is defined as the negative log of the hydrogen ion activity. For purposes of this experiment it will be assumed that the activity and concentration of the hydrogen ion are the same. Equation 22.2 expressed in negative logarithmic terms becomes the following:

$$-\log K_a = pK_a = -\log\left(\frac{[H_3O^+][RCO_2^-]}{[RCO_2H]}\right) \quad \textbf{[22.3]}$$

Simplifying and rearranging:

$$pK_a = pH - \log\left(\frac{[RCO_2^-]}{[RCO_2H]}\right) \quad \textbf{[22.4]}$$

or

$$pH = pK_a + \log\left(\frac{[RCO_2^-]}{[RCO_2H]}\right) \quad \textbf{[22.5]}$$

Using Equation 22.5, it can be shown that the value of the pK_a of an acid is the pH at which the anion and the acid are at the same concentration (that is, the acid is half-neutralized). In other words, if

$$[RCO_2^-] = [RCO_2H]$$

then

$$\log\left(\frac{[RCO_2^-]}{[RCO_2H]}\right) = 0$$

and

$$pK_a = pH$$

The method for determining pK_a by the titration of an acid with a base is illustrated in Figure 22.1. The pH is measured with a pH meter. Note that the equivalent weight of an unknown acid can be determined at the same time if the weight of acid and the concentration of base are known with reasonable accuracy.

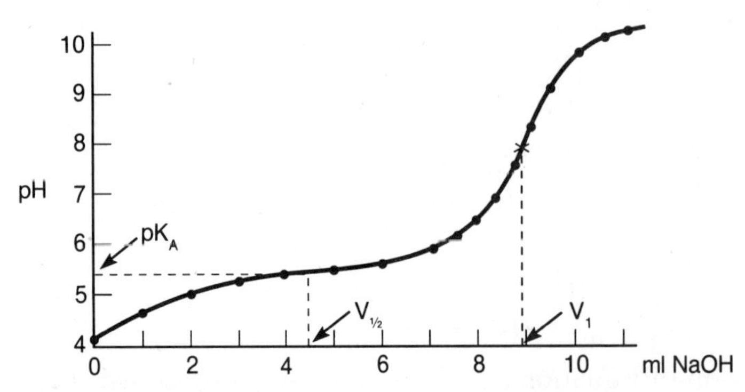

Figure 22.1 pK_a titration curve.

Experiments

A. TITRATION OF AN UNKNOWN ACID (Macroscale)

An unknown aliphatic acid will be provided as an aqueous solution with the concentration in grams per liter noted on the label.

Procedure

The instructor will demonstrate the use of a pH meter. Use them with care.

Rinse a 50-mL buret with small portions of standardized NaOH solution (0.50 *M*), and then fill the buret with the standard solution. Record the molarity of the NaOH and the initial buret reading.

Pipet 10 mL of the unknown aliphatic acid solution into a 250-mL beaker and add 75 mL of distilled water. Position the pH meter electrodes so that they dip sufficiently into the acid solution. Read and record the pH. Add approximately 1 mL of NaOH, swirl the solution gently, then accurately read and record the buret volume and the pH of the mixture being titrated. Repeat the addition of NaOH, swirling, reading, and recording using 1- to 1.5-mL additions of NaOH until the pH begins to change rapidly. At this point, add the NaOH in 3- to 5-drop increments until pH 9 is reached. Then add additional 1-mL increments to continue the titration approximately 5 mL past the end point.

Using phenolphthalein, titrate 75 mL of distilled water as a blank. Subtract this value from the volumes obtained in the titration of the unknown acid. Graph **pH** versus corrected volume. From the graph determine the equivalence point (V_1), which is the point where the curve is the steepest, and the half-equivalence point ($V_{1/2}$) of the titration. Using the half-equivalence point, determine the pK_a of the unknown acid, and from the volume of base at the equivalence point, determine the equivalent weight (neutralization equivalent) of the acid.

After completing duplicate titrations that agree to within ± 2%, empty the buret and rinse it with distilled water, and record your data in a format such as that shown below.

DATA FORMAT

1. Volume of base at the equivalence point ____________ mL
2. Volume of base at half-neutralization ____________ mL
3. pH at half-neutralization ____________
4. pK_a of acid ____________
5. Molarity of base used ____________ *M*
6. Milliequivalents of base used ____________ meq
7. Milliequivalents of acid titrated ____________ meq
8. Milliequivalent weight of acid ____________ g/meq
9. Equivalent weight of acid (neutralization equivalent) ____________ g/eq

B. DETERMINATION OF THE pK_a OF AN UNKNOWN SUBSTITUTED BENZOIC ACID (Microscale)

Procedure

From the sample of the substituted benzoic acid provided, weigh out approximately 20 mg and dissolve it in 25 mL of ethanol in a 100-mL beaker. Dilute the solution with 25 mL of water, and titrate with 0.01 *M* sodium hydroxide solution using a pH meter. Carry out duplicate analyses, and plot the results as shown in Figure 22.1 and described in Part A.

Report the pK_a of your acid to your instructor, who will tabulate the results obtained on all the acids given out. Order this list of acids according to the acidity. (*Note*: Higher acidity = larger K_a = smaller pK_a.) Explain the order and any differences between the effect of *meta*- and *para*-substitution.

QUESTIONS

1. It is not necessary that the pH at half-neutralization be used to calculate the pK_a. Use Equation 22.5 and your titration curve to calculate the pK_a of your acid from the pH of the solution when the acid is 75% neutralized. How does the result compare with the value calculated in the experiment? What does any difference indicate?
2. A solution of acetic acid, pK_a 4.7, has been partially neutralized with base to a pH of 6.0. Calculate the percentage of the acid that is present as the anion.

REFERENCE

King, J. F. In *Techniques of Organic Chemistry*; Bentley, K. W., Ed.; Interscience: New York, 1963; Vol. XI, Part 1; pp. 318–370.

Name ______________________ Section ______________ Date ________

22 Determination of pK_a and Equivalent Weights

PRELABORATORY QUESTIONS

1. List four strong acids and four weak acids.

2. What is the major organic species present in an 0.1 *M* solution of acetic acid: unionized acetic acid or acetate ion?

3. Calculate the pK_a of an acid that has a K_a of 3.5×10^{-5}.

4. What is the K_a of an acid that has a pK_a of 3.6?

5. Which is the *stronger acid*? Acid A, pK_a = 4.67, or acid B, pK_a = 3.45? Explain.

23

Effects of Substituents on Reactivity

In addition to having great utility in the understanding of reaction mechanisms, the study of rate constants also provides an insight into the effects (steric, inductive, or resonance) that substituent groups have on molecular interactions. By taking a particular reaction and by making changes in the nature of the substituents, a correlation of structure and reactivity may be made. Much of our present day understanding of these phenomena is attributed to the pioneering work of such eminent chemists as Brönsted, Bartlett, Hinshelwood, Conant, Norris, and others. A landmark book by L.P. Hammett opened the door to further refinements by Taft, Leffler, and others. The correlation of thermodynamic and kinetic effects has led to a better understanding of molecular dynamics, especially in the field of enzyme-catalyzed reactions. Knowledge of the details of how molecules interact and the effect of a minute change in the structure has led to a wholly new way to approach the fields of medicinal chemistry and pharmacology.

Hammett proposed the correlation of the effect of substituents on the acidity of benzoic acids (see Part B, Chapter 22) with the effect of those substitutents on the rate constants of the hydrolysis of benzoic acid esters. He found a reasonably good correlation of the *para* and *meta* acids and esters. The ortho derivatives, however, did not show the same relationship (the *ortho* effect).

In this experiment, different esters will be hydrolyzed by different students, and the results of all will be compared to see if results similar to those of Hammett can be obtained. If a good selection of methyl benzoates is not available, they may be synthesized by the method described in Part A.

$$C_6H_5C(=O)OMe + OH^- \xrightarrow{k} C_6H_5CO_2^- + MeOH \qquad [23.1]$$

Experiments

A. PREPARATION OF METHYL BENZOATES (Macroscale)

SAFETY NOTE CONCENTRATED SULFURIC ACID CAUSES BURNS ON CONTACT WITH SKIN; RINSE IMMEDIATELY WITH LARGE QUANTITIES OF COLD WATER AND THEN WASH WITH SOAP AND WATER. GET ASSISTANCE FROM A PHYSICIAN IF NECESSARY.

Procedure

Place 5 g of the acid assigned to you in a 100-mL round-bottom flask and add 50 mL of methanol and approximately 1 mL of concentrated sulfuric acid (use a calibrated dropper). Fit a reflux condenser to the flask, add a boiling stone, and reflux the mixture for 1 to 2 hours. The progress of the reaction can be checked at various times (every 20 minutes or so) by TLC on silica gel. The solvent or solvent mixture required for this will depend on the acid being used and will have to be determined individually. Before taking the first sample, find a solvent that barely, but perceptibly, moves the acid from the origin. The ester, being less polar, will have a larger R_f (see Chapter 7) and will move well up the plate.

When the reaction is complete as measured by TLC analysis, cool the solution to room temperature and transfer it to a 250-mL separatory funnel. Add 50 mL of ether and 75 mL of water and shake well. Allow the layers to separate, and discard the aqueous layer. Rinse the organic layer successively with 25 mL of water, twice with 25 mL of 10% $NaHCO_3$ solution, and once with 25 mL of saturated NaCl solution. Dry the organic layer over Na_2SO_4 for 5 to 10 minutes, remove the drying agent by filtration, rinse it on the funnel with ether, and evaporate the ether from the filtrate.

Most of the assigned esters can be obtained as solids and purified by recrystallization from hexane or a mixture of toluene and hexane (exceptions of course, are the liquids methyl benzoate, methyl *m*-toluate, methyl *m*-chlorobenzoate, and methyl *m*-anisate).

Some of the solid esters have relatively low melting points, so their recrystallization must be done at ice bath temperatures. For use in the following experiment, the esters must be completely dried of the recrystallization solvent. Those that are oils should be distilled from a small distilling flask. Because of their high boiling points (190 to 260°C), a condenser will not be needed; the set-up shown in Figure 46.1 will suffice. Report the observed and literature melting (or boiling) range of the ester and the percent yield.

B. RATE MEASUREMENTS

Procedure

Weigh out 0.0100 mole of the known methyl ester supplied by your instructor into a tared 125-mL Erlenmeyer flask; the weight should be measured accurately and should differ from 0.0100 mole by no more than 1%. Accurately weigh 60.0 mL of reagent grade acetone (d = 0.791) into the flask, and swirl the mixture to dissolve the ester completely. Suspend the flask with a buret clamp in an ice-water bath. During the reaction, ice should be added to the bath as necessary to keep the temperature at 0°C.

In a second Erlenmeyer flask accurately weigh out 40.0 mL of 0.25 *M* NaOH solution (d = 1.011) and cool to 0°C. To start the reaction, add this base to the ester solution and mix well by swirling. Note the time, using a watch or clock with a second hand, when the base is added. With a 10-mL pipet, remove 10-mL aliquots of the solution at regular intervals. (The length of the intervals will depend on which ester you are using. See your instructor for directions on this point.) Add each sample to a separate 100-mL flask containing 10 mL (measured by buret) of 0.100 *M* HCl, and note the time when the solution is added. This quenches the reaction.

Add 3 drops of phenolphthalein indicator solution to each flask, and titrate each with 0.010 *M* NaOH solution to a persistent pink end point. For each sample, record the time when it was taken and the volume of titrant required. The acid being titrated is the benzoic acid liberated in the hydrolysis.* Using the volume and molarity of titrant, calculate the number of moles of unreacted ester remaining. The concentration of ester can be calculated simply by dividing the number of moles by the sample volume.

C. CALCULATION OF THE RATE CONSTANT FOR HYDROLYSIS

Procedure

Since the slow (rate-determining) step in the reaction in Equation 23.1 is the attack of OH^- on the ester (Eq. 23.2), the rate of disappearance of ester is proportional to the product of the concentrations of these two species (Eq. 23.3).

---

*This is so because after x percent of the ester has reacted, there are in solution $0.01(1 - x/100)$ mole each of ester and OH^-, and $0.01(x/100)$ mole of benzoate anion. Addition of 10 mL of 0.10 *M* H^+ gives $0.01(1 - x/100)$ mole of H_2O (from OH^-) plus $0.01(x/100)$ mole of the benzoic acid.

$$\mathrm{R{-}C(=O)OMe} + \mathrm{OH^-} \xrightarrow{\text{slow}} \mathrm{R{-}C(O^-)(OMe)(OH)} \xrightarrow{\text{fast}} \mathrm{R{-}C(=O)O^-} + \mathrm{MeOH} \quad [23.2]$$

$$\text{rate} = -\frac{d[\text{ester}]}{dt} = k[\mathrm{OH^-}]\,[\text{ester}] \quad [23.3]$$

In this experiment $[\mathrm{OH^-}]$ = [ester], so it is possible to simplify Equation 23.3 to Equation 23.4

$$\frac{-d[\text{ester}]}{dt} = k[\text{ester}]^2 \quad [23.4]$$

After integration of this differential form of the rate equation, Equation 23.5 is obtained: $[\text{ester}]_t$ is the concentration of ester after t seconds of reaction.

$$\frac{1}{[\text{ester}]_t} = kt + \frac{1}{[\text{ester}]_0} \quad [23.5]$$

This relationship is true because after x percent of the ester has reacted, there are in solution $0.01(1-x/100)$ mole each of ester and $\mathrm{OH^-}$, and $0.01(x/100)$ mole of benzoate anion. Addition of 10 mL of 0.10 M $\mathrm{H^+}$ gives $0.01(1-x/100)$ mole of H_2O (from $\mathrm{OH^-}$) plus $0.01(x/100)$ mole of the benzoic acid. The value of k can be obtained by substituting values of $[\text{ester}]_t$, t, and $[\text{ester}]_0$ (which equals 0.1 M) into Equation 23.5. However, it is simpler and more accurate to treat Equation 23.5 as that of a straight-line plot of $[\text{ester}]_t^{-1}$ versus t, whose slope is k. Therefore, to obtain k for your ester, plot the values of $[\text{ester}]_t^{-1}$ (Column 7 in Table 23.1) versus time (Column 2) on graph paper with appropriate scales, and then with a straightedge, draw a line that passes as close as possible to all of the points. (Because of experimental errors, you should not expect that the line will go through each of the points.) Using two well-separated points on the line (not necessarily data points), (X_1, Y_1) and (X_2, Y_2), calculate the slope of the line $k = (Y_2 - Y_1)/(X_2 - X_1)$. Since the ordinate ($Y$) is in units of

Table 23.1 Data Format

Sample Number	Time Taken(X)	Vol of 0.01M NaOH	mmoles of Acid	mmoles of Ester	Concentration of Ester(M)	(Concentration of Ester)$^{-1}$ Y

M^{-1} (liters/mole) and the abscissa (X) is in units of sec, the rate constant has units of M^{-1} sec^{-1}.

After everyone in the class has calculated a value of k for a different ester, the results can be collected and compared. For this purpose, it is useful to prepare a table of esters and rate constants in order of increasing values of k. How does this list compare with the corresponding list of acids and pK_a's from the previous experiment? Why? If a more quantitative comparison of the two sets of quantities seems justified, you can make a graph of the values of log k for the esters versus the pK_a's ($-$ log K_a's) of the corresponding acids.

QUESTIONS

1. How many moles of benzoic acid are present in a sample that requires 30 ml of 0.200 M NaOH for titration?

2. How many grams of benzoic acid are present in a sample that requires 20 ml of 0.500 M NaOH for titration?

3. a. How many moles of terephthalic acid are present in a 1.66-g sample?

b. How many equivalent weights of terephthalic acid are present in a 1.66-g sample?

4. What is the equivalent weight of an acid that requires 20.0 mL of 0.25 M NaOH to titrate an 0.83-g sample?

5. Calculate the pH of a 0.1 M solution of glycolic acid ($HOCH_2COOH$) that has a value of $K_a = 1.5 \times 10^{-4}$.

6. How would you expect the acid strength (as reflected in K_a values) of acetic acid to compare with fluoroacetic acid. Explain.

REFERENCES

Hammett, L.P. *Chem. Rev.* **1935**, *17*, 125.

Moore J. W.; Pearson, R. G. *Kinetics and Mechanism* 3rd ed.; John Wiley & Sons: New York, 1981; pp 12–26.

Name ________________________ Section ______________ Date ________

23 Effects of Substituents on Reactivity

PRELABORATORY QUESTIONS

1. Write an equation for the basic hydrolysis (saponification) of methyl benzoate.

2. List three electron-donating groups and three electron-withdrawing groups.

3. Draw the structure of the following substances:

a. methyl *para*-nitrobenzoate

b. methyl *para*-toluate

4. What is the formula of the product of the reaction of excess water with sodium sulfate?

5. Write the equation for a second-order reaction.

24

The Friedel–Crafts Reaction

The attachment of alkyl groups to an aromatic ring is an extremely important reaction, and it is carried out on an industrial scale to produce compounds such as cumene (isopropylbenzene) in multimillion-pound amounts each year. Alkylation involves the attack of an aromatic ring by an electrophilic carbon species, the choices of reagent and conditions depending on the alkyl group and the reactivity of the ring.

The Friedel–Crafts reaction using an alkyl halide and aluminum halide is a general method for such aromatic alkylation. Other important variations of this reaction involve the generation of the electrophile from an alcohol or alkene. However, there are some practical limitations. Since these reactions involve the development of a positive charge on the alkyl carbon atom, carbocation rearrangements may occur. Furthermore, the ring becomes more susceptible to substitution (more nucleophilic) as alkyl groups are introduced, and there is thus a tendency for polysubstitution. Finally, "scrambling" of ring-attached alkyl groups may occur, when a thermodynamically more stable product can arise from one that is formed initially in a faster reaction.

The first experiment in this chapter involves the macroscale preparation of an antioxidant. Such a compound inhibits the process of autoxidation, which is the reaction of organic compounds with oxygen in the air. Autoxidation leads to the incorporation of oxygen, oxidative degradation, and cross-linking and is the cause of deterioration of foods, plastics, rubber, and many other materials.

The major reactions in autoxidation involve free radical chains and include the steps shown in Equations 24.1 to 24.4.

Initiation:

$$RH + X\cdot \rightarrow R\cdot + HX \quad \textbf{[24.1]}$$

$$R\cdot + O_2 \rightarrow RO_2\cdot \quad \textbf{[24.2]}$$

Propagation:

$$RO_2\cdot + RH \rightarrow RO_2H + R\cdot \qquad [24.3]$$

$$RO_2H \rightarrow RO\cdot + OH\cdot \rightarrow \text{oxidation and degradation products} \qquad [24.4]$$

Important intermediates in the process are the peroxy radical $RO_2\cdot$ and the hydroperoxide RO_2H. Breakdown of these intermediates can provide further radicals for the initiation step (Eq. 24.1), or it can lead to various end products (Eq. 24.4). In the case of some solvents such as ether, hydroperoxides can accumulate and act as highly hazardous contaminants (Chapter 1).

Unsaturated materials, such as edible fats and oils with one or two double bonds in the fatty acid chain, are particularly susceptible to autoxidation because they form relatively stable allylic radicals (Eq. 24.5).

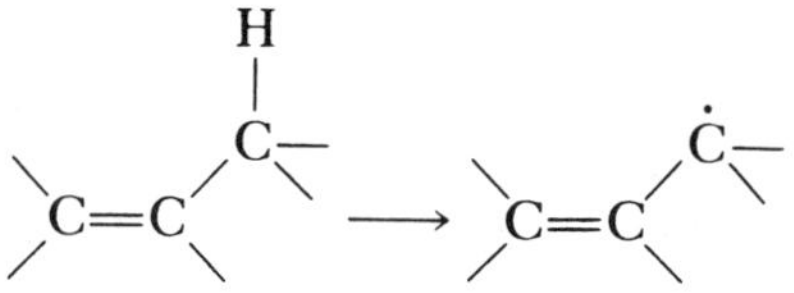

H

C—

C=C ⟶ C=C C̣— [24.5]

In foods, these reactions lead to off-flavor and rancidity. One situation in which autoxidation is a desirable process is in drying oils such as linseed or tung oil, which are used in paints, varnishes, and printing inks. In this case the very highly unsaturated chains undergo cross-linking and polymerization by radical attack, forming a hard protective coating.

To slow down autoxidation, inhibitors have been developed that trap peroxy radicals in the propagation step of the radical chain (Eq. 24.2). Since the process has a long chain length, one molecule of inhibitor can interrupt a cycle that would otherwise lead to many molecues of hydroperoxide. Among the most effective antioxidants are phenols with a *para* substituent and one or two bulky alkyl groups at the *ortho* positions. These substituents stabilize the initially formed aryl radical until it can react with a second radical to form a stable end product.

Two phenols that are nontoxic and are permitted for use as food additives are 2-*t*-butyl-4-methoxyphenol and 2,6-di-*t*-butyl-4-methylphenol. When used as antioxidants these compounds are called **BHA** and **BHT**, respectively, abbreviations for the very unsystematic names "butylated hydroxyanisole" and "butylated hydroxytoluene."

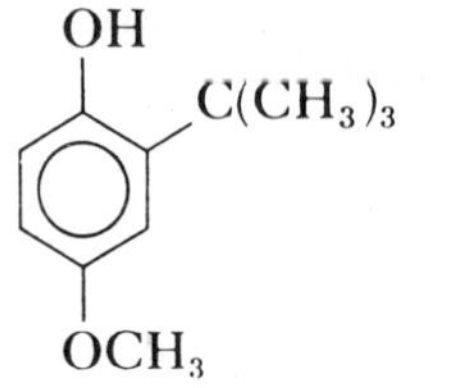

OH, $C(CH_3)_3$, OCH_3

BHA

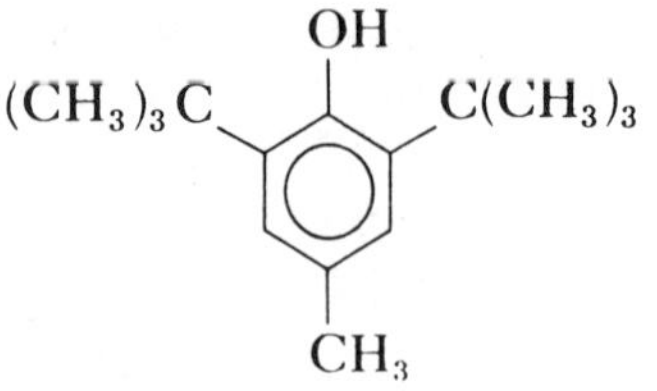

OH, $(CH_3)_3C$, $C(CH_3)_3$, CH_3

BHT

The first experiment involves the preparation of the antioxidant BHT (2,6-di-*t*-butyl-4-methylphenol) on a macroscale by alkylation of the highly activated ring in *p*-cresol with *tert*-butyl alcohol and sulfuric acid. In the industrial process for this compound, the alkylation is carried out with isobutylene and boron trifluoride.

$$p\text{-}CH_3C_6H_4OH + 2(CH_3)_3COH \xrightarrow{H_2SO_4} 2,6\text{-}[(CH_3)_3C]_2\text{-}4\text{-}CH_3C_6H_2OH + 2H_2O$$

d 0.79

BHT, mp 70°C

The conditions and molar ratio of reactants are rather critical in this reaction, since byproducts can be formed that interfere with the isolation of the di-*t*-butyl-*p*-cresol. One obvious possibility is the monosubstituted compound, 2-*t*-butyl-*p*-cresol, which becomes the major product if the concentration of the sulfuric acid is reduced from 96% to 75%. Disubstitution is favored by high acid strength and use of excess *t*-butyl alcohol. However, the excess alcohol can lead to further complications caused by dehydration and the formation of diisobutylene.

The second experiment describes the preparation of *p*-di-*t*-butylbenzene on a microscale by the alkylation of *t*-butylbenzene with *t*-butyl chloride in the presence of aluminum chloride. This procedure parallels the method originally discovered by chemists Charles Friedel and James Crafts in 1877 at the Sorbonne University in Paris. Alkyl halides are insufficiently electrophilic by themselves to react with aromatic rings, but in the presence of Lewis acids, they produce highly reactive carbocations, which readily undergo electrophilic substitution reactions with activated rings. As in other such substitutions, the rate is enhanced by electron-donating substituents and retarded by those that are electron withdrawing. In fact, the reaction fails with ring substituents more highly deactivating than the halogens, and this constitutes yet another limitation. Nevertheless, despite these drawbacks, the Friedel–Crafts reaction continues to be a highly useful synthetic method.

Experiments

A. PREPARATION OF BHT (Macroscale)

SAFETY NOTE IN THIS EXPERIMENT, *p*-CRESOL AND CONCENTRATED SULFURIC ACID, ESPECIALLY THE LATTER, WILL CAUSE BURNS IN CONTACT WITH THE SKIN. IF EITHER IS SPILLED ON THE HAND, RINSE THOROUGHLY WITH WATER AND WASH WITH SOAP AND WATER.

Procedure

In a dry 50-mL Erlenmeyer flask, place 2.16 g of *p*-cresol, 1 mL of glacial acetic acid, and 5.6 mL of *tert*-butyl alcohol. (The alcohol has a mp of 26°C and should be warmed to between 30 and 35°C before pouring or pipeting; it also helps to warm the graduated cylinder slightly to avoid solidification during transfer.) When the cresol has dissolved, cool the solution to 0°C in an ice bath. If a magnetic stirring apparatus is available, add a stirring bar and place the ice bath (use a beaker) on the stirring plate.

Obtain 5.0 mL of concentrated sulfuric acid in a dry 10-mL graduated cylinder, and stand the cylinder in a small beaker to keep it from tipping over. With a pipet and bulb, add the sulfuric acid to the cresol–alcohol solution in the ice bath, swirling thoroughly or stirring to mix and disperse the acid as it is added. If a pink color develops, stop the addition until the color fades. Addition of the acid should require 4 to 5 minutes, and the color should remain pale yellow; an oily layer will separate.

When addition of the acid is completed, continue the stirring or intermittent shaking and swirling in the ice bath for 20 minutes; record any changes in appearance. Remove from the bath, add a few pieces of ice, and then fill the flask with water. Pour the mixture into a separatory funnel, rinse the flask with 30 mL of ether, and add the ether solution to the funnel. Shake vigorously for 1 to 2 minutes, venting the funnel cautiously several times. Remove the lower aqueous layer, and wash the ether twice more with 10-mL portions of water and finally with 10 mL of 0.5 *M* (2%) aqueous KOH. Dry the ether solution with Na_2SO_4, filter through a cotton plug, and evaporate to approximately a 10-mL volume. Transfer the solution to a 25 × 100-mm test tube, rinsing with a little ether, and then evaporate as completely as possible, finally under aspirator vacuum to remove diisobutylene (bp 101 to 105°C). Some of this byproduct will condense as a liquid in the upper end of the test tube and can be removed by wiping with a paper towel.

Cool the solution to room temperature, scratch to initiate crystallization, and then swirl to promote crystal growth in the viscous mixture. Complete the crystallization process by placing the tube in an ice–salt or ice–HCl bath for a few minutes. Collect the crystals on a Hirsch funnel, and press the mixture of solid and oil to squeeze out as much of the oily mother liquor as possible. Weigh the crude moist solid and recrystallize it from methanol, using approximately 2 mL of solvent per gram of solid. Collect the resulting crystals, weigh, and determine the mp. If the mp is low, recrystallize again from methanol and determine the final yield and mp.

Antioxidant Properties

A quick and direct way to evaluate antioxidant effectiveness is based on the drying properties of linseed oil. This oil, extracted from flax seed, contains a high proportion of linolenic acid, with the nonconjugated triene unit $—(CH{=}CHCH_2)_3—$. For most applications, such as in varnishes and printing inks where rapid drying is desired, the oil is heated with a small amount of lead oxide. This treatment causes isomerization of the double bonds and preliminary polymerization. The resulting "boiled" oil dries in approximately one sixth of the time required for the "raw" oil. Drying occurs by autoxidation, with the formation of C—O and C—C bonds between low molecular weight polymer units. In the presence of an effective autoxidation inhibitor, the drying process is retarded.

Procedure

In each of four small test tubes labeled 1 through 4, place approximately 0.5 mL of boiled linseed oil (simply divide 2 mL of oil evenly among the four tubes). Prepare 1% solutions of your di-*t*-butyl-*p*-cresol and of *p*-cresol in acetone (e.g., 50 mg in 5 mL of acetone). Dilute the linseed oil samples as follows:

Tube 1: Add 0.5 mL of acetone (blank).

Tube 2: Add 0.25 mL of di-*t*-butyl-*p*-cresol solution and 0.25 mL of acetone (0.5% BHT).

Tube 3: Add 0.5 mL of di-*t*-butyl-*p*-cresol solution (1% BHT).

Tube 4: Add 0.5 mL of *p*-cresol solution (1% *p*-cresol).

Place an open-end capillary tube in each solution and stir well until the solutions are mixed, and then place 2 to 3 drops from each tube on microscope slides (four samples can be placed on two slides). Place the slides on a paper towel, number them for identification, and place them on a hot plate at low heat or in an oven (70 to 80°C). Examine the slides after 20 minutes, compare the fluidity or stickiness of the four samples, and record the results.

If it is more convenient, the slides may be allowed to stand at room temperature and compared after several hours or on the following day.

B. PREPARATION OF *p*-DI-*t*-BUTYLBENZENE (Microscale)

SAFETY NOTE ALUMINUM CHLORIDE IS EXTREMELY HYGROSCOPIC AND PRODUCES HCl IN THE PRESENCE OF MOISTURE. AVOID BREATHING THE DUST AND IMMEDIATELY WASH OFF ANY MATERIAL WHICH MAY GET ON THE SKIN.

Procedure

In a thoroughly dry 5-mL round-bottom flask, place 1.0 mL of *t*-butyl chloride, 0.5 mL of *t*-butyl benzene, and a small stirring bar, if available. Place the reaction vessel in an ice bath (a 100-mL beaker or a small

$$C_6H_5C(CH_3)_3 + (CH_3)_3C\text{—}Cl \xrightarrow{AlCl_3} p\text{-}(CH_3)_3C\text{—}C_6H_4\text{—}C(CH_3)_3 + HCl$$

t-butylbenzene d 0.87 + *t*-butyl chloride d 0.85 → *p*-di-*t*-butylbenzene mp 78–79°C

crystallizing dish may be used for the bath), and place it on a stirring plate. HCl gas is produced in this reaction, and it should be trapped using one of the following two methods. A convenient HCl trap, which can also be used for larger quantities of evolved gas, consists of a funnel inverted over a beaker half filled with water (Figure 24.1A). Gaseous HCl is *very* soluble in water and care must be taken to prevent water from backing up in the system. This is prevented by suspending the funnel so that its edge is *just* below the surface of the water. This is a simple and safe way to handle the acid fumes; the system is sealed with water to prevent the escape of HCl, but the water cannot back up into the reaction flask because of the large empty volume of the funnel.

An alternate procedure for trapping the HCl gas, which is suitable for small volumes (as is the case here), consists of passing the gas from the reaction vessel through a tube into a beaker (approximately 100 mL) containing water-moistened cotton (Figure 24.1B). The HCl is readily absorbed by the wet cotton, which can then be disposed of by removing with forceps and rinsing under tap water.

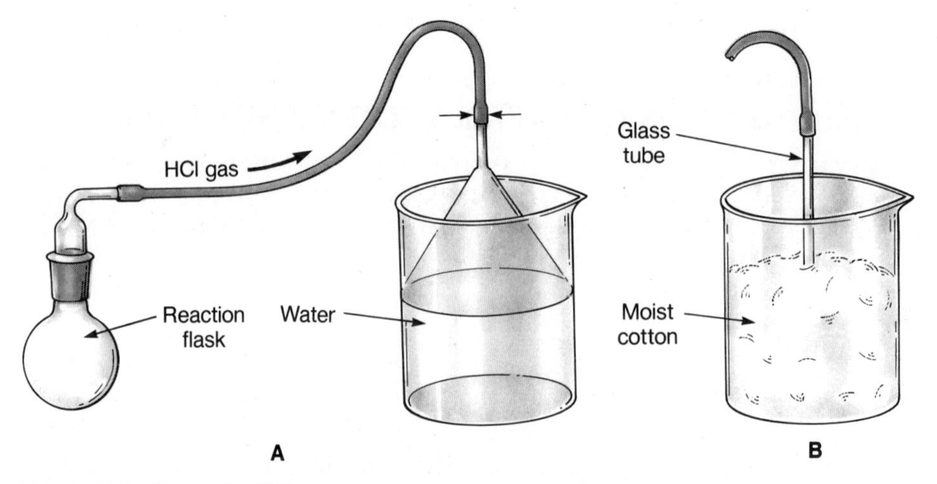

Figure 24.1 Traps for HCl gas.

Weigh 0.05 g of anhydrous aluminum chloride and store in a capped vial until you are ready to use it. It may be difficult to weigh this material accurately because it reacts readily with the moisture in the air. To minimize exposure to air, keep the reagent bottle closed as much of the time as possible, have the weighing paper tared beforehand, and weigh and transfer the material as quickly as possible.

Add one half of the aluminum chloride to the chilled *t*-butyl benzene/*t*-butyl chloride mixture. Recap immediately, and if a stirring bar is not being used, gently shake to effect mixing.* A vigorous reaction should ensue, and after 5 to 6 minutes, add the remainder of the aluminum chloride. When the reaction begins to subside—a white product should be formed, although this may be difficult to detect—remove the flask from the ice bath and allow the mixture to warm to room temperature.

Add 1 mL of ice-cold water (from the bath), followed by 1 mL of ether. Swirl, and with a pipet transfer the contents to a centrifuge tube. Remove the ether layer with a clean pipet, and wash the remaining aqueous layer with two 1-mL portions of ether, adding these washes to the original ether layer. Dry the combined ether layers by filtering through a small column of $MgSO_4$ prepared in a pipet with a cotton plug at the bottom to contain the drying agent. Collect the filtrate in a small Erlenmeyer flask, evaporate the solvent over a steam bath, and recrystallize the residue from 1 mL of methanol. Collect the product, on a Hirsch funnel, let dry, weigh and determine its melting point. If required, the product can again be recrystallized from methanol.

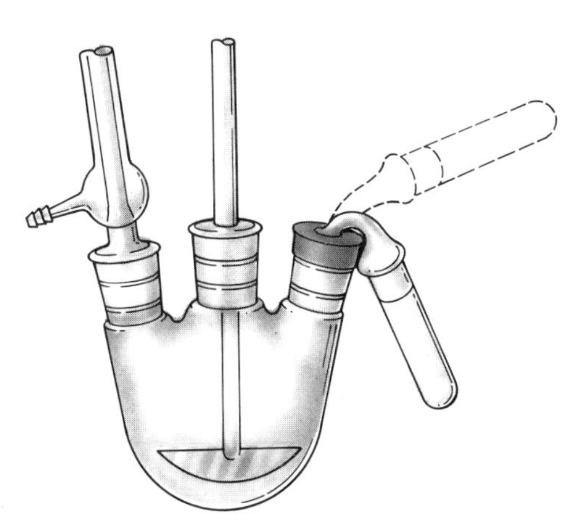

Figure 24.2 Technique for adding a solid to a moisture-sensitive reaction.

*In a larger-scale reaction, it would be quite undesirable to open the system repeatedly to add the $AlCl_3$. A very satisfactory procedure for adding a solid reagent in portions to a moisture-sensitive reaction is shown in Figure 24.2. A wide test tube containing the solid is connected to a neck of the flask by a short length of Gooch crucible tubing. The solid can be sealed from the reaction by collapsing the tubing, and then added simply by tipping up the test tube and shaking in the reagent.

QUESTIONS

Preparation of BHT

1. In an attempt to improve the conditions for dialkylation of *p*-cresol and reduce the amount of monosubstituted product, it was decided to use 100% sulfuric acid, which is prepared from concentrated H_2SO_4 (96%) and fuming H_2SO_4 (contains 30% SO_3 by weight). Calculate the volumes of the two acids that must be mixed to prepare 50 mL of 100% H_2SO_4. (The densities of concentrated H_2SO_4 and 30% fuming H_2SO_4 are 1.84 and 1.95, respectively.)
2. Diisobutylene is a mixture of two isomers, C_8H_{16}, formed in the reaction of *t*-butyl alcohol with concentrated H_2SO_4. Outline the steps in this reaction, showing the structures of the two C_8H_{16} isomers.
3. The NMR spectra of BHT and of the mother liquor from the crystallization are given in Figures 24.3 and 24.4, respectively. Assign as completely as possible the peaks in both spectra to protons in BHT and other compounds that are present in the mother liquor.
4. Account for the results observed in Tubes 1 to 4 in the antioxidant testing.

Preparation of *p*-Di-*t*-Butylbenzene

5. What monoalkylated product(s) would one expect to obtain if *n*-butyl chloride were used instead of *t*-butyl chloride?
6. Other than comparison of melting points, how could one prove that the isolated product is *p*-di-*t*-butylbenzene and not the *ortho* or *meta* isosmer?
7. A byproduct in this reaction arises from dialkylation to give a tri-*t*-butylbenzene. How can the experimental conditions be changed to

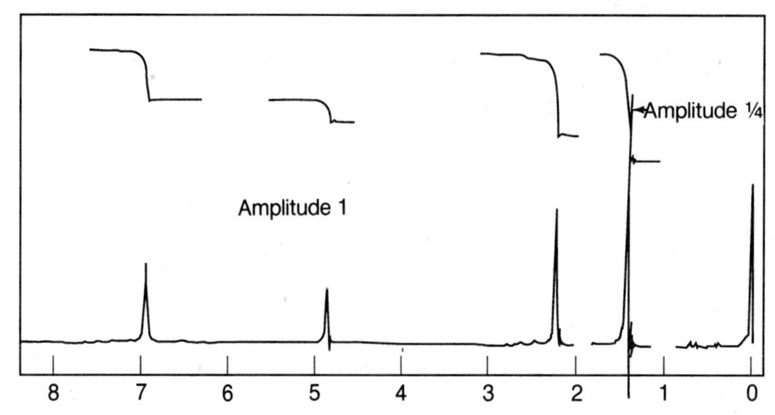

Figure 24.3 The NMR spectrum of BHT.

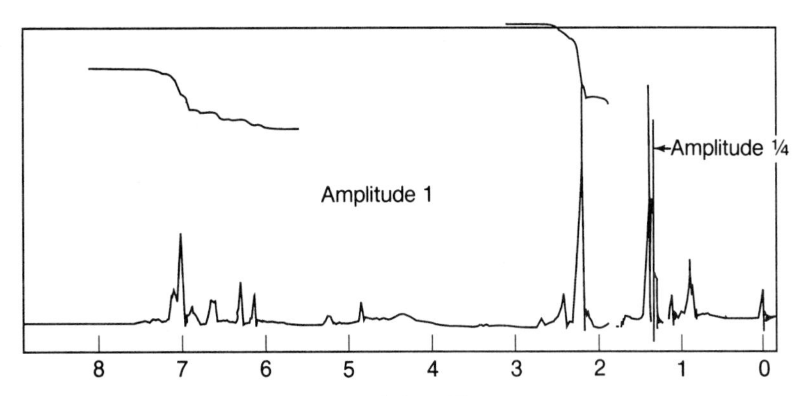

Figure 24.4 The NMR spectrum of the BHT recrystallization mother liquor.

make this the major product? Predict the thermodynamically most stable tri-*t*-butylbenzene.

8. Suggest two starting compounds other than *t*-butyl chloride that do not use aluminum chloride as a catalyst which might be used to prepare *p*-di-*t*-butylbenzene from *t*-butylbenzene. What catalyst(s) do these methods use?

9. A Friedel–Crafts "acylation" reaction utilizes an acid chloride instead of an alkyl chloride. The catalyst is again aluminum chloride. Give a stepwise mechanism for the acylation of benzene with acetyl chloride in the presence of aluminum chloride.

10. Is the introduced ring substituent in Question 9 activating or deactivating toward further electrophilic substitution? Does this favor or disfavor further acylation of the aromatic ring?

Name ______________________ Section ____________ Date ________

24 The Friedel–Crafts Reaction

PRELABORATORY QUESTIONS

General

1. The Friedel–Crafts reaction involves the electrophilic attack of an aromatic ring by a carbocation. List three unrelated ways that carbocations can be generated for use in this reaction.

2. An aromatic ring more highly deactivated than a monohalobenzene cannot be alkylated by the Friedel–Crafts procedure. Which of the following *will not* undergo the Friedel–Crafts alkylation? acetanilide, nitrobenzene, *p*-xylene, anisole, benzamide.

3. The fact that deactivating groups suppress Friedel–Crafts alkylations (Question 2) constitutes a practical limitation for its use in syntheses. List two other aspects of this reaction which might limit its synthetic use.

Preparation of BHT

4. BHA and BHT are trade names for antioxidants used as food additives, and so forth. What do the letters stand for? Give the systematic names for these compounds.

5. In an autoxidation process, a hydrogen atom (H·) is abstracted from a C—H bond to form a carbon free radical. In the system

$$-\underset{H_a}{\underset{|}{C}}=\underset{H_b}{\underset{|}{C}}-\underset{H_c}{\underset{|}{\overset{|}{C}}}-$$

which hydrogen atom is most easily removed in such a process? Explain why.

6. Calculate the molar concentrations of the organic reactants used in the BHT experiment. Which one (if any) is in excess? Is there a reason for this?

Preparation of *p*-Di-*t*-butylbenzene

7. Write an equation showing how $AlCl_3$ participates in this reaction. Aluminum chloride acts as a catalyst in that only small amounts are required. Explain why.

8. Why must moisture be excluded from the reaction vessel?

9. Is the reaction exothermic, or endothermic? How do you know?

25

Aromatic Nitration

Nitration is one of the most important and thoroughly studied examples of electrophilic aromatic substitution. Benzene and benzene derivatives of widely different reactivities can be nitrated under suitable conditions; mononitro compounds can usually be obtained in good yields because the first nitro group deactivates the ring toward further substitution. Nitro compounds are useful as explosives (TNT), but their greater interest for organic chemists is in their usefulness as synthetic intermediates. The nitro group may be reduced to a nitroso, hydroxylamine, or amine function, and via the diazotization reaction amines can be converted to many other functional groups (—OH, —F, —I, —CN, —N=N—Ar, etc).

Nitration may be carried out by using fuming nitric acid, mixtures of concentrated nitric acid and sulfuric acid, and nitric acid in glacial acetic acid. Selection of the nitrating agent and the reaction conditions will depend upon the reactivity of the compound to be nitrated; however, nitration is most frequently carried out with a mixture of concentrated nitric acid and concentrated sulfuric acid. In the HNO_3/H_2SO_4 mixture (called mixed acid or nitrating acid), the electrophile is the nitronium ion, NO_2^+, which is generated by protonation of the nitric acid followed by the loss of water.

$$HONO_2 + H_2SO_4 \rightleftarrows H_2\overset{+}{O}NO_2 + HSO_4^-$$

$$H_2\overset{+}{O}NO_2 \rightarrow NO_2^+ + H_2O$$

$$H_2O + H_2SO_4 \rightarrow H_3O^+ + HSO_4^-$$

Attack on the aromatic ring by the nitronium ion, usually the rate-determining step, is then followed by the rapid loss of a proton to give the nitro derivative.

$$NO_2^+ + C_6H_6 \longrightarrow [C_6H_6NO_2]^+ \xrightarrow{-H^+} C_6H_5NO_2$$

The ease or difficulty of nitration depends upon the nature of the substituents already present. Electron-donating groups (—OH, —OR, —NHCOR, —CH_3) facilitate the nitration, but electron-withdrawing groups (—NO_2, —COOH, —COR) retard the reaction. With substituted benzenes, the nitro group may be directed to the *ortho* and *para* positions (by —OH, —NHCOR, —Cl, —Br) or to the *meta* position (by —NO_2, —COOR).

Experiments

In the experiments in this chapter, we illustrate how nitration can be controlled by the proper choice of conditions with substrates of significantly different reactivities. A variety of methods for the separation and purification of the products are also illustrated in the procedures. The first two nitrations are done on a macroscale while the second two are carried out on a microscale.

SAFETY NOTE CONCENTRATED SULFURIC ACID AND CONCENTRATED NITRIC ACID CAN CAUSE BURNS ON CONTACT WITH THE SKIN. PARTICULAR CARE MUST BE TAKEN IN MEASURING AND HANDLING THESE REAGENTS. CLEAN UP ANY SPILLS IMMEDIATELY (CONSULT YOUR INSTRUCTOR). SHOULD ANY ACID GET ON YOU, WASH QUICKLY AND THOROUGHLY WITH COLD WATER AND THEN WITH SOAP AND WATER. WEAR SAFETY GOGGLES! AROMATIC NITRO COMPOUNDS ARE GENERALLY TOXIC, AND SOME MAY CAUSE CONTACT DERMATITIS; AVOID CONTACT WITH ANY OF THE PRODUCTS.

A. PREPARATION AND SEPARATION OF *o*- AND *p*-NITROANILINE (Macroscale)

The —NH_2 group activates the benzene ring so strongly for electrophilic substitution that direct nitration of aniline cannot be controlled. For the preparation of mononitroanilines, therefore, the reactivity is moderated by acetylation of the amine; the acetyl group is subsequently removed by acid hydrolysis. Even with the acetyl "protecting group" the ring is strongly activated, and it is difficult to avoid the formation of some dinitro products.

O
‖
$NHCCH_3$

1) HNO/H_2SO_4
2) H_3O^+
3) OH^-

NH_2, NO_2
o-nitroaniline
mp 72°C

NH_2, NO_2
p-nitroaniline
mp 147°C

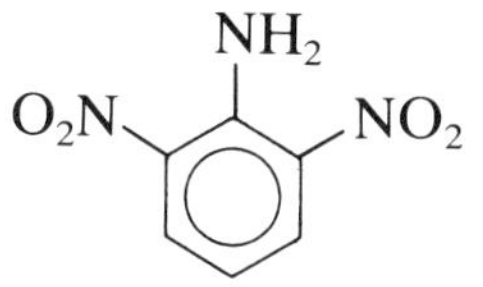

NH_2 / O_2N / NO_2

2,6-dinitroaniline
mp 142°C

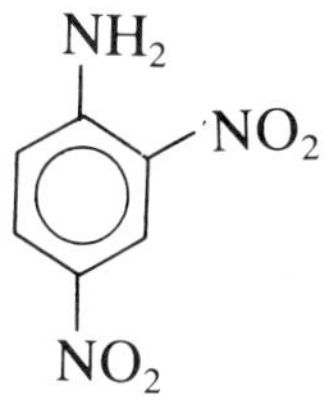

NH_2 / NO_2 / NO_2

2,4-dinitroaniline
mp 188°C

The *para* isomer is the major product, and it can be purified readily by recrystallization. A small amount of *o*-nitroaniline and traces of the 2,4- and 2,6-dinitro compounds can be isolated or detected by chromatography.

Procedure

In a 125-mL Erlenmeyer flask, place 2.7 g of acetanilide. Obtain 9 mL of concentrated sulfuric acid in a graduated cylinder, and pour approximately half of the acid into the flask. Swirl and stir the mixture until all but traces of the acetanilide dissolve (a small amount of remaining solid will subsequently dissolve). Cool the solution in an ice bath. With a buret or graduated pipet, measure out 1.5 mL of concentrated nitric acid and add it to the remaining sulfuric acid. Mix the acids by drawing up samples from the bottom with a transfer pipet and bulb and emptying them on top; complete the mixing by discharging a stream of bubbles from the empty pipet at the bottom of the cylinder.

Using the pipet and bulb, add the mixed acid in small portions (approximately 0.5 mL) to the cooled sulfuric acid solution of acetanilide. Swirl the flask in the ice bath after each portion is added. The flask should not become perceptibly warm to the touch; the addition requires 10 to 15 minutes. After 20 minutes, including addition time, add 25 mL of a mixture of ice and water (loosely pack a 16 $\times$ 150-mm test tube with ice and fill with water) to the reaction mixture.

The resulting suspension of nitroacetanilides is now hydrolyzed in the same flask, using the aqueous sulfuric acid. Add a boiling stone and heat the flask on a hot plate or wire gauze with a burner flame. Watch carefully as the color darkens and the solid begins to dissolve, and remove the heat source if necessary to avoid excessively vigorous boiling and foaming. Continue heating at a gentle boil for 15 minutes, and then cool the flask and contents in an ice bath. Place the bath and flask in the hood, and add, in five or six portions, 25 mL of concentrated aqueous ammonium hydroxide; swirl in the ice bath after each addition.

Collect the precipitated nitroaniline in a Büchner funnel by suction filtration. Rinse the flask with 3 mL of water, disconnect the suction tubing, pour the rinsing liquid into the funnel, and then suck the water through the precipitate; press the solid, and allow air to be drawn through it for 2 or 3 minutes.

Transfer the yellow crystals, which can be peeled cleanly from the filter paper, to a 25 × 100-mm test tube and add 4 mL of ethanol. Heat and stir the mixture on the steam bath until the alcohol boils and most of the solid dissolves, and then cool in an ice bath for 10 to 15 minutes to crystallize the *p*-nitroaniline. Collect the crystals by suction filtration, wash with a minimum volume of ethanol, and dry on glassine paper. Save the mother liquor for the examination of minor products. Determine the melting point and weight, and calculate the percent yield of crude *p*-nitroaniline. Run a TLC on samples of the solid and mother liquor, using methylene chloride for development.

For crystallization, dissolve the crude product in a minimum volume of hot ethanol. If the crude nitroaniline is very dark, add a small amount of charcoal and filter through fluted paper. Cool, collect the crystals, and determine the melting point.

For examination of the minor products, evaporate the mother liquor from the *para* isomer to a thick syrup and filter off any additional solid that crystallizes. Prepare a chromatography column (see Fig. 7.5) with 15 g of alumina, using methylene chloride as the solvent. A 10 × 300-mm chromatography tube or a 25-mL pinch-cock buret can be used for the column.

Dissolve the nitroaniline residue in a minimum volume of methylene chloride; 1 mL should suffice. Apply the sample to the adsorbent, and then develop and elute the column with methylene chloride. Collect the eluate in test tubes in 5-mL fractions, beginning with the first colored solution that emerges from the column. After five 5-mL fractions, collect one final large fraction until either all of the colored material is eluted or the volume of this fraction reaches 25 mL. Examine each fraction by TLC (three fractions per plate), and pool fractions on the basis of TLC behavior; if two fractions show the same TLC spot(s), combine them.

Evaporate and weigh any fractions or groups of fractions that contain a single component. Compare the TLC with authentic samples of the possible minor products; determine the melting point and weight of the crystalline residues. Record the yield of *o*-nitroaniline and any other compounds isolated from the chromatogram.

B. PREPARATION AND SEPARATION OF *o*- AND *p*-NITROPHENOL (Macroscale)

OH (phenol) $\xrightarrow{HNO_3}$ OH, NO_2 (*o*-nitrophenol) + OH, NO_2 (*p*-nitrophenol)

o-nitrophenol
mp 45°C

p-nitrophenol
mp 114°C

Electrophilic substitution in the highly activated phenol ring occurs under very mild conditions, and mononitration must be carried out with dilute aqueous nitric acid. The usual nitric acid/sulfuric acid mixture gives a complex mixture of polynitro compounds and oxidation products.

Separation of *o*- and *p*-nitrophenol can be accomplished by taking advantage of the strong intramolecular hydrogen bonding in the *ortho* isomer (I). In the *para* isomer (II), the hydrogen bonding is intermolecular, and these attractive forces between the molecules lead to a lower vapor pressure. On steam distillation of the mixture, the *ortho* isomer is obtained in pure form in the distillate. The *para* isomer can then be isolated from the nonvolatile residue.

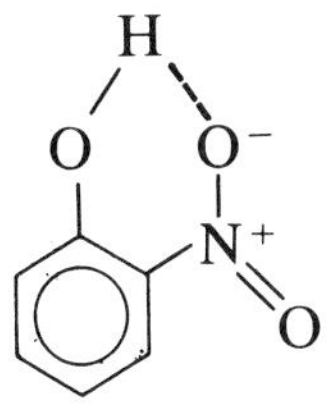

(I) 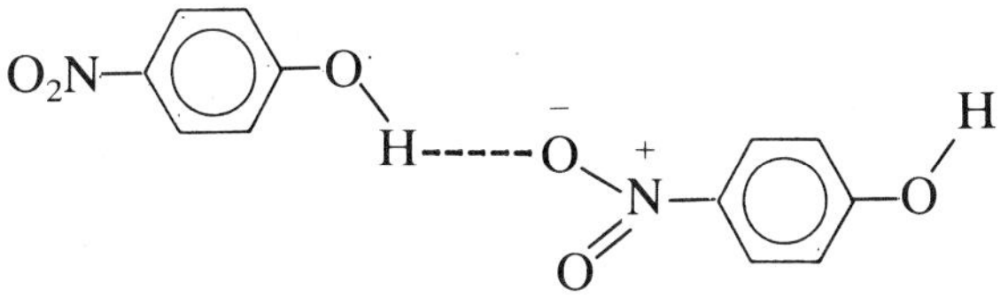 (II)

Procedure

In a 250-mL Erlenmeyer flask, add 10 mL of concentrated nitric acid to 35 mL of water. Weigh out 9.0 g of "liquid phenol" (approximately 90% phenol in water; density 1.06 g/mL) in a 50-mL beaker (**CAUTION:** phenol is corrosive; avoid contact with skin). With a disposable pipet, add 1- or 2-mL portions of the phenol to the nitric acid and cool as necessary by swirling the flask in a pan of cold water to keep the temperature of the reaction mixture at 40 to 50°C. After all of the phenol has been added (rinse the beaker with 1 mL of cold water), swirl the flask intermittently for 5 to 10 minutes while the contents cool to room temperature. Meanwhile assemble the apparatus for a steam distillation (see Fig. 8.2).

Transfer the reaction mixture to a separatory funnel, and drain the oily organic layer into a 500-mL round-bottom flask. (If the mixture is so dark that separate layers are not evident, add 10 mL of water.) Add 200 mL of water, and then carry out a steam distillation until no further *o*-nitrophenol appears in the distillate. Collect the *o*-nitrophenol by filtration, allow it to air dry, then determine yield and melting point.

For the isolation of the *p*-nitrophenol, adjust the total volume of the distillation residue to approximately 200 mL by adding more water or removing water by distillation. Decant the hot mixture through a coarse fluted filter or cotton plug, and add approximately 1 g of charcoal and 2 mL of concentrated, hydrochloric acid to the hot filtrate, heat again to boiling, and filter through fluted filter paper to remove the charcoal. Chill a 500-mL Erlenmeyer flask in ice, and pour a small portion of the hot *p*-nitrophenol solution into it to promote rapid crystallization and prevent the separation of the product as an oil. Add the remainder of the solution in small portions so that each is quickly chilled. Collect the crystals by vacuum filtration, air dry, and determine the yield and melting point.

Submit both isomers to your instructor in appropriately labeled containers.

C. NITRATION OF NITROBENZENE (Microscale)

$$C_6H_5NO_2 \xrightarrow{HNO_3/H_2SO_4} m\text{-}C_6H_4(NO_2)_2$$

mp 90°C

In a 25-mL Erlenmeyer flask, mix 1 mL of concentrated sulfuric acid and 1 mL of nitric acid. (*Note*: With the small quantities of sulfuric acid, nitric acid, and aromatic compounds used in this experiment, no difficulty with overheating will occur on mixing. These reactions can be highly exothermic, however, and the reader is cautioned not to attempt nitration reactions on a larger scale without consulting the literature and making provision for the mixing of chemicals in small portions so that temperature control can be maintained.) To the mixed acid, add 0.20 mL (0.24 g, 0.0019 mole) of nitrobenzene. Place the flask in a boiling water bath for 20 minutes with frequent swirling, and then pour the reaction mixture into 20 mL of ice water in a 50-mL beaker. Stir the mixture well, collect the crude *m*-dinitrobenzene by vacuum filtration on a small Hirsch funnel, and wash the product with two or three 3-mL portions of water. For most effective washing, release the suction, drip the water over all of the filter cake, and apply suction again to remove the water.

Recrystallize the product using ethanol, collect the purified product by vacuum filtration, and wash it with two 1-mL portions of cold ethanol. Dry the crystals, determine the melting point and weight of the purified product. Calculate the percent yield, and submit the product to your instructor in an appropriately labeled container.

D. NITRATION OF METHYL BENZOATE (Microscale)

$$C_6H_5C(=O)OCH_3 \xrightarrow{HNO_3/H_2SO_4} m\text{-}O_2NC_6H_4C(=O)OCH_3$$

mp 79°C

Place approximately 1 mL of concentrated sulfuric acid in a small test tube, and chill it in an ice bath. While the sulfuric acid is cooling, place 0.250 mL (0.272 g, 0.002 mole) of methyl benzoate in a clean, dry 13 × 100-mL test tube. Add the chilled sulfuric acid dropwise to the methyl benzoate, and cool the mixture in the ice bath if it begins to warm up noticeably. After the sulfuric acid has been added, add 0.3 mL of concentrated nitric acid dropwise using a graduated pipet, again chilling the reaction mixture if it warms up noticeably. After the addition of the nitric acid is complete, allow the reaction mixture to stand at room temperature for 10 minutes.

Add 5 mL of ice water to the reaction mixture, and stir the mixture thoroughly in an ice bath until the mixture is chilled and crystallization is complete. Filter the precipitate using vacuum filtration, wash the filter cake with two 2-mL portions of ice water, and suck the filter cake as dry as possible.

Air dry the solid (or oven dry it at a temperature below 50°C for 30 minutes). Determine the weight and melting point of the crude product, and recrystallize it from methanol. Determine the weight, melting point, and the percent yield of the purified material. Submit the product to your instructor in an appropriately labeled container.

QUESTIONS

1. Which compound in each of the following pairs is more reactive towards aromatic nitration? Explain your answers.
   a. phenol or nitrobenzene
   b. methyl benzoate or phenol
   c. nitrobenzene or methyl benzoate
   d. benzene or toluene
2. If all the methyl benzoate is not nitrated in Part D, where is the unreacted starting material removed in the purification procedure?
3. Show all possible mononitration products of toluene. Do you think it would be easy to isolate and identify all of them if each is produced in some amount?
4. In the nitration of nitrobenzene, why would you expect little (if any) trinitrobenzene to be produced (dinitration of the starting material)? If the temperature were not regulated, and some were produced, what isomer would it be?
5. Why is steam distillation, as opposed to simple distillation of the oily product mixture, preferred for separation of the nitrophenols?
6. The solubility of phenol in water is 9 g/100 mL at room temperature; explain how 8 g of phenol and 2 mL of water can form a homogeneous solution.

REFERENCE

de la Mare, P. B.; Ridd, J. H. *Aromatic Substitution: Nitration and Halogenation*; Academic Press: New York, 1959.

Name ______________________ Section ____________ Date ________

25 Aromatic Nitration

PRELABORATORY QUESTIONS

1. a. What mixture of reagents is used to nitrate nitrobenzene and methyl benzoate?

b. What should you do if either of these reagents is spilled on your hands?

2. The nitration of acetanilide yields two mononitration products. Draw their structures.

3. What product would you expect if methyl benzoate underwent dinitration?

4. Explain in your own words why *o*-nitrophenol and *p*-nitrophenol can be separated by steam distillation.

26

Aromatic Nucleophilic Substitution

Chlorobenzene and aryl chlorides generally are extremely unreactive toward nucleophilic substitution. Displacement reactions that occur under drastic conditions, as in the preparation of phenol from chlorobenzene with NaOH at 250°C, often occur by elimination to give an aryne (Chapter 35) followed by addition.

With nitro groups *ortho* or *para* to the halogen, however, nucleophilic substitution becomes a facile and useful reaction. The preparation of 2,4-dinitrophenol is an example. This compound cannot be obtained satisfactorily by the nitration of phenol or mononitrophenols because of the susceptibility of the ring to oxidation, but displacement of halogen in 2,4-dinitrochlorobenzene to give the phenol can be carried out in good yield in boiling aqueous base. The reaction occurs by addition of the nucleophile to give an intermediate with an sp^3 carbon, followed by loss of halide anion.

Cl, NO_2, NO_2 $\xrightarrow{OH^-}$ HO Cl, NO_2, NO_2 $\longrightarrow$ OH, NO_2, NO_2 + Cl^-

This addition–elimination sequence leads to two characteristic differences between aromatic nucleophilic substitution and S_N2 reactions of alkyl halides. In the latter case, the reactivity sequence within a series of halides is I>Br>Cl>F, corresponding to the progressively greater strength of the C—X bond, which is broken in the rate-determining step of the reaction. In aromatic substitution this reactivity difference is not observed, and the fluoro compound is in fact the most reactive, since the highly electronegative fluorine atom increases the stabilization of the transition state for the addition step.

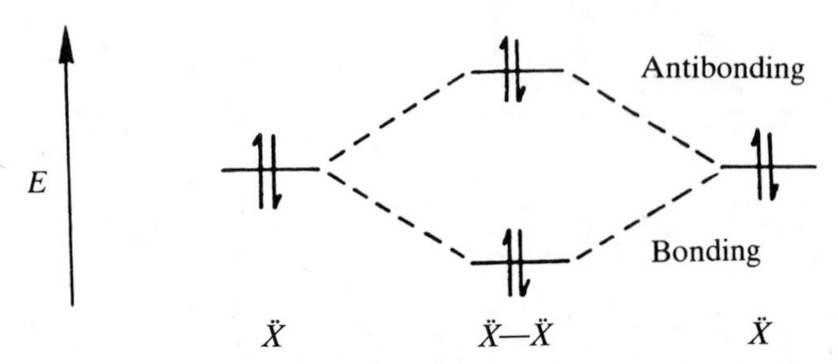

Figure 26.1 Interaction of adjacent electron pairs in α-nucleophiles. The Ẍ denotes an atom with one or more unshared pairs of electrons (e.g., N, O, Cl).

A second distinctive feature of aromatic nucleophilic substitution as contrasted to aliphatic S_N2 reactions is the **α-effect**. This term denotes a large acceleration of the rate with nucleophiles such as NH_2NH_2, NH_2OH, ClO^-, or HO_2^- in which an electronegative atom with an unshared electron pair is located adjacent (i.e., α) to the nucleophilic atom. This effect is also seen in reactions of esters and other carbonyl compounds with α-nucleophiles. It has been suggested that the α-effect arises from the interaction of the orbitals containing the unshared electron pairs on adjacent atoms, leading to a situation in which both bonding and antibonding molecular orbitals are occupied (Fig. 26.1). The electrons in the antibonding orbital are at a higher energy than the unshared pairs in simple nucleophiles and are more readily donated to an unsaturated electrophile.

The α-effect in reactions of 2,4-dinitrochlorobenzene is a very large one, the rate constant k_2 for reaction with hydroperoxide ion being nearly 10^5 times greater than that for hydroxide ion. The consequences of such an increase in rate can be translated in practice to faster reactions at lower temperatures. In the preparation of 2,4-dinitrophenol, for example, a traditional procedure requires reaction with aqueous base at 130 to 200°C for several hours. As will be demonstrated in the first experiment in this chapter, the phenol can be obtained in the presence of hydroperoxide in a few minutes at 50 to 60°C. In this procedure the product formed initially is probably the aryl hydroperoxide, which rapidly undergoes attack by hydroxide to give the phenoxide anion.

Cl, NO_2, NO_2 $\xrightarrow{^-OOH}$ [O—OH, NO_2, NO_2] $\xrightarrow{\ddot{O}H^-}$ O^-, NO_2, NO_2 + H_2O_2

Likewise, differences in reactivity due to the α-effect and also due to the nature of the halogen can be seen in the reactions of dinitrohalobenzene with hydrazine and with ammonia.

$$\text{2,4-}(NO_2)_2C_6H_3NH_2 \xleftarrow{NH_3} \text{2,4-}(NO_2)_2C_6H_3X \xrightarrow{NH_2NH_2} \text{2,4-}(NO_2)_2C_6H_3NHNH_2$$

2,4-dinitroaniline
mp 188°C

2,4-dinitrophenylhydrazine
(2,4-DNPH) mp 197°C

The compound 2,4-dinitroaniline can be obtained more economically by another route, but the reaction with hydrazine is the only practical process for preparing the very useful reagent 2,4-dinitrophenylhydrazine (2,4-DNPH). The reaction of 2,4-dinitrofluorobenzene (Sanger's reagent) with amines has an important application in determining the sequence of amino acids in peptides.

Experiments

A. PREPARATION OF 2,4-DINITROPHENOL (Microscale)

SAFETY NOTE 2,4-DINITROCHLOROBENZENE IS VERY IRRITATING TO THE SKIN; AVOID ANY CONTACT WITH THE HANDS. DINITROPHENOL IS HIGHLY TOXIC.

In a 25-mL round-bottom flask with a ground glass joint, place 0.2 g of 2,4-dinitrochlorobenzene, 4.0 mL of 2 *M* aqueous KOH solution, and 0.5 mL of methanol. Warm the mixture on the steam bath with swirling until the solid melts. Obtain 1.0 mL of 3% H_2O_2 or 0.5 mL of 6% H_2O_2 (30% solution diluted 1:4), and add about one fourth of the peroxide solution to the flask. Place a ground glass stopper in the flask, hold the stopper in place with your hand, and shake the flask vigorously for 3 to 4 minutes; then remove the stopper. (**CAUTION:** Handle the stopper with a paper towel and avoid contact of the solution with the skin; rinse thoroughly with water if contact occurs.)

Warm the flask to about 50°C (as warm as can be comfortably held in the hand), add one fourth to one third of the peroxide solution, re-stopper, and shake again for several minutes. When solid begins to appear in the dark solution, remove the stopper and warm the flask on the steam bath until the solid redissolves. Add the remaining peroxide, and swirl the flask on the steam bath until no more oily droplets can be seen on the bottom of the flask (approximately 5 minutes' heating at 60 to 65°C should suffice).

Pour the solution into another Erlenmeyer flask, rinse with a few tenths mL of water, and cool the solution to 5 to 10°C in an ice bath. Mix 0.8 mL of concentrated HCl with approximately 1 mL of crushed ice, and add this dilute acid to the solution until the mixture is a light yellow color. Stir the mixture for 2 minutes, and rub the solid against the wall to insure that all of the phenoxide is acidified. Collect the solid on a small Büchner funnel, wash with a few tenths mL of dilute HCl, followed by water, and dry thoroughly (60°C oven for 30 minutes or longer at room temperature). Determine the melting point (lit. mp 114–115°C) and calculate the percent yield. Retain the product for use in Chapter 28 or submit it to your instructor as directed.

B. REACTION OF 2,4-DINITROHALOBENZENES WITH AMMONIA AND HYDRAZINE (Microscale)

In a 10 × 75-mm test tube (Tube 1) dissolve 0.06 g of 2,4-dinitrochlorobenzene in 1.0 mL of ethanol. Transfer half of the solution to a second test tube (Tube 2). In a third tube (3) dissolve 0.04 mL of 2,4-dinitrofluorobenzene (d = 1.47 g/mL) in 1.0 mL of ethanol and transfer half of this solution to a fourth tube (4).

To the first tube, add 0.02 mL of 85% hydrazine hydrate (suspected carcinogen) ($N_2H_4 \cdot H_2O$; 0.55 g N_2H_4/mL), and shake the tube briefly to mix the solutions. Note any changes and the time required. In the same way, add 0.02 mL of hydrazine solution to Tube 3, and note changes.

To Tubes 2 and 4, add 0.02 mL of concentrated ammonium hydroxide (0.26 g NH_3/mL), mix well, and record any changes.

Collect any crystalline products that were obtained, and wash with a few mL of alcohol. Dry the products, weigh, and calculate percent yields. Place the filtrates and any unreacted solutions in the waste solvent receptacle. Label the products and retain for later use or submit them as directed.

Compare the results in the four tubes and explain the reactivity order observed.

Both Parts A and B can be carried out on a **macroscale** using ten times the quantities specified without difficulty.

QUESTIONS

1. The reaction of a polypeptide with 2,4-dinitrofluorobenzene is used in an analytical procedure for identifying the terminal amino acid in the chain. After reaction, the amide bonds are hydrolyzed, liberating all but the N-terminal amino acid (containing R_1).

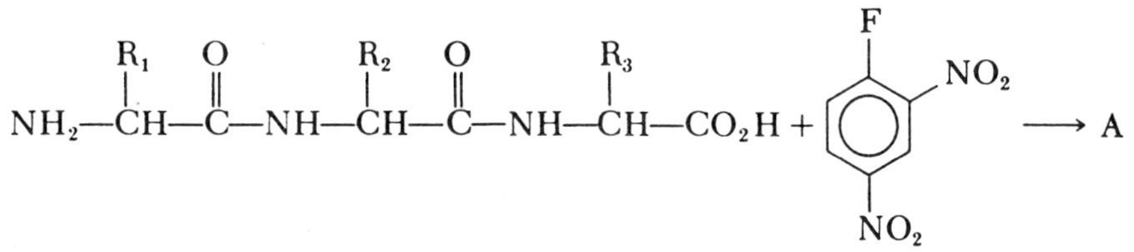

$$NH_2{-}\underset{\displaystyle\vert}{\overset{R_1}{C}}H{-}\overset{O}{\overset{\Vert}{C}}{-}NH{-}\overset{R_2}{\underset{}{C}}H{-}\overset{O}{\overset{\Vert}{C}}{-}NH{-}\overset{R_3}{C}H{-}CO_2H + \text{(1-F-2,4-}(NO_2)_2C_6H_3) \longrightarrow A$$

$$A \xrightarrow{H_3O^+} B + NH_2{-}\overset{R_2}{C}H{-}CO_2H + NH_2{-}\overset{R_3}{C}H{-}CO_2H$$

Write structures for compounds A and B. Suggest how this reaction sequence permits the identification of the amino acid containing R_1.

2. Groups that are deactivating in electrophilic substitution reactions (e.g., —NO_2) become activating groups in nucleophilic substitution reactions. Explain.

3. Smiles rearrangements are rearrangements that follow an intramolecular nucleophilic substitution course, as exemplified by the following reaction.

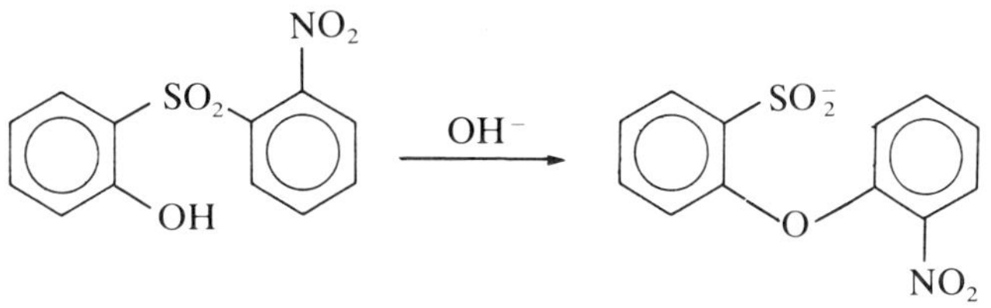

$$\text{2-HO-}C_6H_4{-}SO_2{-}C_6H_4\text{-2-}NO_2 \xrightarrow{OH^-} \text{2-}(SO_2^-){-}C_6H_4{-}O{-}C_6H_4\text{-2-}NO_2$$

Write the mechanism by which this rearrangement takes place.

REFERENCES

Bernasconi, C. F. *Accounts of Chemical Research* **1978**, *11*, 147.
Bunnett, J. F. *J. Chem. Education* **1974**, *51*, 312.
Miller, J. *Aromatic Nucleophilic Substitution*; American Elsevier: New York, 1968.

Name ______________________ Section ______________ Date ________

26 Aromatic Nucleophilic Substitution

PRELABORATORY QUESTIONS

1. Define the α-effect and what is the requirement of the nucleophile that brings it about?

2. What is Sanger's reagent?

3. Draw the structural formula for the product of the nucleophilic substitution reaction of 2,4-dinitrochlorobenzene with methyl amine.

4. When ethyl picrate (2,4,6-trinitroethoxybenzene) is heated with an excess of sodium methoxide, the product formed is 2,4,6-trinitroanisole (2,4,6-trinitromethoxybenzene). Give a plausible mechanism for this reaction.

27

Properties of Amines

Amines have an unshared pair of electrons on nitrogen and, therefore, act as bases, nucleophiles, and compounds that may be oxidized. As Lewis bases, they form salts with acids and form coordination complexes with metal cations. As nucleophiles, they displace halogen from alkyl halides and acid chlorides to give more highly alkylated amines and amides, respectively. They may be oxidized by a variety of oxidizing agents including oxygen, permanganate ion, hydrogen peroxide, and nitrous acid.

The basicity of an amine is influenced by the number and type of groups attached to the nitrogen atom. Aliphatic amines are stronger bases than ammonia because the alkyl groups are electron donors relative to hydrogen. Aromatic amines are weaker bases than ammonia because delocalization of the unshared electron pair on nitrogen into the ring lowers the electron density on nitrogen.

Amines up to approximately five carbon atoms are soluble in water. Higher molecular weight amines that are water insoluble dissolve in aqueous acid through the formation of salts. This provides a convenient method for separating such amines from water insoluble neutral and acidic compounds (see Chapter 4, Part B).

Many amines, as well as other nitrogen compounds, are physiologically active and should be handled with caution. Amines that occur naturally in plants are called alkaloids; the potency of many of these such as morphine, cocaine, and nicotine is widely known.

Experiments

SAFETY NOTE MANY AMINES ARE TOXIC SUBSTANCES; INHALATION OF THEIR VAPORS OR CONTACT WITH THE SKIN SHOULD BE AVOIDED. IF ANY SHOULD CONTACT THE SKIN, WASH WITH WATER, THEN WITH 10% ACETIC ACID SOLUTION, AND AGAIN WITH WATER. BENZENESULFONYL CHLORIDE IS CORROSIVE AND GIVES OFF IRRITATING VAPORS.

A. SOLUBILITY IN WATER

Place 1 mL of water in a 13 × 100-mm test tube, add 1 drop of cyclohexylamine, and shake the tube to mix the two materials. Note whether the amine is soluble; if it is, determine the solubility of a second and third drop. Save the mixture for use in Part B.

Repeat this solubility test with aniline, pyridine, and triethylamine. Also save the mixtures for Part B.

B. BASICITY

1. Determine the pH of each of the mixtures from Part A by placing a drop of the aqueous phase on a piece of universal pH paper.
2. To those mixtures from Part A in which the amine is insoluble, add 10% HCl dropwise with stirring. Does the amine dissolve?

C. HINSBERG TEST

Amines react as nucleophiles toward acid chlorides and acid anhydrides to yield amides. Many of these amides, such as those of acetic, benzoic, and benzenesulfonic acid, are crystalline solids that may be used as derivatives to identify the amine.

The reaction of an amine with benzenesulfonyl chloride can be useful in determining whether the amine is primary, secondary, or tertiary. The sulfonamide from a primary amine is acidic and dissolves in sodium hydroxide solution, but the sulfonamide of a secondary amine lacks the amide hydrogen and is insoluble. Tertiary amines do not form amides. This overall process is known as the **Hinsberg test**.

$$C_6H_5\text{—}SO_2Cl + RNH_2 \xrightarrow{NaOH} C_6H_5\text{—}SO_2\text{-}NHR + NaCl + H_2O$$

$$\text{HCl} \uparrow \downarrow \text{NaOH}$$

$$C_6H_5\text{—}SO_2\text{-}\ddot{N}R^{-} \quad Na^+$$

Place two drops of amine, 4 drops of benzenesulfonyl chloride, and 4 mL of 10% sodium hydroxide in a 13 × 100-mm test tube. Stopper the tube and shake it for approximately 5 minutes. If a clear or nearly clear solution is obtained, a primary amine is indicated; acidification with hydrochloric acid should precipitate the sulfonamide. The presence of a solid or liquid residue in the original reaction mixture indicates a secondary or tertiary amine. Separate the residue and determine its solubility in 10% HCl. A tertiary amine will dissolve; if the residue does not dissolve, then it is a sulfonamide of a secondary amine. Perform the Hinsberg test on aniline, *N*-methylaniline, and triethylamine, and then determine whether an unknown is a primary, secondary, or tertiary amine.

D. REACTIONS OF DIAZONIUM SALTS (Microscale)

Amines undergo a variety of reactions with nitrous acid depending on whether the amine is a primary, secondary, or tertiary amine and whether it is aliphatic or aromatic. The reaction of primary aromatic amines with nitrous acid is unique and provides a very versatile synthetic intermediate. The process is called **diazotization**, and the intermediates formed are called diazonium salts. Aqueous solutions of diazonium salts are stable at 0 to 5°C for several hours; however, they lose nitrogen on warming and form phenols. Solid diazonium salts sometimes crystallize from solution but should *never* be separated by filtration as they can decompose explosively.

$$C_6H_5{-}NH_2 \xrightarrow[\text{HCl, } 0\text{–}5°C]{\text{HONO}} C_6H_5{-}\overset{+}{N}{\equiv}N\ Cl^-$$

Diazonium salts react with nucleophiles to liberate N_2 and form phenols, aryl halides, and nitriles with such nucleophiles as water, halide ion, and cyanide ion, respectively. They react with phenols and tertiary aromatic amines (aromatic compounds particularly reactive towards electrophilic aromatic substitution) to give azo compounds by a reaction known as azo coupling. In the following experiment, *p*-nitroaniline will be reacted with nitrous acid to form the diazonium salt, *p*-nitrobenzenediazonium chloride, which will then be used to prepare *p*-iodonitrobenzene and/or the dye Para Red. If both products are to be prepared, the diazonium salt preparation should be doubled.

1. Preparation of the Diazonium Salt

$$O_2N{-}C_6H_4{-}NH_2 \xrightarrow[\text{HCl, } 0\text{–}5°C]{\text{HONO}} O_2N{-}C_6H_4{-}N_2^+\ Cl^-$$

In a 25-mL Erlenmeyer flask, place 5 mL of water, 1 mL of 20% hydrochloric acid, and 0.276 g (0.002 mole) of *p*-nitroaniline. Warm the mixture, stir until the amine has dissolved, and then cool the solution to 0 to 5°C. Add approximately 1 g of chipped ice and keep the mixture at 0 to 5°C in an ice bath.

Next prepare a solution of 0.138 g (0.002 mole) of sodium nitrite in 1 mL of cold water. Add the sodium nitrite solution to the solution of amine hydrochloride while swirling the latter vigorously. Continue swirling for 3 to 5 minutes, and then keep the solution at ice temperature until ready for use in Part 2 or Part 3 below. If appreciable solid remains, filter the cold solution through a small cotton plug.

The diazonium solution may be tested for nitrous acid with potassium iodide–starch paper (a purple to black coloration indicates HNO_2; the test paper must darken immediately) and for the actual formation of a diazonium salt (streak a solution of 2-naphthol in aqueous NaOH across a piece of filter paper, and then cross it with a streak of the reaction solution; an orange to red coloration at the junction indicates the presence of a diazonium salt).

2. Preparation of *p*-Iodonitrobenzene

$$O_2N{-}C_6H_4{-}N_2^+ + I^- \longrightarrow O_2N{-}C_6H_4{-}I + N_2\uparrow$$

Dissolve 0.6 g of potassium iodide in 2 mL of water and cool the solution to 0 to 5°C. Add it, with swirling, to the diazonium salt solution at such a rate that the temperature of the latter is maintained below 5°C. When the addition is complete, allow the mixture to stand in an ice bath for 3 to 5 minutes with occasional swirling. Collect the product by vacuum filtration, dry it, and recrystallize it from ethanol. Determine the weight, melting point, and percent yield.

3. Preparation of Para Red

$$O_2N{-}C_6H_4{-}N_2^+ + \text{2-naphthol (HO)} \longrightarrow O_2N{-}C_6H_4{-}N{=}N{-}C_{10}H_6{-}OH$$

In a 50-mL Erlenmeyer flask, dissolve 0.288 g of 2-naphthol (0.002 mole) in a mixture of 3 mL of 10% sodium hydroxide and 20 mL of water. Chill this solution in an ice bath until the temperature is below 5°C (a few chips of ice may be added to lower the temperature). Pour the diazonium solution all at once into the cold solution of 2-naphthol, and swirl the mixture for 5 to 10 minutes. Add 1 mL of concentrated hydrochloric acid, heat the mixture with stirring for 20 to 30 minutes on a steam bath, and filter the product. Allow the dye to dry and weigh the crude material. Do not determine a melting point or percent yield since the product contains inorganic salts.

QUESTIONS

1. Write an equation for the reaction of butylamine with one mole of HCl; for nicotine with one mole of HCl.
2. Write equations for the reaction of aniline with
   a. acetyl chloride
   b. benzoyl chloride
   c. acetic anhydride
3. How might one distinguish between the following pairs of substances using reactions described in this chapter?
   a. butylamine and dibutylamine
   b. cyclohexylamine and aniline
4. Write equations to illustrate the diazotization of aniline and subsequent reaction with phenol; with dimethylaniline.

Name ______________________ Section ____________ Date ________

27 Properties of Amines

PRELABORATORY QUESTIONS

1. a. Would you expect butylamine, aniline, and octylamine to be water soluble? Explain.

   b. If you wanted to get an insoluble amine to dissolve in an aqueous solution, what would you do?

2. Would either butylamine or diethylamine give a benzenesulfonamide (Hinsberg test) that is soluble in NaOH? Explain.

3. Write an equation for the reaction of aniline with benzenesulfonyl chloride.

4. Show the structure of the diazonium salt formed when aniline is diazotized in HCl solution.

28

Chemiluminescence

Chemiluminescence is the phenomenon in which visible light is emitted from a chemical reaction that occurs at or near room temperature. The most familiar examples of chemiluminescence are those that occur in fireflies and several marine organisms. These bioluminescent processes are enzymatic reactions of molecular oxygen with rather complex heterocyclic substrates, as exemplified by firefly luciferin. The overall reaction is shown in the following equation:

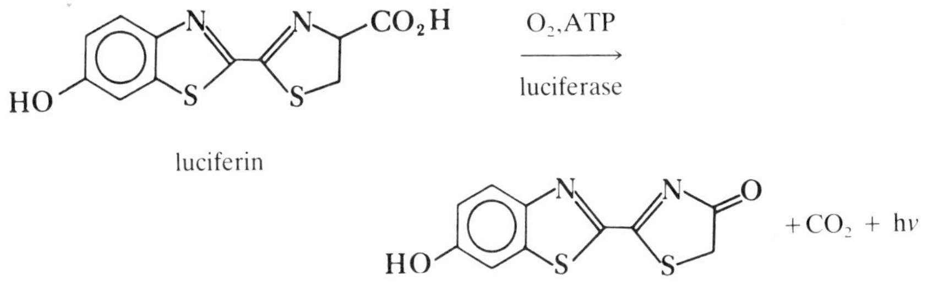

luciferin $\xrightarrow[\text{luciferase}]{O_2,\text{ATP}}$ (oxidized luciferin) $+ CO_2 + h\nu$

Chemiluminescence occurs by emission of energy in the visible region of the electromagnetic spectrum from a molecule in an electronically excited state. The process generally involves three steps: (1) formation of an energy-rich intermediate, I, from the reactants; (2) fragmentation of the intermediate to form products, of which one, P^*, is in an electronically excited state; and (3) emission of a quantum of light from P^* as it decays to its ground state.

$$R \xrightarrow{1} I \xrightarrow{2} P^* \xrightarrow{3} P + h\nu$$

The overall efficiency or quantum yield in chemiluminescence, as in any chemical process, is the product of the yields in Steps 1 through 3. The limiting factor is usually Step 2, in which a molecule in an excited state is produced. This reaction must be sufficiently exothermic (50 to 80 kcal/mole) to provide the energy for excitation. Even when this degree of exothermicity is available, significant chemiluminescence is only seldom observed because most of the energy is dissipated as vibrational, rather than electronic, excitation of the products.

A few systems have been discovered, however, in which the number of vibrational modes is reduced by a rigid structure, and the overall efficiency of chemiluminescence is nearly 25%. This value can be compared with an efficiency of approximately 3% for the conversion of energy to light by an incandescent lamp. The most efficient nonenzymatic chemiluminescence is produced by the reaction of diaryl oxalate esters with the highly reactive nucleophile hydrogen peroxide. Devices based on this oxalate–peroxide reaction are used in commercially available Cyalume emergency lights, in which a glass capsule containing one of the reactants is broken in a tube containing the other to activate the light.

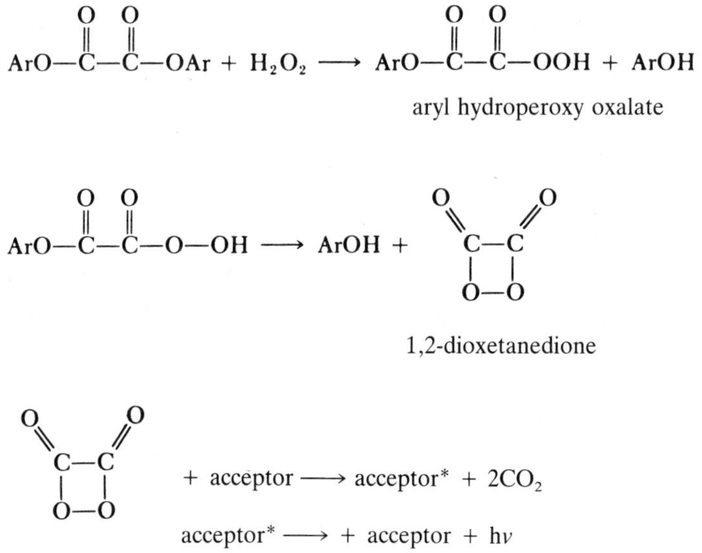

$$ArO{-}\overset{\overset{O}{\|}}{C}{-}\overset{\overset{O}{\|}}{C}{-}OAr + H_2O_2 \longrightarrow ArO{-}\overset{\overset{O}{\|}}{C}{-}\overset{\overset{O}{\|}}{C}{-}OOH + ArOH$$

aryl hydroperoxy oxalate

$$ArO{-}\overset{\overset{O}{\|}}{C}{-}\overset{\overset{O}{\|}}{C}{-}O{-}OH \longrightarrow ArOH + \text{(cyclic } C_2O_4\text{: O=C–C=O with ring O–O)}$$

1,2-dioxetanedione

$$\text{1,2-dioxetanedione} + \text{acceptor} \longrightarrow \text{acceptor}^* + 2CO_2$$

$$\text{acceptor}^* \longrightarrow + \text{acceptor} + h\nu$$

The energy-rich intermediate in this reaction has been suggested to be the strained peroxide, dioxetanedione, produced by cyclization of the hydroperoxy oxalate. Decomposition of dioxetanedione to two molecules of the very stable compound carbon dioxide can be estimated to be exothermic by at least 100 kcal/mole. The energy liberated in this reaction is not released directly as light, but it is first transferred to an acceptor molecule. The excited state of the acceptor is the species that emits light.

The final emission step is identical to that observed in fluorescence, in which radiation absorbed at one energy is reemitted at a lower energy. In chemiluminescence, the energy of a chemical reaction is used instead of radiation to pump the acceptor to an excited state. The acceptors used in peroxyoxalate chemiluminescence can be any compounds that undergo efficient visible fluorescence. The characteristic feature of most fluorescent compounds is a rigid polycyclic aromatic structure. As the conjugated system becomes more extended, the wavelength of the fluorescent emission becomes longer.

One group of compounds that are efficient acceptors for chemiluminescence are linear polycyclic hydrocarbons and phenyl derivatives of these structures, such as 9,10-diphenylanthracene, naphthacene, and rubrene.

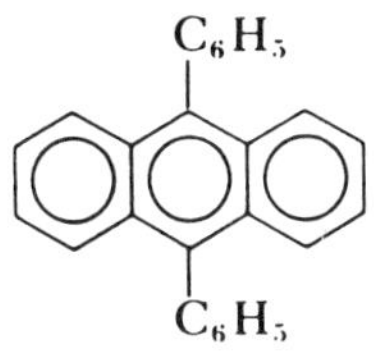

9,10-diphenylanthracene

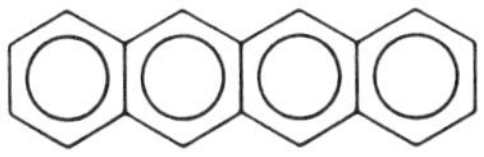

naphthacene

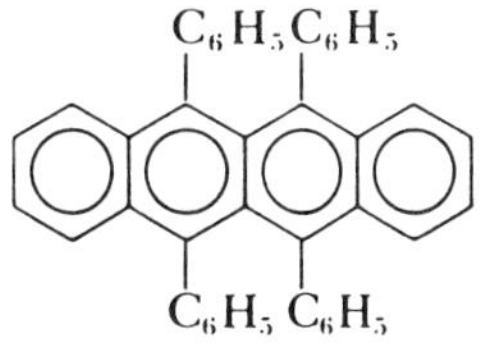

rubrene (5,6,11,12-tetraphenylnaphthacene)

Another type of acceptor is based on the highly fluorescent xanthene dyes, such as fluorescein and the rhodamines. The latter dyes are used as pigments in fluorescent signs, tape, and fabric. Minor changes in structure—for example, the addition of alkyl groups or the substitution of an ester for an acid substituent—cause shifts in the wavelength of the fluorescent emission and a corresponding change in the color of the chemiluminescence.

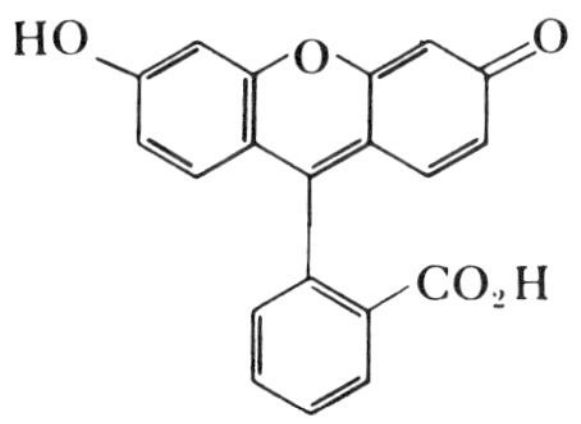

fluorescein

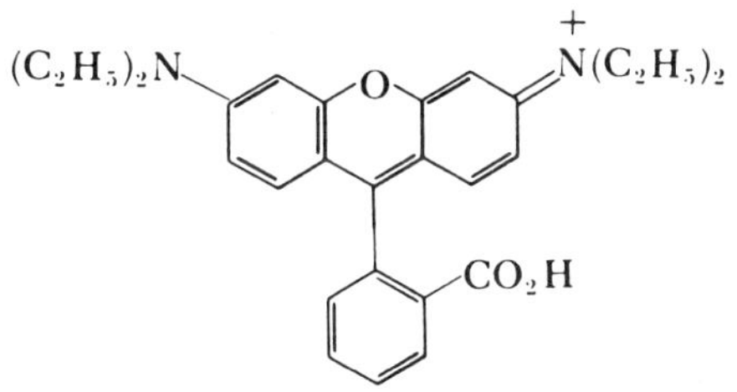

rhodamine B

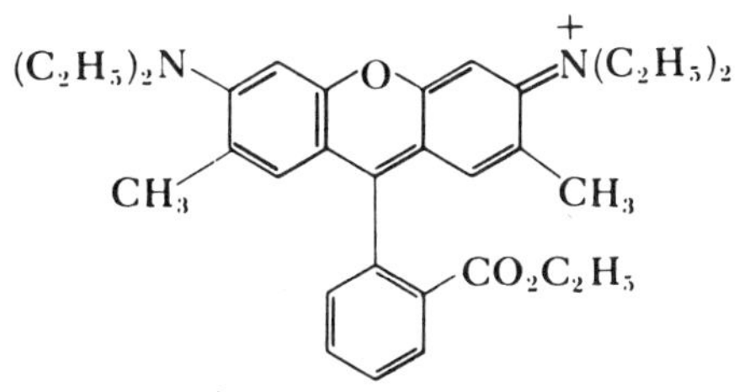

rhodamine 6G

Experiments

A. BIS-(2,4-DINITROPHENYL) OXALATE (Microscale)

The reaction used in this preparation is a common one—esterification with an acid chloride in the presence of a tertiary amine. The ester is quite susceptible to hydrolysis, and contact with water must be avoided. It is possible to separate it from the triethylammonium chloride by washing with chloroform since it is less soluble than is the chloride. However, the washing invariably dissolves appreciable amounts of ester as well. Since the triethylammonium chloride by-product does not interfere with the subsequent chemiluminescence reactions, the ester can be used without further purification.

$$2\ \text{2,4-}(NO_2)_2C_6H_3OH + ClCOCOCl + 2(C_2H_5)_3N \rightarrow [2,4\text{-}(NO_2)_2C_6H_3O\text{-}C(=O)]_2$$

2,4-dinitrophenol | oxalyl chloride | triethylamine | bis(2,4-dinitrophenyl) oxalate, mp 185°

$$+\ 2\ (C_2H_5)_3\overset{+}{N}HCl^-$$

triethylammonium chloride

SAFETY NOTE OXALYL CHLORIDE IS TOXIC AND HYDROLYZES IN AIR TO PRODUCE HCl. WORK WITH IT IN A WELL-VENTILATED HOOD. 2,4-DINITROPHENOL CAN CAUSE SKIN BURNS AND WILL STAIN.

Procedure

Add 0.12 g of *dry* 2,4-dinitrophenol to a graduated centrifuge tube (or distillation receiver) and dissolve the phenol in 0.6 mL (estimated in the centrifuge tube) of reagent-grade *dry* acetone. Add 0.1 mL of triethylamine, and cool the solution to below room temperature. In the hood, add 0.1 mL of a 30% (v/v) solution of oxalyl chloride in ether (0.03 mL or 0.045 g of oxalyl chloride) *dropwise* to the acetone solution, and constantly stir the mixture until it is a thick cream-colored suspension. Back-titrate by adding 1 drop of triethylamine. If a yellow/orange color develops, add an additional 0.05 mL (1 or 2 drops) of oxalyl chloride until the loss of the yellow/orange color.

After 10 minutes, warm the mixture on a steam or hot water bath and stir to evaporate most of the acetone. When the mixture is a thick pasty consistency, cool in an ice bath, and remove the remaining solvent under reduced pressure, shaking the receiver by hand or by using a rotary evaporator. Have the system close to vertical because some bumping and spattering may occur. After most of the remaining acetone has been evaporated, warm the evacuated tube to room temperature, and release the vacuum. Remove the product from the centrifuge tube with a microspatula and let dry. Use this material without further purification in Part B.

B. CHEMILUMINESCENCE (Microscale)

Procedure

The chemiluminescent reaction can be carried out in several ways. A simple and effective procedure is to add the solid bis-(2,4-dinitrophenyl) oxalate to a solution of the fluorescent acceptor and hydrogen peroxide. One or more of the hydrocarbon or xanthene emitters will be available as a 0.002 *M* solution in dimethyl phthalate.

Label test tubes with the fluorescent emitters to be used, and obtain 1 mL of each solution. To each tube, add 1 mL of a 0.2 *M* solution of hydrogen peroxide in dimethyl phthalate/*t*-butyl alcohol (2 mL of 30% H_2O_2 in 20 mL *t*-butyl alcohol plus 80 mL of dimethyl phthalate). Mix the solutions well and take the tubes, your notebook, and your vial of bis-(dinitrophenyl) oxalate to a dark room or a shielded box in the laboratory. Add a few crystals of the oxalate to each tube and record your observations.

To observe the effect of added base on the reaction, mix 2 mL each of 0.002 *M* 9,10-diphenylanthracene solution and 0.2 *M* hydrogen peroxide and divide the mixture between two tubes. Dissolve a few mg of your oxalate in 2 mL of dimethyl phthalate and obtain a few drops of a 10% solution of triethylamine in dimethyl phthalate. In the dark area, add 1 mL of the oxalate solution to each of the tubes of fluorescer, shake the tubes, and then add a drop of the triethylamine solution to one tube and compare the two reactions.

To observe the effect of temperature, place the other tube in an ice bath and then rewarm it.

QUESTIONS

1. Bis-(2,4-dinitrophenyl) oxalate and the 2,4,5-trichlorophenyl ester give much higher light yields than diphenyl oxalate in peroxy–oxalate chemiluminescence. Suggest a reason in terms of the mechanism of the process.
2. It has been reported that when hydrogen peroxide is added to a solution

2. It has been reported that when hydrogen peroxide is added to a solution of bis-(dinitrophenyl) oxalate, and a stream of nitrogen is bubbled through the solution and into a solution of rubrene, light is emitted from the latter. Suggest an explanation for this observation.
3. Firefly luminescence requires "high-energy" phosphate in the form of ATP, and the process has been suggested to involve formation of structure *A* followed by further steps. Suggest the subsequent steps and a plausible high-energy intermediate *I*.

CO_2H $\xrightarrow[O_2]{ATP}$ [$C(=O)-O\overline{P}O_3H$, $O-OH$] *A* $\longrightarrow I \longrightarrow$ [O] $+ CO_2$

REFERENCES

Adam, W. *J. Chem. Educ.* **1975**, *52*, 138.
Burr, J. G. Ed.; *Chemi- and Bioluminescence*; Marcel Decker: New York, 1985.
Mohan, A. G.; Turro, N. J. *J. Chem. Educ.* **1974**, *51*, 528.
Rauhut, M. M. *Accounts of Chemical Research* **1969**, *2*, 80.

Name ______________________ Section ____________ Date ________

28 Chemiluminescence

PRELABORATORY QUESTIONS

1. Define "chemiluminescence."

2. How does chemiluminescence compare with fluorescence?

3. How is the efficiency of a chemiluminescent reaction measured?

4. What role do acceptors play in chemiluminescent reactions?

5. Draw the structures of the following:

a. firefly luciferin

b. the high energy intermediate in the oxalate–peroxide reaction sequence

c. a xanthene dye chemiluminescence energy acceptor

29

Photochemical Reactions

Photochemical reactions, involving molecules in high-energy excited states, provide synthetic pathways to otherwise inaccessible compounds. A simple illustration is the photoisomerization of 1,3-cycloheptadiene to bicyclo[3.2.0]-6-heptene

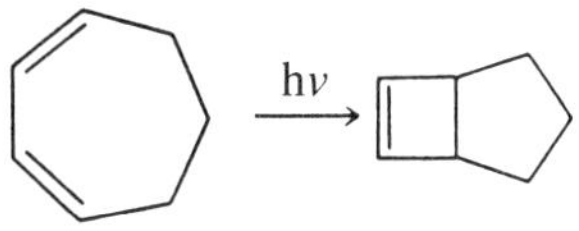

$$\text{1,3-cycloheptadiene} \xrightarrow{h\nu} \text{bicyclo[3.2.0]-6-heptene}$$

Other types of photochemical reactions include dimerizations (cycloadditions) and oxidation–reduction processes. Examples of these are given in the following two experiments.

PHOTOCHEMICAL DIMERIZATION OF *TRANS*-CINNAMIC ACID

This reaction is a classic example of the photochemical cycloaddition of two alkenes to form a cyclobutane. For unsymmetrically substituted alkenes such as cinnamic acid (**1**), two modes of reaction are possible, leading to head-to-head (**2**) or head-to-tail (**3**) dimers named truxinic and truxillic acids, respectively.

$$2\ C_6H_5{-}CH{=}CH{-}CO_2H \xrightarrow{h\nu}$$

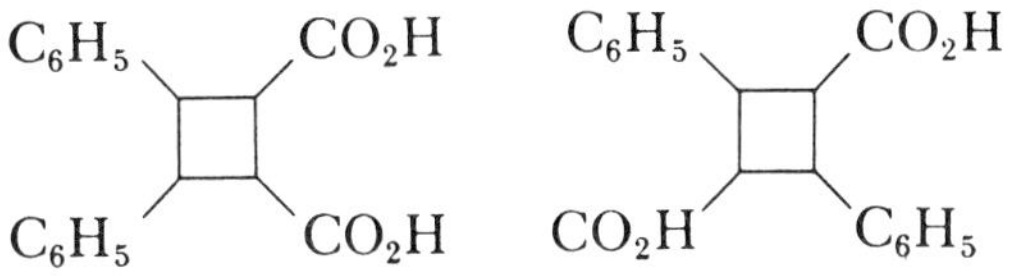

1 cinnamic acid

2 truxinic acids (cyclobutane with C_6H_5, C_6H_5 on one side and CO_2H, CO_2H on the other)

3 truxillic acids (cyclobutane with C_6H_5 and CO_2H alternating)

These dicarboxylic acids were originally isolated along with cocaine and other alkaloids from coca leaves growing near Truxillo, Peru. The compounds are prepared by photodimerization of cinnamic acid in solution or in the crystalline state and are formed by this reaction in the plant.

The stereochemistry of the product poses a complex problem, since there are six geometrical isomers of the truxinic acids and five of the truxillic acids. The product obtained by photodimerization depends on the crystalline form of cinnamic acid that is irradiated. This experiment illustrates how the structure of a given isomer can be elucidated, using NMR spectra and chemical information to narrow the eleven possibilities to one.

The stereochemistry of the isomers was established long before the discovery of NMR spectroscopy and required a much more elaborate chemical analysis. It was based primarily on the knowledge that certain of the dicarboxylic acids readily form cyclic anhydrides (**4** or **5**) on heating with acetic anhydride, indicating that in these isomers the carboxyl groups are situated *cis* to each other. The other isomers, with *trans* CO_2H groups, do not give cyclic anhydrides under these conditions. However, these *trans* isomers are transformed to cyclic anhydrides when sodium acetate is added to the reaction mixture. In this case, sodium acetate acts as a base to invert the configuration (epimerize) at one of the carboxyl-substituted carbons, giving *cis* CO_2H groups that can cyclize to an anhydride. (In the diagram below a wavy bond indicates unspecified stereochemistry.)

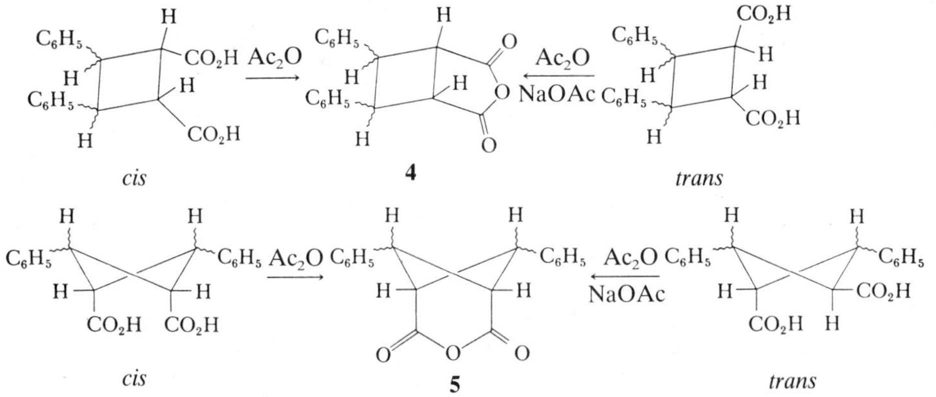

cis —Ac_2O→ **4** ←Ac_2O, NaOAc— *trans*

cis —Ac_2O→ **5** ←Ac_2O, NaOAc— *trans*

Data on the six geometrical isomers of the truxinic and five isomers of the truxillic acids, their anhydrides, and the dimethyl esters are listed in Table 29.1. As indicated in the third column, six of these acids give cyclic anhydrides and five do not. In the last column the number of methoxyl proton peaks in the NMR spectrum of each isomeric dimethyl ester is shown. If the two methoxycarbonyl groups are in equivalent chemical environments, for example, both *cis* to one phenyl group and *trans* to the other as in **6**, only one OCH_3 group is seen in the spectrum. If the methoxycarbonyl groups are nonequivalent, as in **7**, two peaks are present in the OCH_3 region.

Table 29.1 Data on Isomers of Truxinic and Truxillic Acids

ISOMER	MP OF ACID (°C)	MP OF ANHYDRIDE* (°C)	MP OF DIMETHYL ESTER (°C)	MeO-PROTON PEAKS IN NMR OF ESTER
1	286	—	174	1
2	285	—	112	2
3	266	287	104	1
4	245	Unknown	133	1
5	239	150	116	2
6	228	191	127	1
7	210	116	76	1
8	209	—	127	2
9	196	—	199	1
10	192	Unknown	64	1
11	175	—	77	1

*A dash (—) indicates that the acid cannot form a cyclic anhydride; "unknown" means that the anhydride presumably can exist but has not been reported.

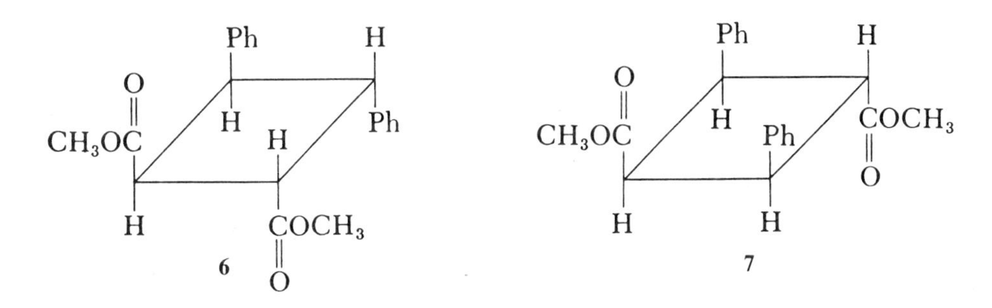

In this experiment, *trans*-cinnamic acid will be photodimerized and the dimeric acid converted to derivatives whose NMR spectra permit the assignment of the structure of the dimer. These NMR spectra are given in Figures 29.1 to 29.3. One derivative is the dimethyl ester (Fig. 29.1), obtained by treatment with methanolic acid. Another derivative is the anhydride (Fig. 29.2), obtained in the reaction with acetic anhydride and sodium acetate. (The acid obtained in this experiment does not form a cyclic anhydride with acetic anhydride alone, but does give an anhydride in the presence of sodium acetate.) Finally, the anhydride obtained by the acetic anhydride/sodium acetate reaction is converted to the dimethyl ester of the new acid (Fig. 29.3).

The objectives of this experiment are twofold. First, you are to determine which of the isomers in Table 29.1 is the original dimer that you obtain in the photoreaction and which of the isomers corresponds to the anhydride and dimethyl ester obtained after acetic anhydride/sodium acetate treat-

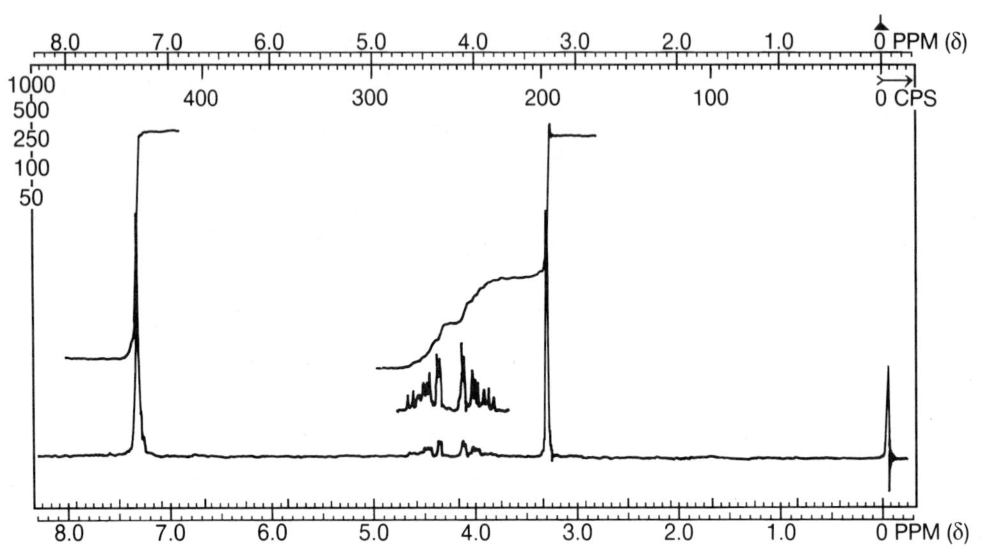

Figure 29.1 Proton NMR spectrum of the cinnamic acid dimer diester.

ment. Second, you are to determine the structure of each of these acids based on the NMR spectra given in Figures 29.1 to 29.3 and the analysis outlined in Questions 1 through 5. For the second objective, the stereochemical possibilities and their consequences must be fully understood. To do this, answer Questions 1 through 5 at the end of the experiments, and then answer Question 6.

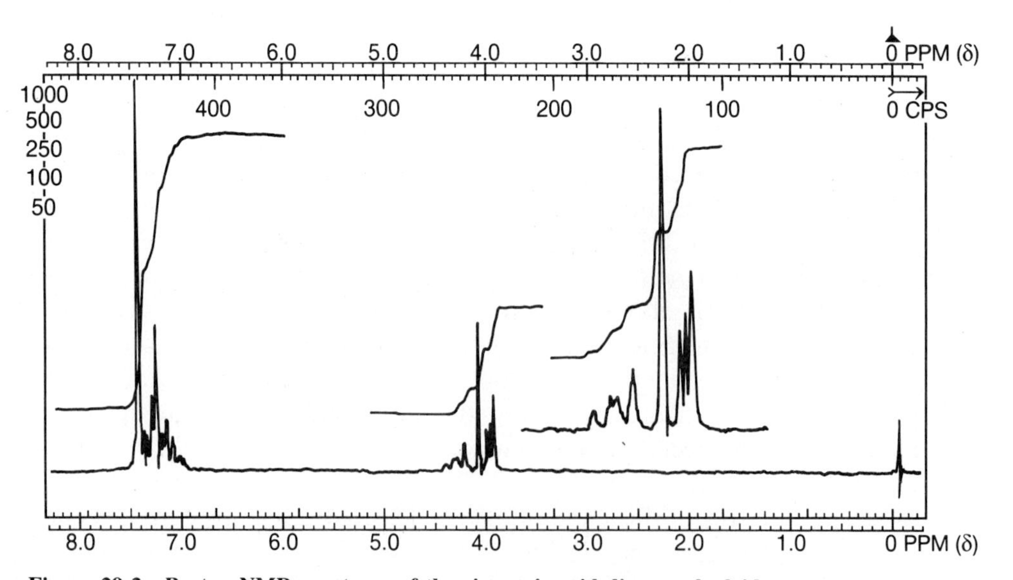

Figure 29.2 Proton NMR spectrum of the cinnamic acid dimer anhydride.

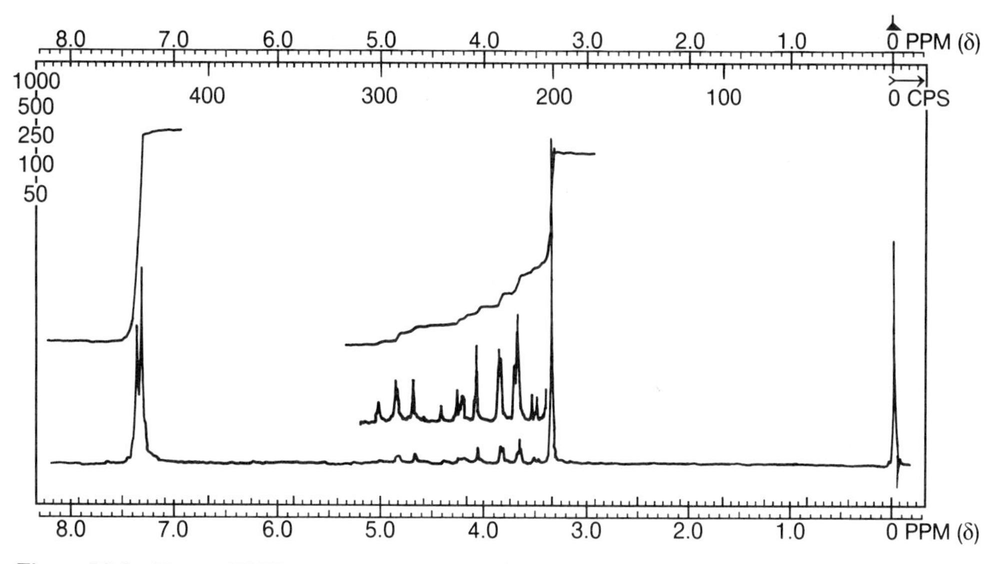

Figure 29.3 Proton NMR spectrum of the isomeric cinnamic acid dimer diester.

Experiments

SAFETY NOTE THERE ARE NO PARTICULARLY HAZARDOUS STEPS OR REAGENTS IN THESE EXPERIMENTS EXCEPT CONCENTRATED SULFURIC ACID WHICH WILL CAUSE BURNS ON THE SKIN; WASH IMMEDIATELY WITH COLD WATER IF A SPILL OCCURS.

A. DIMERIZATION OF CINNAMIC ACID (Macroscale)

Procedure

Weigh out 1.5 g of *trans*-cinnamic acid and transfer it to a 125-mL Erlenmeyer flask. By heating on a steam bath, dissolve the acid in approximately 2 mL of tetrahydrofuran (THF). Remove the flask from the bath, and while the solution is still hot, rotate the flask to coat the walls with the crystallizing acid. If a very uneven coating is obtained, reheat to redissolve the acid and repeat the process. When the residue is sufficiently dry, clamp the flask in an inverted position for 30 minutes to permit all solvent vapor to flow out. Stopper the flask with a cork, label it with your name, and place it, cork down, in a beaker on a window ledge (with southern exposure if possible) for irradiation by the sunlight. After 1 week, rotate the flask 180 degrees to expose the opposite side.

After the second week of irradiation, transfer the solid to a 25 × 100-

mm test tube, add 15 mL of toluene, stir, and warm to 40°C to dissolve the cinnamic acid remaining. Collect the product in a Hirsch funnel, and wash it in the funnel with 10 mL of additional toluene. Air-dry and weigh the product, and record the percent yield and melting point.

Proceed with one of the two following experiments as directed by your instructor; it is advantageous for students to pair up, with each doing one procedure, and then to exchange results.

B. ESTERIFICATION OF CINNAMIC ACID DIMER

Procedure

Transfer the acid from the photoreaction to a 25-mL round-bottom flask and add 10 mL of methanol, 2 to 3 drops of concentrated sulfuric acid, and a boiling stone. Reflux the mixture gently on a steam bath for 1 hour, cool the solution, add 25 mL each of water and ether, shake well, and separate the aqueous layer. Extract the ether layer with $NaHCO_3$ solution, and dry ($MgSO_4$). Filter and concentrate the solution to small volume, transfer to a 25 × 100-mm test tube or small Erlenmeyer flask, evaporate to dryness, and recrystallize the residue from a minimum amount of hexane or methanol. Collect and air-dry the crystallized diester and report the yield and melting point. The proton NMR spectrum of the diester is given in Figure 29.1.

C. EPIMERIZATION AND DEHYDRATION OF CINNAMIC ACID DIMER

Procedure

Place the acid from the photoreaction in a 10-mL flask with 0.3 g of sodium acetate and 2 mL of acetic anhydride and reflux the mixture gently with a heating mantle or sand bath for 10 minutes. (If necessary, the reaction can be carried out *in the hood* in a test tube without a reflux condenser.) Add 1 mL of water dropwise to the warm mixture; after the exothermic hydrolysis of the excess acetic anhydride is complete, add another 5 mL of water and transfer the mixture to a small separatory funnel. Rinse the test tube with 20 mL of water and 10 mL of methylene chloride, shake well, and separate the layers. After washing the organic layer with $NaHCO_3$ solution, dry over $MgSO_4$, and concentrate the solution to approximately 2 mL on the steam bath (use a boiling stone). Add ethanol to the refluxing solution (2 to 4 mL) until crystals begin to form. Cool and collect the anhydride. Rinse the crystals with 2 to 3 mL of ethanol and air-dry; then weigh, and determine the melting point. A second crop may be obtained by concentrating the filtrate. The NMR spectrum of this anhydride is shown in Figure 29.2.

This anhydride is converted to the corresponding dimethyl ester using the procedure given for the esterification of the cinnamic acid dimer. If desired, the progress of the reaction can be followed conveniently by TLC on silica gel with chloroform (see Question 7). Crystallize the product from hexane and determine the melting point. The NMR spectrum of the diester formed is shown in Figure 29.3.

From the melting points of the original acid from the photoreaction and its dimethyl ester, and the melting points of the anhydride and dimethyl ester from the $Ac_2O/NaOAc$ reaction, determine to which of the acids in Table 29.1 these products correspond.

PHOTOCHEMICAL PREPARATION OF BENZOPINACOL (Microscale)

This experiment provides an exceedingly simple and efficient example of a photochemical reduction of a diaryl ketone to a pinacol, using a secondary alcohol as the hydrogen donor.

$$2\ (C_6H_5)_2C{=}O + R_2CHOH \longrightarrow C_6H_5{-}C(OH)(C_6H_5){-}C(OH)(C_6H_5){-}C_6H_5 + R_2C{=}O$$

Experiments

Procedure

In a 10 × 75-mm test tube, place 0.2 g of benzophenone. Dissolve the ketone in approximately 1 mL of isopropyl alcohol with gentle warming, add 1 drop of glacial acetic acid, and then fill the tube just to the top with more isopropyl alcohol, so that air is excluded during the reaction. Stopper the tube tightly with a cork of the proper size. Hold the cork in place with a strip of plastic tape. Invert the tube in a small beaker to present the maximum surface to the light, label with your name, and place on the window sill for 1 week. After several sunny days, the reaction should be complete, with a heavy deposit of large crystals. Collect the product on a Hirsch funnel, label with melting point and yield, and submit it to your instructor.

QUESTIONS

Photochemical Dimerization of *Trans*-Cinnamic Acid

1. Write out the structures of the six geometric isomers of the truxinic and five geometric isomers of the truxillic acids, showing the stereochemistry as illustrated in structures 6 and 7. (*Hint*: Do not confuse geometric isomers and enantiomers; write only one enantiomer of each acid. In the photodimerization, any asymmetric dimers are obtained as racemates.)
2. Which of the acids in Question 1 can form cyclic anhydrides with acetic anhydride alone?
3. Which of the acids give methyl esters with identical methoxyl groups by NMR?
4. For each acid that cannot form an anhydride with acetic anhydride alone, show the anhydride that would be formed with acetic anhydride/sodium acetate by epimerization of one of the carboxy-substituted carbons.
5. Which of the truxillic or truxinic acids is isomer number 5 in Table 29.1; that is, what is the stereochemistry of isomer 5?
6. From your analysis in Questions 1 through 5 and the NMR spectra of the products, determine the structures of the compounds formed by photochemical dimerization and the epimerization–cyclization sequence.
7. Account for the TLC observations in the anhydride-to-diester conversion; write equations for the reactions.
8. Assign the peaks in the NMR spectra (Figs. 29.1 to 29.3) as completely as you can.

Photochemical Preparation of Benzopinacol

9. The photochemical pinacol reduction begins with the excitation of benzophenone, followed by abstraction of a hydrogen atom from 2-propanol to give two radicals. The reaction occurs less efficiently with ethanol and not at all with *t*-butyl alcohol. In the second step of the reaction, a hydrogen atom is transferred from the radical derived from 2-propanol to a second molecule of benzophenone. The combination of two radicals derived from benzophenone gives the pinacol. Write equations showing these three steps, with the overall formation of benzopinacol and acetone.
10. The acetic acid is added to suppress a subsequent reaction that can occur with benzopinacol and traces of base. This reaction involves cleavage to give one molecule of benzophenone. Write an equation for this reaction, showing the other product(s) and the mechanism.

11. When the benzopinacol is warmed with acetic acid, rearrangement occurs to give an isomeric product, benzopinacolone. Write the structure of this product and indicate the mechanism by which it is formed.

Reference

Coyle, J.D. *Introduction to Organic Photochemistry*; Wiley: New York, 1986.

Name ______________________ Section ____________ Date ________

29 Photochemical Reactions

PRELABORATORY QUESTIONS

Fill in the blanks with the right word(s).

1. The dicarboxylic truxinic and truxillic acids were first isolated from ____________ leaves growing near Truxillo, ____________.

2. These acids are possible products in the experiment described in this chapter that involves the irradiation of ____________________ with sunlight.

3. In the experiment, the truxinic or truxillic acid product is converted to the anhydride using ____________________.

4. The spectroscopic technique used to identify some of the products is ____________________.

5. The esterification of the dimer is carried out using ________________ alcohol and ____________________ acid.

30

The Diels–Alder Reaction

The Diels–Alder reaction is an important synthetic tool for building cyclic systems. The reaction comprises the cycloaddition of a conjugated diene and another unsaturated compound, the dienophile. The dienophile usually contains a double or triple bond conjugated with an electron-withdrawing group or groups. Occasionally an acid catalyst is found to increase the rate of reaction, presumably by making the dienophile more electrophilic (dienophilic). A typical synthetic application of the Diels–Alder reaction is the condensation of butadiene and 4-methoxy-2,5-toluquinone; this is the first step in a classic synthesis of the steroid hormones.

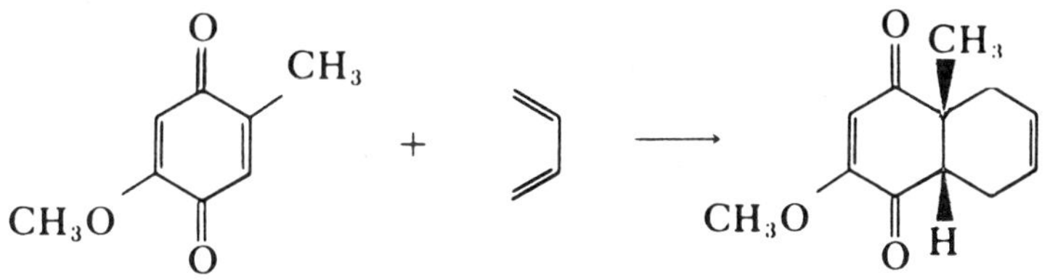

O, CH_3, CH_3O, O + ⟶ O, CH_3, CH_3O, O, H

The Diels–Alder reaction occurs by a concerted cycloaddition process in which bonds are broken and formed in a continuous change from reactants to products, without ionic or free-radical intermediates. The transition state for such a process requires a well-defined orientation of the reactants, and this in turn leads to high stereospecificity. Thus, if groups in the dienophile are *cis*, they remain *cis* in the product.

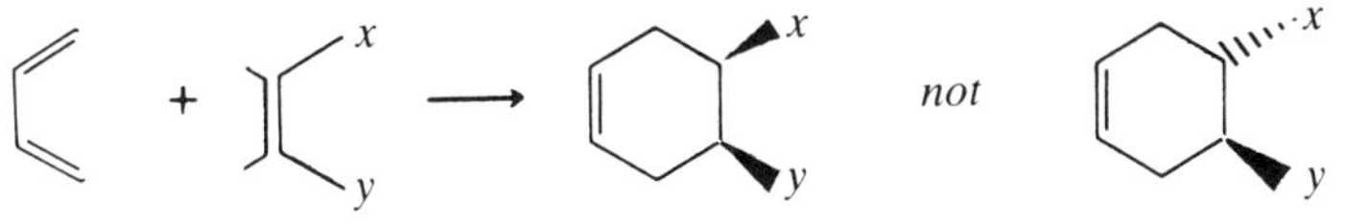

+ *x*, *y* ⟶ *x*, *y* *not* *x*, *y*

A second "rule" governing the stereochemistry of the cycloaddition is that in the formation of a bridged bicyclic product from a cyclic diene, the reaction proceeds with the orientation that places the new double bond closer to the substituents in the dienophile. This rule is illustrated by the reaction to be carried out in this chapter, in which cyclopentadiene and maleic anhydride react to give *endo*-norbornene-5,6-*cis*-dicarboxylic anhydride.

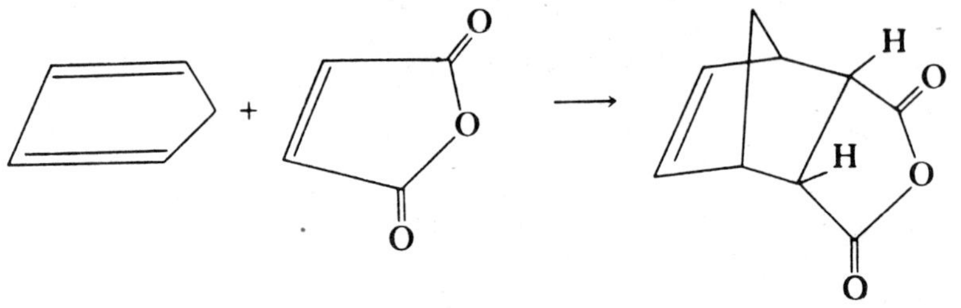

cyclopentadiene
bp 41°C, d 0.80

maleic
anhydride
mp 60°C

endo-norbornene-5,6-*cis*-
dicarboxylic anhydride
mp 165°C

The preference for the *endo* orientation is suggested to arise from the more complete overlap of *p* orbitals in the transition state, compared with that for the *exo* product. This stereospecificity is noteworthy, since in this case the *exo* anhydride is sterically less hindered and is the lower-energy product.

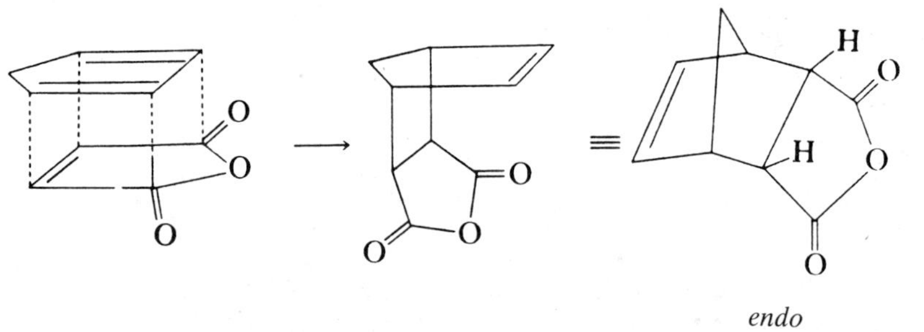

endo

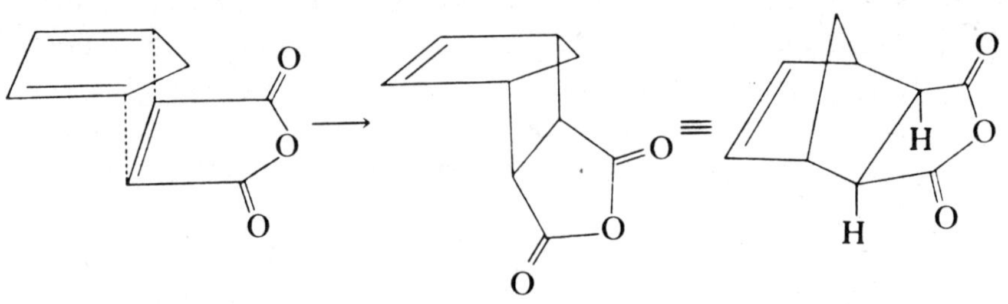

exo

Diels–Alder cycloadditions are reversible reactions, and the adducts can often be converted back to the diene and dienophile by heating. The reverse process is illustrated in this experiment by using the reaction to obtain the diene employed in the synthesis. Cyclopentadiene is extremely reactive as a diene component, and on standing it dimerizes by utilizing one mole as a diene and another as dienophile. The compound is therefore stored and sold as the dimer, called dicyclopentadiene. The reverse reaction is carried out simply by heating the dimer (cracking) and distilling out the monomeric cyclopentadiene as it is formed.

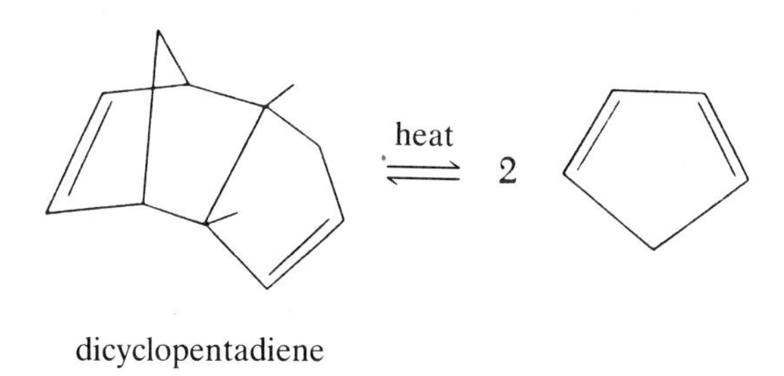

Experiment

ENDO-NORBORNENE-5,6-*CIS*-DICARBOXYLIC ANHYDRIDE (Macroscale)

SAFETY NOTE CYCLOPENTADIENE AND MALEIC ANHYDRIDE ARE MILDLY TOXIC, AND MALEIC ANHYDRIDE DUST IS IRRITATING. CYCLOPENTADIENE AND HEXANE ARE EXTREMELY FLAMMABLE; IF A BURNER IS USED FOR THE DISTILLATION, BE PARTICULARLY CAREFUL THAT JOINTS ARE TIGHT AND THAT NO VAPOR CAN ESCAPE FROM THE RECEIVER. IT IS BEST TO CARRY OUT THE MACROSCALE PROCEDURE IN A HOOD SINCE SOME MAY FIND THE SMELL OF CYCLOPENTADIENE ANNOYING.

Procedure

In a 50-mL round-bottom flask, obtain 20 mL of dicyclopentadiene. Set up an apparatus for fractional distillation with thermometer, packed column, and vacuum take-off adapter. As a receiver, use a 10 or 25-mL round-bottom flask attached to the adapter and immersed in a beaker of ice water. The side arm of the adapter should be fitted with a small calcium chloride drying tube to prevent condensation of moisture in the cold receiver.

Turn on the condenser water, add a boiling stone, check that joints are tight, and heat the flask until boiling begins and the monomeric diene distills. Continue heating until approximately 5 mL of liquid remains in the flask. Record the boiling point of the product; keep the distillate in an ice bath until it is used. If it is cloudy at the end of the distillation, add a few grains of calcium chloride to absorb the moisture.

While the distillation is in progress, place 4 g of maleic anhydride in a 125-mL Erlenmeyer flask and dissolve it in 15 mL of ethyl acetate by warming on the steam bath. Add 15 mL of hexane (or petroleum ether), and then cool the solution in an ice bath.

Measure out 4 mL of the cold cyclopentadiene in a 10-mL graduated cylinder, add it to the maleic anhydride solution, and swirl the solution to effect mixing. Wait until the product crystallizes from solution, then heat it on the steam bath to redissolve, and allow it to recrystallize. Collect the product by suction filtration, and record the melting point, weight, and percentage yield.

ENDO-NORBORNENE-5,6-*CIS*-DICARBOXYLIC ANHYDRIDE (Microscale)

Procedure

Fill a small ceramic heating mantle with sand, and allow the sand to come to a temperature of 200 to 250°C. Assemble an apparatus for cracking dicyclopentadiene as shown in Figure 30.1. This consists of a 5-mL round-bottom flask fitted with a distillation head with a septum at the top, and a condenser attached to the side arm. A vacuum take-off adapter is fitted to the condenser, to which a distillation receiver and a drying tube are attached.

In the 5-mL round-bottom flask, place 2 to 3 mL of mineral oil, check that joints are tight, turn on the condenser water, and allow the mineral oil to come to temperature. During this time, place 0.4 g of maleic anhydride in a 25-mL Erlenmeyer flask and dissolve it in 1.5 mL of ethyl acetate by warming on the steam bath. Add 1.5 mL of hexane, and then cool the solution in an ice bath.

Fill a 5-mL syringe with 1.0 mL of dicyclopentadiene, and when the sand bath has reached a temperature of 200°C, begin to add the dicyclopentadiene drop by drop through the septum. Cracking—which you will see as vaporization of the drop upon impact with the hot oil—will occur immediately. The entire procedure should take less than 15 minutes. When cracking has been completed, add a few grains of calcium chloride to the distillate to absorb residual moisture if it appears cloudy. Remove 0.4 mL of the distillate and add this to the cold maleic anhydride solution. Swirl the solution to cause mixing, replace it in the ice bath, and scratch the side of the flask to induce crystallization. Once the product crystallizes, heat the mixture on the steam bath to redissolve the solid and recrystallize it in the

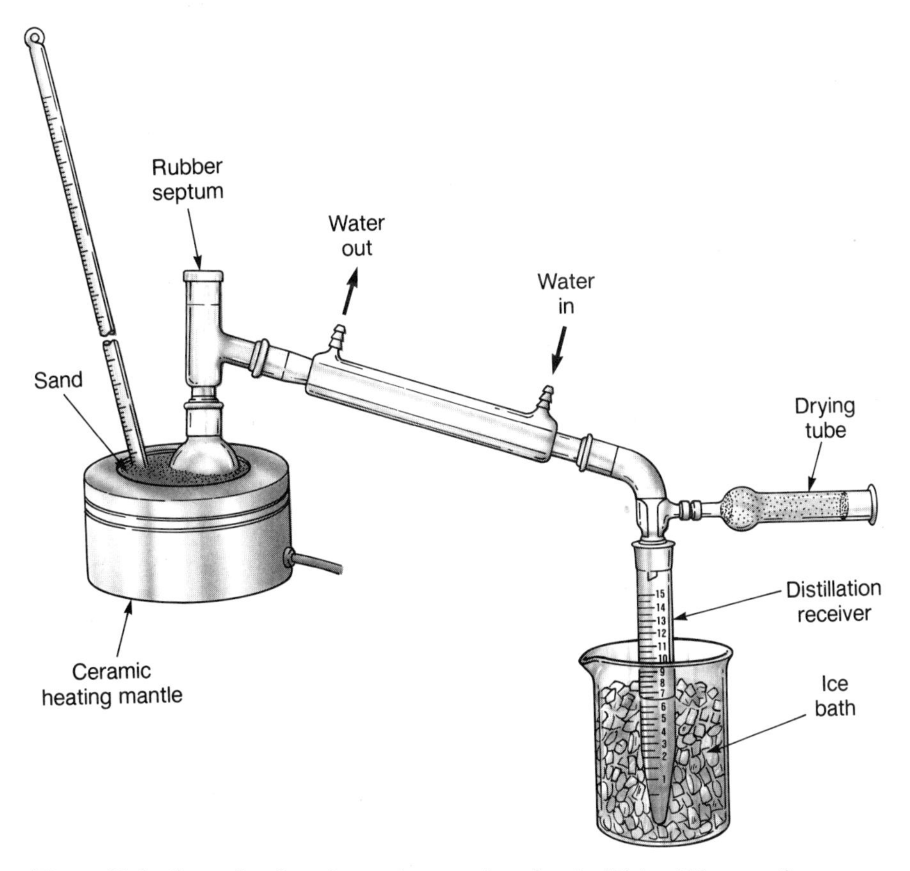

Figure 30.1 Setup for the microscale procedure for the Diels–Alder reaction.

same container. Collect the product by suction filtration on a Hirsch funnel, record the melting point of the dry crystals, and calculate the percent yield.

QUESTIONS

1. Write the structures of the products that would be obtained in the following Diels–Alder reactions:

a. + O O O

b. + $CH_2{=}CHCO_2Me$

c. 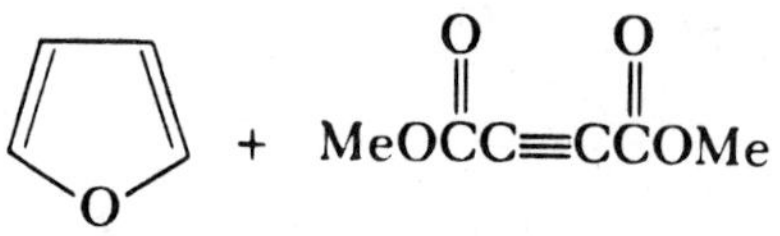 (furan) + $MeO\overset{O}{\overset{\|}{C}}C{\equiv}C\overset{O}{\overset{\|}{C}}OMe$

d. 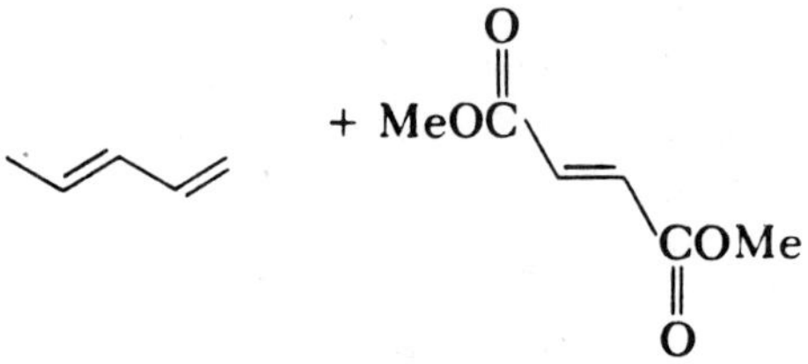 (1,3-butadiene) + $MeO\overset{O}{\overset{\|}{C}}CH{=}CH\underset{\underset{O}{\|}}{C}OMe$

e. 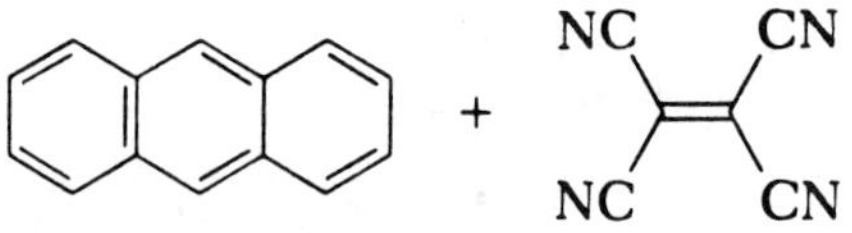 (anthracene) + $(NC)_2C{=}C(CN)_2$

2. Write the products that would be obtained by the thermal reverse Diels–Alder reaction of the following:

a. 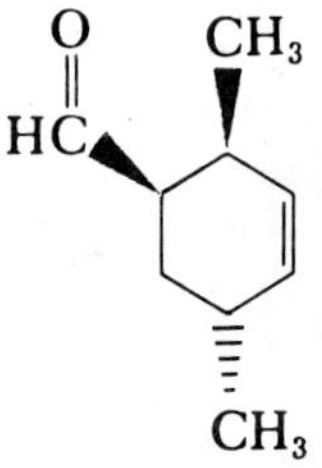 (cyclohexene bearing $\overset{O}{\overset{\|}{HC}}$, CH_3 and CH_3 substituents)

b. 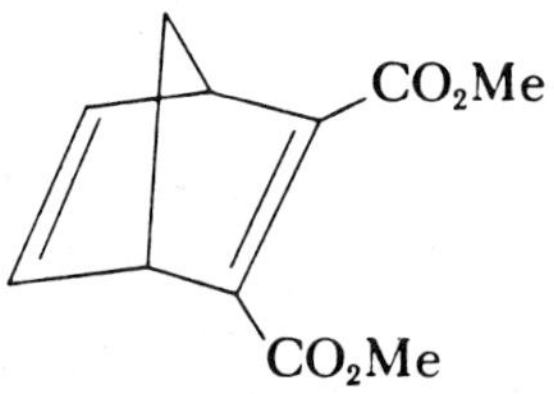 (bicyclic diene bearing two CO_2Me groups)

c. 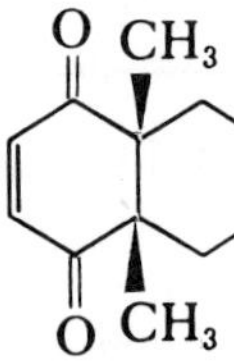 (bicyclic dione bearing two CH_3 groups)

d.  (cyclic sulfone, SO_2)

3. A student neglected to attach a drying tube in the preparation of cyclopentadiene and used his cloudy distillate directly for the Diels–Alder

reaction. On heating the mixture after reaction to redissolve and recrystallize it, some insoluble solid remained undissolved. Suggest what this compound was.

4. After collecting a first crop of norbornene-5,6-dicarboxylic anhydride, TLC indicated the presence of further product in the mother liquor. The solution was concentrated on the steam bath to isolate a second crop of product, but the residue proved to be an oily liquid, and the desired product could not be crystallized. Suggest the reason for the problem. What was the oily liquid?

Name ______________________ Section ____________ Date ________

30 The Diels–Alder Reaction

PRELABORATORY QUESTIONS

1. What is the product of the following Diels–Alder reaction? (Show stereochemistry where applicable.)

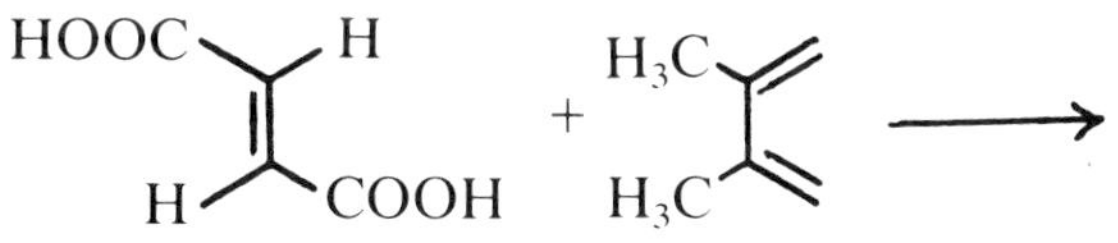

HOOC, H / H, COOH + H_3C, H_3C ⟶

2. The Diels–Alder reaction is said to be "stereospecific." What does that mean in terms of the product shown for the answer to Question 1?

3. Why cannot cyclopentadiene be purchased directly from a supplier?

4. How is cyclopentadiene made in this experiment?

5. How are "exo" and "endo" defined?

31

Phase Transfer Catalysis

The traditional approach to bringing about a chemical reaction between two substances of different types, such as an organic compound and an inorganic salt, is to use a solvent in which both reactants are at least partially soluble. In some cases, alcohols can be used; however, dipolar protic solvents such as dimethylsulfoxide (DMSO) or dimethylformamide (DMF) are often more suitable. A different and very general way to effect such reactions is to use a **phase transfer catalyst**. In this method the reactants may be in two different liquid phases. The catalyst is a quaternary ammonium salt, $R_4N^+ \ X^-$, in which long alkyl groups provide solubility in organic solvents and the cation is the hydrophilic portion.

Many nucleophilic reactions can be enhanced by the use of phase transfer catalysts. Typically, an S_N2 reaction of an alkyl halide (RX_{org}, soluble in organic phases) with a nucleophilic anion (Nuc_{aq}^-, soluble in polar phases) can be carried out using a catalyst such as benzyltributylammonium chloride.

$$RX_{org} + Nuc_{aq}^- \longrightarrow RNuc_{org} + X_{aq}^-$$

The catalyst transfers the nucleophile to the organic phase, thus setting up an equilibrium cycle between the two phases as illustrated in the following diagram:

$$Na^+X^- + (R'_4N)^+Nuc^- \rightleftharpoons Na^+Nuc^- + (R'_4N)^+X^-$$

Aqueous Phase

interface ——————————————————————

Organic Phase

$$RX + (R'_4N)^+Nuc^- \rightleftharpoons RNuc + (R'_4N)^+X^-$$

The reaction to be studied involves the preparation of a naphthyl ether from an active halide and a naphtholate ion that have very different solubility characteristics. The phase transfer catalyst is benzyltributylammonium chloride, an organic ion that is soluble in organic solvents because of the alkyl and aryl groups and, at the same time, is slightly soluble in aqueous media because of its ionic nature. The overall reaction is as follows:

$$p\text{-}NO_2C_6H_4CH_2Cl + C_{10}H_7O^- \longrightarrow p\text{-}NO_2C_6H_4CH_2\text{-}O\text{-}C_{10}H_7$$

p-nitrobenzyl chloride naphtholate ion *p*-nitrobenzyl naphthyl ether

The naphtholate ion is soluble in water but almost insoluble in the organic phase; the *p*-nitrobenzyl chloride is soluble in the organic phase but almost insoluble in water.

Experiment

SAFETY NOTE *CAUTION*: *p*-NITROBENZYL CHLORIDE IS A VERY CORROSIVE LACHRYMATOR AND β-NAPHTHOL IS AN IRRITANT. HANDLE THESE MATERIALS WITH CARE.

PREPARATION OF *p*-NITROBENZYL NAPHTHYL ETHER (Microscale)

Procedure

Place a small magnetic stirring bar in a 50-mL Erlenmeyer flask and add 10 mL of 0.1 *M* NaOH and 0.1 g of β-naphthol. Add 0.05 g of benzyltributylammonium chloride and stir the mixture thoroughly until all components are fully dispersed.

In a separate container, dissolve 0.06 g of *p*-nitrobenzyl chloride (dispense in a hood) in 10 mL of methylene chloride. Add this solution to the naphtholate mixture and allow it to stir for 1 hour. Separate the two layers using a small separatory funnel, and wash the lower (organic) layer with 5 mL of water. Collect the lower layer in a small Erlenmeyer flask and add a small amount of Na_2SO_4 to remove traces of water. Using a Pasteur pipet with some cotton in the tip to act as a filter, transfer the solution to a tared 25 × 100-mm test tube and carefully evaporate the methylene chloride under aspirator vacuum (see Fig. 4.5). Weigh the test tube and contents; the solid should be reasonably crystalline. If it is still moist, remove the remaining solvent by gently warming the test tube under vacuum on a steam bath. Recrystallize the product from ethanol, determine the melting point, and calculate the percent yield.

QUESTIONS

1. What effect would doubling the concentration of the naphtholate ion have on the reaction rate?
2. Sketch a possible transition state for the S_N2 reaction of *p*-nitrobenzyl chloride and β-naphtholate ion.
3. Give an example of a compound other than benzyltributylammonium chloride that can be used as a phase-transfer agent. (Consult your textbook if necessary.) What are the necessary features of the compound?
4. Give two solvent pairs, other than water and methylene chloride, that would make convenient solvent systems for phase transfer catalysis.

Name ______________________ Section ____________ Date ________

31 Phase Transfer Catalysis

PRELABORATORY QUESTIONS

1. What is meant by the term "lachrymator"?

2. Give three examples of other types of compounds that could be used as nucleophiles in this experiment.

3. Draw the structures of dimethylsulfoxide (DMSO), dimethylformamide (DMF), and β-naphtholate ion.

4. Draw the structural formula of the "phase transfer" agent used in this experiment.

5. What other types of compounds are used as phase transfer agents?

6. How does a phase transfer agent differ from a detergent?

32

Derivatives of Carboxylic Acids: Synthesis of a Plant Hormone

Plant hormones are substances synthesized in the cells of a plant that regulate growth and physiological function. One of the major plant hormones is indoleacetic acid, found in several parts of the plant. This compound and other plant hormones that promote and control growth are called **auxins**. In addition, several structural analogs of indoleacetic acid elicit the same growth response in plants. Among these synthetic auxins are indolebutyric acid, 1-naphthaleneacetic acid, and 2,4-dichlorophenoxyacetic acid (2,4-D). The application of auxins at very low concentrations promotes stem growth and stimulates rooting. High concentrations can cause destruction of the plant, as in the use of 2,4-D as a weed killer.

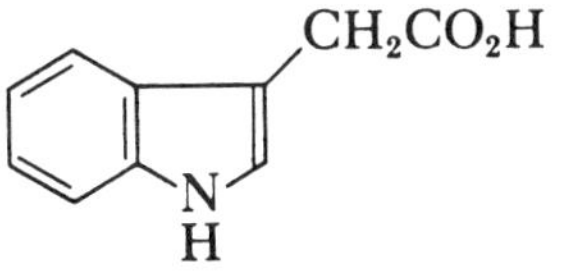

CH_2CO_2H N H

indoleacetic acid

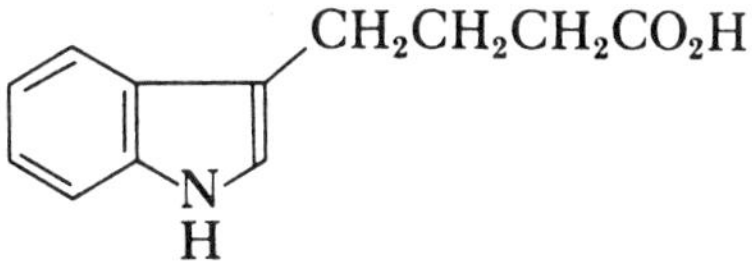

$CH_2CH_2CH_2CO_2H$ N H

indolebutyric acid

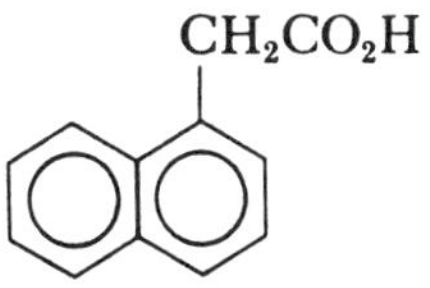

CH_2CO_2H

naphthaleneacetic acid

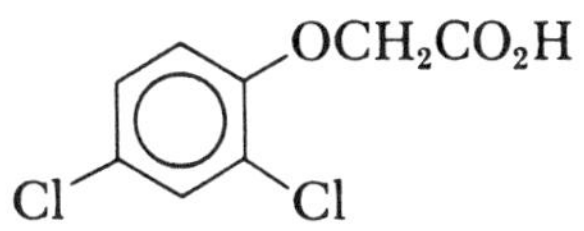

OCH_2CO_2H Cl Cl

2,4-dichlorophenoxyacetic acid

One of the important uses of auxins is in the propagation of plant cuttings. For this and several other horticultural purposes, the preferred auxin is the amide of naphthaleneacetic acid. This compound probably acts by the same mechanism as the acid but is more effective. Commercial preparations such as Transplantone® or Rootone® contain 1-naphthalene-acetamide as the principal active ingredient (less than 0.1%) together with fertilizer or inert carrier. Fungicides are often added to protect the developing roots from fungal attack. In this experiment, 1-naphthaleneacetamide will be synthesized by two different methods and compared by TLC with commercial material.

SYNTHESES OF 1-NAPHTHALENEACETAMIDE

One method of synthesis illustrates a general approach to several types of carboxylic acid derivatives, namely the preparation and reaction of an acid chloride with a nucleophile, in this case ammonia. The second method involves the conversion of a nitrile to an amide by hydration.

CH_2CO_2H $\xrightarrow[DMF]{SOCl_2}$ $CH_2\overset{O}{\overset{\|}{C}}Cl$

$\downarrow NH_3$

$CH_2C{\equiv}N$ $\xrightarrow[H_2SO_4]{H_2O}$ $CH_2\overset{O}{\overset{\|}{C}}NH_2$

naphthaleneacetamide
mp 183–184°C

The most commonly used reagent for converting an acid to the acid chloride is thionyl chloride, $SOCl_2$; the reaction is convenient because the by-products SO_2 and HCl are gases and are readily removed. A complication, however, is the possibility of chlorination at the α-position of the acid on heating with thionyl chloride, particularly in the case of an acid of the type $ArCH_2CO_2H$, in which the α-hydrogens are readily enolized. An effective way to circumvent this problem is to use dimethylformamide (DMF) as a catalyst. The DMF is converted to dimethylformamide chloride, which then reacts with the acid at room temperature to give the acid chloride:

$$\mathrm{H\overset{O}{\overset{\|}{C}}N(CH_3)_2 + SOCl_2 \longrightarrow H{-}\overset{Cl}{\overset{|}{C}}{=}\overset{+}{N}(CH_3)_2Cl^- + SO_2}$$

$$\mathrm{RCO_2H + H\overset{Cl}{\overset{|}{C}}{=}\overset{+}{N}(CH_3)_2Cl^- \longrightarrow R\overset{O}{\overset{\|}{C}}Cl + HCl + H\overset{O}{\overset{\|}{C}}N(CH_3)_2}$$

In the second synthesis, hydration of the nitrile is carried out in 67% (v/v) sulfuric acid solution. Strong acid protonates the nitrile, and water then attacks the nitrilium ion to give the imidic acid, which isomerizes to the amide:

$$\mathrm{RC{\equiv}N \xrightarrow{H^+} \underset{\text{nitrilium ion}}{RC{\equiv}\overset{+}{N}H} \xrightarrow[-H^+]{H_2O} \underset{\text{imidic acid}}{R\overset{OH}{\overset{|}{C}}{=}NH} \longrightarrow R\overset{O}{\overset{\|}{C}}NH_2}$$

Since hydrolysis of the amide to the carboxylic acid can occur, the hydration must be carried out under conditions that are as mild as possible.

Experiments

A. ACID CHLORIDE METHOD (Microscale)

SAFETY NOTE THIONYL CHLORIDE IS A FUMING, CORROSIVE LIQUID AND REACTS VIOLENTLY WITH WATER. CONCENTRATED AQUEOUS AMMONIA HAS CHOKING FUMES. HANDLE ALL CHEMICALS IN THE HOOD AND AVOID BREATHING VAPORS; IF MATERIALS ARE SPILLED ON THE SKIN, WASH THOROUGHLY WITH WATER.

Procedure

In a dry, heavy-wall 10-mL Erlenmeyer flask or 13 × 100-mm test tube, place 0.47 g of 1-naphthaleneacetic acid (1-naphthylacetic acid). In the hood, obtain 0.25 mL of thionyl chloride (graduated pipet) and add it all at once to the acid. Cool the mixture of acid and thionyl chloride in an ice bath, and then remove from the bath. Add 0.1 mL of dimethylformamide, and swirl the contents (note the temperature of the container with your hand). Loosely cork the flask or test tube, label it with your name, and allow it to stand in the hood. After 1 hour, connect the flask or test tube to an aspirator and evaporate the excess thionyl chloride by keeping it under aspirator vacuum for about 2 minutes and warming to 30°C.

Mix 1.3 mL of concentrated aqueous ammonia and approximately 3 mL of crushed ice in a graduated cylinder, and pour the mixture into the flask containing the acid chloride. Stir with a rod or spatula until all of the oil is converted to solid. Collect the solid by vacuum filtration, wash with 1 mL of cold water, and press out excess water. Recrystallize the solid from 1 to 2 mL of hot ethanol. Save a sample of the mother liquor for TLC. Dry the product and determine its weight, melting point, and the yield.

B. NITRILE METHOD (Microscale)

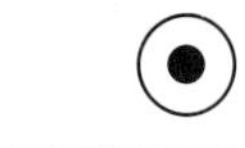

SAFETY NOTE CONCENTRATED SULFURIC ACID CAUSES BURNS ON CONTACT WITH SKIN. IF ANY IS SPILLED ON THE SKIN, WASH IMMEDIATELY AND THOROUGHLY WITH WATER, FOLLOWED BY SOAP AND WATER.

Procedure

In a 10-mL Erlenmeyer flask, place 1 mL of water and then slowly add 2 mL of concentrated sulfuric acid. (**CAUTION**: The flask will become very hot.) Add 0.5 g of naphthalene-1-acetonitrile (1-naphthylacetonitrile) and place the flask on a steam bath. Using a strip of paper towel in a loop to protect your hand, swirl the mixture intermittently to keep the oily layer suspended. When all of the oil dissolves (approximately 15 to 20 minutes), allow the hot solution to stand for an additional 10 minutes and then cool. Add approximately 4 g of ice, and stir the oil until it solidifies. Collect the solid by vacuum filtration, press out excess water, and recrystallize from 1 to 2 mL of hot ethanol. Save a sample of the mother liquor for TLC. Dry the product and determine the weight, melting point, and yield.

C. DETECTION OF NAPHTHALENEACETAMIDE IN ROOTONE®

Place 1 g of Rootone® in an 18 × 150-mL test tube and add 3 mL of ethanol. Heat the mixture on a steam bath, and after boiling for 2 to 3 minutes, filter through a small fluted filter paper into another test tube and rinse the insoluble residue with 1 mL of alcohol. Evaporate the filtrate almost to dryness by warming it under aspirator vacuum. The residue may contain yellow crystals of fungicide bis-(dimethylthiocarbamyl) disulfide, $(CH_3)_2NCS_2CS_2N(CH_3)_2$. With a pipet remove approximately 0.1 mL of the oil and transfer to a spotting capillary for TLC.

On a 3 × 10-cm strip of fluorescent silica gel TLC sheet, spot a sample of recrystallized 1-naphthaleneacetamide from your preparation along with a sample of the recrystallization mother liquor and the Rootone extract. Make several applications of Rootone extract. Develop the TLC with methylene chloride–ethanol (19:1), and visualize the chromatogram under UV light.

QUESTIONS

1. Calculate the molar amounts of acid and thionyl chloride (d = 1.63 g/mL) used in the acid chloride preparation.
2. A student absentmindedly omitted the aqueous ammonia in his preparation of the amide; ice was added to the acid chloride, and the oil was rubbed and stirred to give a white solid, which was then collected, washed with water, and recrystallized. What was this product?
3. The 1989 prices of 1-naphthaleneacetic acid, the nitrile, and the amide were as follows: acid, $9.00/100 g; nitrile, $13.80/100 g; amide, $31.60/100 g. Ignoring the cost of solvents, other reagents, etc, calculate the following:
   a. the cost of 100 g of amide based on these starting material costs and your yields in the preparation of amide
   b. your profit or loss in both preparations compared with purchasing the amide
4. A 0.40-oz packet (11 g) of commercial rooting hormone containing 0.067% of 1-naphthaleneacetamide costs $1.99. Using the price for the amide from Question 3, calculate the cost of 1-naphthaleneacetamide in the packet.

Name ______________________ Section ______________ Date ________

32 Derivatives of Carboxylic Acids: Synthesis of a Plant Hormone

PRELABORATORY QUESTIONS

1. Write an equation to show the reaction of benzoic acid with thionyl chloride to give the acid chloride.

2. Write an equation to show the reaction of the product formed in Question 1 above with concentrated aqueous ammonia.

3. What properties of thionyl chloride make it a reagent to be used with caution? Why is it prudent to use it in a hood?

4. Write the structural formula of the product to be synthesized in this experiment; then, show the product expected from its complete hydrolysis.

33

Active Methylene Condensation, Ester Hydrolysis, and Decarboxylation: Synthesis of Coumarin

A major process in synthetic organic chemistry is the condensation of an enolate with a carbonyl group to form a carbon–carbon bond. The simplest form of the reaction is the aldol condensation, in which an aldehyde or ketone is converted by a base to the enolate, which then adds to the carbonyl group of another mole of aldehyde or ketone to give a β-hydroxy carbonyl compound. Dehydration to the unsaturated carbonyl compound frequently occurs as a second step:

$$\mathrm{RCH_2\overset{\overset{O}{\|}}{C}R'} \xrightarrow{\mathrm{B:^-}} \mathrm{R\overset{\overline{\bullet\bullet}}{C}H\overset{\overset{O}{\|}}{C}R'}$$

$$\mathrm{R''\overset{\overset{O}{\|}}{C}H} + \mathrm{R\overset{\overline{\bullet\bullet}}{C}H\overset{\overset{O}{\|}}{C}R'} \longrightarrow \mathrm{R''\overset{\overset{O^-}{|}}{C}H\underset{\underset{R}{|}}{C}H\overset{\overset{O}{\|}}{C}R'}$$

$$\xrightarrow{+\mathrm{H^+}} \mathrm{R''\overset{\overset{OH}{|}}{C}H\underset{\underset{R}{|}}{C}H\overset{\overset{O}{\|}}{C}R'} \xrightarrow{-\mathrm{H_2O}} \mathrm{R''CH{=}\underset{\underset{R}{|}}{C}\overset{\overset{O}{\|}}{C}R'}$$

Several methods can be used to apply this condensation to the preparation of an unsaturated acid or ester.

1. The **Perkin reaction** utilizes the condensation of an aldehyde and the acid anhydride in the presence of the sodium salt of the acid. The method generally requires temperatures of 160 to 200°C and rather long reaction times.

$$R'CH_2\overset{O}{\overset{\|}{C}}O\overset{O}{\overset{\|}{C}}CH_2R' \xrightarrow{NaO\overset{O}{\overset{\|}{C}}CH_2R'} R'\overset{\bar{\cdot\cdot}}{C}H\overset{O}{\overset{\|}{C}}O\overset{O}{\overset{\|}{C}}CH_2R'$$

$$R\overset{O}{\overset{\|}{C}}H + R'\overset{\bar{\cdot\cdot}}{C}H\overset{O}{\overset{\|}{C}}O\overset{O}{\overset{\|}{C}}CH_2R' \longrightarrow R\overset{O^-}{\overset{|}{C}}H\underset{R'}{\underset{|}{C}}H\overset{O}{\overset{\|}{C}}O\overset{O}{\overset{\|}{C}}CH_2R'$$

$$\downarrow \text{several steps}$$

$$RCH{=}\underset{R'}{\underset{|}{C}}CO_2H \xleftarrow{H_2O} RCH{=}\underset{R'}{\underset{|}{C}}\overset{O}{\overset{\|}{C}}O\overset{O}{\overset{\|}{C}}CH_2R'$$

2. The **Wittig–Horner reaction** involves condensation of the stabilized anion obtained from a dialkylphosphonate ester, which must be prepared in a separate step.

$$P(OEt)_3 + XCH_2CO_2R' \longrightarrow (EtO)_2\overset{O}{\overset{\|}{P}}CH_2CO_2R' + EtX$$

$$\downarrow NaOR'$$

$$R{-}\overset{O^-}{\underset{H}{C}}{-}\underset{CO_2R'}{\overset{P(=O)(OEt)_2}{CH}} \xleftarrow{R\overset{O}{\overset{\|}{C}}H} (EtO)_2\overset{O}{\overset{\|}{P}}\overset{\bar{\cdot\cdot}}{C}H{-}CO_2R'$$

$$\downarrow$$

$$RCH{=}CHCO_2R' + (EtO)_2PO_2^-\ Na^+$$

3. The **Knoevenagel condensation** takes advantage of the greater acidity of the α-hydrogens in a compound with two activating groups such as RCO, CO_2R, or CN. A weak base such as an amine can be used, and the reactions usually proceed rapidly under mild conditions to give the unsaturated product in high yield.

$$\mathrm{CH_2(COOR')_2} \xrightarrow{\text{B:}} {}^{-}\!:\mathrm{CH(COOR')_2}$$

$$\mathrm{RCH{=}O} + {}^{-}\!:\mathrm{CH(COOR')_2} \longrightarrow \mathrm{RCH(O^-)CH(COOR')_2} \xrightarrow{+\mathrm{H^+}} \mathrm{RCH(OH)CH(COOR')_2}$$

$$\mathrm{RCH(OH)CH(COOR')_2} \xrightarrow{-\mathrm{H_2O}} \mathrm{RCH{=}CH(COOR')_2}$$

To obtain a monocarboxylic acid, the condensation can be carried out with malonic acid, and one of the CO_2H groups subsequently is removed by decarboxylation. Alternatively, the less expensive malonic ester can be used, followed by hydrolyses of the ester groups and decarboxylation in separate steps:

$$\mathrm{RCH{=}C(CO_2R')_2} \xrightarrow[\text{2) H}^+]{\text{1) 2 KOH}} \mathrm{RCH{=}C(CO_2H)_2} \xrightarrow[\Delta]{} \mathrm{RCH{=}CHCO_2H} + \mathrm{CO_2}$$

SYNTHESIS OF COUMARIN

In the sequence of experiments in this chapter, the Knoevenagel condensation, ester hydrolysis, and decarboxylation are illustrated by a classic synthesis of the aromatic lactone, coumarin. This compound occurs in nature in many plants, including sweet clover, the herb woodruff, and the tonka bean, from a South American plant that was for many years the principal source. Coumarin has a strong, sweet odor of newmown hay and is widely used in perfumery. It was formerly used with vanilla as a flavoring agent, but because of its toxicity, coumarin is now prohibited for use in human food. It has been classified as a suspected carcinogen, but the hazards of coumarin to humans are low.

The synthesis begins with the Knoevenagel condensation of salicylaldehyde and diethyl malonate using piperidine and acetic acid as the catalyst. In this step, the C═C bond is formed and the lactone ring is closed by transesterification involving the *o*–hydroxy group and one of the carboxylate esters. The second step is a simple hydrolysis of the other ester group with KOH, and the final step involves the removal of the CO_2H group by decarboxylation. The first two steps can be carried out in one 3-hour laboratory period and the last reaction in a second period. The coumarin is isolated by short-path distillation in a setup in which the distillate is collected as a solid on a cold-finger condenser.

CHO, OH + H_2C ($CO_2C_2H_5$, $CO_2C_2H_2$) → (piperidine, N–H; CH_3CO_2H) $CO_2C_2H_5$, O, O

salicylaldehyde — diethyl malonate — ethyl coumarin-3-carboxylate mp 92–93°C

1) KOH
2) HCl

CO_2H, O, O ← ($-CO_2$, Δ) O, O

coumarin mp 69–70°C — coumarin-3-carboxylic acid mp 191–192°C

Experiments

SAFETY NOTE NONE OF THE STARTING MATERIALS IN THIS SEQUENCE IS PARTICULARLY TOXIC OR HAZARDOUS. PIPERIDINE HAS A VERY UNPLEASANT ODOR AND SHOULD BE TRANSFERRED IN A HOOD. BOTH KOH AND HCl ARE CORROSIVE AND SHOULD BE HANDLED WITH CARE.

A. ETHYL COUMARIN-3-CARBOXYLATE (Microscale)

Procedure

In a 5-mL round-bottom flask, place 0.27 mL (0.3 g) of salicylaldehyde, 0.48 mL (0.5 g) diethyl malonate, and 1.2 mL of absolute ethanol. With a graduated micropipet, add 0.03 mL of piperidine, and then add one drop of glacial acetic acid. Add a boiling stone, fit the flask with a reflux condenser, and heat the solution at reflux for approximately 1 hour. If desired, samples (1 to 2 drops) of the solution can be removed before starting and at 30 minute intervals to follow the progress of the reaction by TLC (use fluorescent silica gel, develop with CH_2Cl_2, and visualize with a UV lamp).

While the solution is refluxing, prepare a setup for suction filtration with a Hirsch funnel. For the next step, prepare a solution of 0.40 g of KOH in 2.0 mL of water and 0.8 mL of ethanol unless this solution has already been prepared in volume for your use.

When the condensation reaction has refluxed to completion—confirmed by TLC analysis—cool the solution to approximately 60°C, add 1.5 mL of warm water, and then chill the solution in an ice bath. When the mixture is cold, collect the crystals and wash with a small amount of chilled 50% ethanol–water. Press the crystals thoroughly on the funnel and then spread them out on paper. Remove a small sample, allow it to dry completely, and then record the melting point. Meanwhile, continue directly to Part B.

B. COUMARIN-3-CARBOXYLIC ACID

Since this reaction is carried out in aqueous ethanol solution, there is no point in completely drying the ester from the preceding step. Place the damp crystals in a 5-mL round-bottom flask, and add 2.8 mL of the aqueous alcoholic KOH solution prepared above and a boiling stone. Reflux the solution for 30 minutes. While the solution is refluxing, obtain 1.2 mL of concentrated HCl, mix with 1.2 mL of water, and cool in an ice bath.

After the solution has refluxed for 30 minutes, cool in an ice bath and add the cold HCl solution all at once. Collect the crystals of the acid on a Hirsch funnel, and wash them thoroughly with water. Press the solid on the funnel, and spread it out to dry until the next laboratory period. Record the melting point and weight, and calculate the yield for the two-step sequence.

C. COUMARIN

SAFETY NOTE IN THIS PART, YOU WILL BE CARRYING OUT A DISTILLATION AT REDUCED PRESSURE. BE SURE TO KEEP SAFETY GOGGLES ON AT ALL TIMES. COUMARIN IS ON THE LIST OF SUSPECTED CARCINOGENS.

In a 50-mL heavy-wall side-arm filter flask, place 0.2 g of the dried carboxylic acid from Part B. Clamp the flask about 1 cm deep in a sand bath with a thermometer in the bath. Place a rubber stopper in the neck of the flask. Connect a piece of rubber tubing to the side arm, attach a short length of glass tubing or a pipet to the other end of the rubber tubing, and place the glass tip in a test tube containing 2 to 3 mL of water so that bubbling can be observed. (**CAUTION**: Arrange the tube with the tip *just* below the surface of the water so that water cannot be sucked back into the side-arm flask if the system is cooled.) Heat the sand bath to 220°C, and then maintain the temperature between 220 and 240°C until CO_2 evolution (bubbling) ceases; this usually requires 20 to 25 minutes. Tap the rubber stopper occasionally to dislodge droplets of distillate from the walls.

When gas evolution has stopped, remove the heat source, raise the flask out of the bath, disconnect the tubing, and allow the flask to cool, first in air and then in cold water. (Be careful not to use the cold water too quickly to avoid cracking the flask.) Remove the stopper, scrape off any oil or solid, and return it to the flask. Place a cold-finger condenser (a 16 × 150-mm test tube fitted through a rubber cone or a rubber stopper containing a wide hole; see Fig. 18.2) in the flask with the test tube approximately 2 cm above the bottom of the flask.

Connect the side arm of the flask to the aspirator, evacuate the flask, and pack the cold finger with small pieces of ice. (Ice can be crushed finely for use in the cold finger by crushing it in a towel.) Be sure that you have and maintain vacuum! Clamp the flask in the sand bath and heat to 200°C. Maintain the temperature at 210 to 220°C for 25 minutes. Every few minutes, remove water from the cold finger with a pipet and bulb and replace it with more ice. Take care not to get water inside the flask. Some splattering usually occurs during the distillation, and it may be difficult to see inside the flask. Most of the coumarin distills in the first 20 minutes; a total time of 25 to 30 minutes at 220°C is usually sufficient to complete the distillation.

Raise the flask out of the bath and allow it to cool while still under aspirator vacuum, first in air and then in water. Pour out the ice water in the cold-finger test tube (to avoid condensation of moisture), and then remove the aspirator tube. Carefully lift out the cold finger, and scrape the distilled coumarin onto a piece of weighing paper. The material is usually a mixture of brown and white solid at this point. With the spatula, scrape the white nodules of distillate from the walls of the flask and add them to the material from the cold finger. Record the weight of the solid and take a melting point of the crude sample.

Clean out the flask (use a few mL of acetone to loosen the residue) and dry it with aspirator suction. Place the distillate back in the flask, replace the cold finger, fill with ice, connect the flask to the aspirator, evacuate, and distill the crude coumarin again at a bath temperature of approximately 140 to 160°C. Remove the product as described above, weigh, and report the yield and melting point. Compare the melting point with that of the crude coumarin.

QUESTIONS

1. Show the mechanistic steps involved in the base-catalyzed aldol condensation of propionaldehyde.
2. Write a reaction to illustrate the preparation of ethyl cinnamate using the Wittig–Horner reaction.
3. The first synthesis of coumarin in the nineteenth century was carried out by Perkin to establish the structure, using the reaction that now bears his name. Write reactions to illustrate this synthesis.
4. Calculate the molar amounts of the starting materials in the Knoevenagel condensation leading to ethyl coumarin-3-carboxylate. Which compound is used in excess?

REFERENCE

Jones, G. *Organic Reactions*: The Knoevenagel Reaction; John Wiley & Sons: New York, 1967; Vol. 15, p. 204.

Name ______________________________ Section ______________ Date ________

33 Active Methylene Condensation, Ester Hydrolysis, and Decarboxylation: Synthesis of Coumarin

PRELABORATORY QUESTIONS

1. Write an equation to show the Knoevenagel reaction of benzaldehyde with diethyl malonate.

2. a. Write an equation to show the product that would form if salicylaldehyde were reacted with diethyl malonate involving only the Knoevenagel reaction comparable to that which took place in Question 1.

b. Using the product from Question 2a, write an equation to show the compound that would form on reaction of the phenolic OH group with one of the carboxylate ester groups (transesterification).

3. Decarboxylation of β-keto acids leads to ketones. With this in mind, show the product of the decarboxylation of benzoylacetic acid, $C_6H_5COCH_2COOH$.

34

Synthesis of a Citrus Bioregulator

The word **bioregulator** is a broad term referring to any substance that affects the growth or development of a plant. Some bioregulators are natural plant hormones (e.g., auxins) or structurally similar analogs that when applied to a plant in amounts greater than normal affect the development in some way. Regulators of this type may in fact be lethal, as with broad-leaf weed killers such as 2,4-dichlorophenoxyacetic acid, which is a highly active auxin. Other plant growth regulators are used as yield-enhancing agents, since they alter the plant physiology to increase the amount of fruit or seed.

An important type of plant bioregulator that acts directly on the genetic mechanism has been developed to enhance the color and vitamin content of citrus fruit. Oranges and other citrus crops contain carotenes as natural pigments; certain carotenes are biochemical precursors of vitamin A, which makes citrus fruit so important in the diet. Several groups of compounds, when applied to mature fruit, cause "depression" (i.e., stimulation) of specific genes responsible for carotene production and also prevent loss of accumulated carotenes by further metabolic steps. As a result, more deeply pigmented fruit and higher provitamin A levels can be achieved even after the fruit has been harvested.

One type of regulator has the general structure shown in Formula A, a diaryl unsaturated ketone (chalcone) with a diethylaminoethoxy group. These regulators cause a large increase in the bright red carotenoid lycopene, which is the principal pigment in tomatoes. A second type of regulator, shown in Formula B, with the diethylamino group attached to an eight- or nine-carbon alkyl chain, causes a buildup in provitamin A carotenoids such as β-carotene. It is noteworthy that both types of regulator contain the diethylamino group present in so many biologically active compounds.*

*We thank Dr. Henry Yokoyama of the US Department of Agriculture for numerous helpful suggestions concerning this experiment.

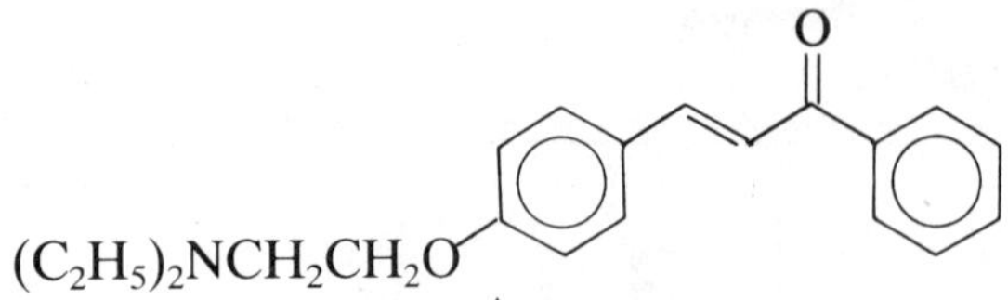

$(C_2H_5)_2NCH_2CH_2O$

A

l-phenyl-3-(*p*-diethylamino-ethoxyphenyl)-2-propen-1-one

$CH_3(CH_2)_nN(CH_2CH_3)_2$
n = 7,8

B

l-diethylaminoalkane

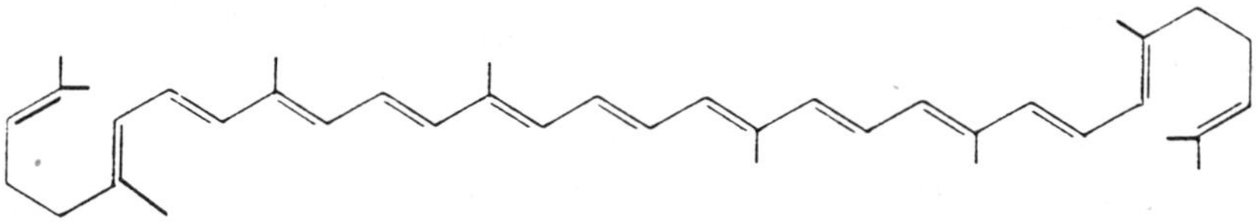

lycopene

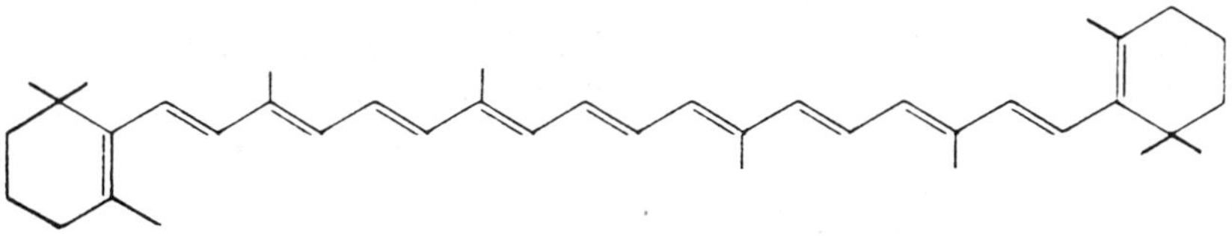

β-carotene

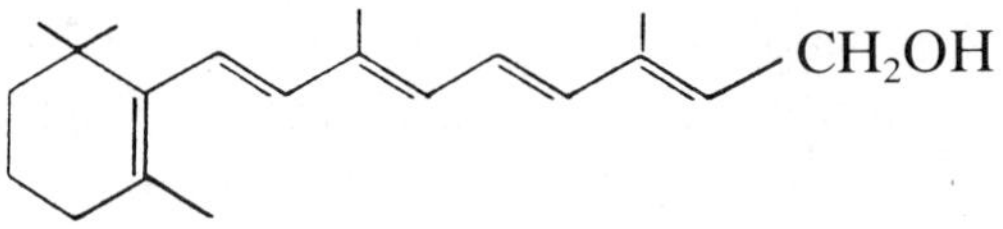

CH_2OH

vitamin A

SYNTHESIS OF THE BIOREGULATOR

The structure of the chalcone-type regulators can be dissected into three main parts, and the synthesis involves the assembly of these subunits in two steps. A diarylpropenone (chalcone) can be readily obtained by utilizing the aldol condensation. Alkyl aryl ethers are best prepared by alkylation, that is, formation of the O—alkyl bond in a nucleophilic substitution reaction (Chapter 14).

The sequence of the two steps is not crucial in this synthesis, and the choice depends mainly on convenience. In the procedure used in this experiment, the aldol condensation is carried out first, and the alkylation is the second step. The aldol reaction is carried out with a very large excess of base, because the aldehyde contains an acidic OH group and the con-

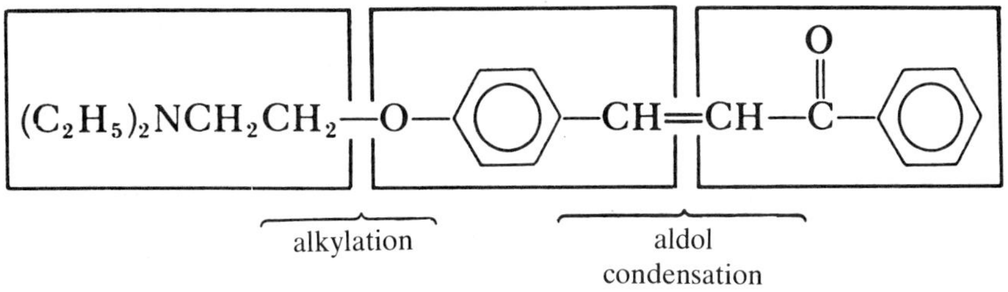

$(C_2H_5)_2NCH_2CH_2$—O—C_6H_4—CH=CH—C(=O)—C_6H_5

alkylation

aldol condensation

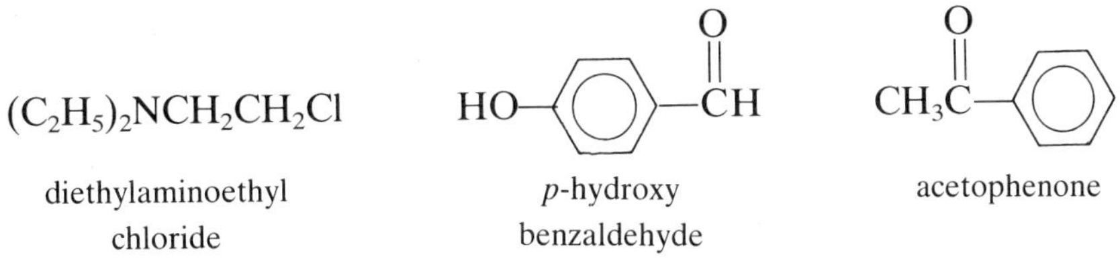

$(C_2H_5)_2NCH_2CH_2Cl$ — diethylaminoethyl chloride

HO—C_6H_4—CHO — *p*-hydroxy benzaldehyde

$CH_3C(=O)$—C_6H_5 — acetophenone

densation therefore requires the combination of two anions. The condensation product is isolated as the crude phenoxide salt, which can be used directly in the next step.

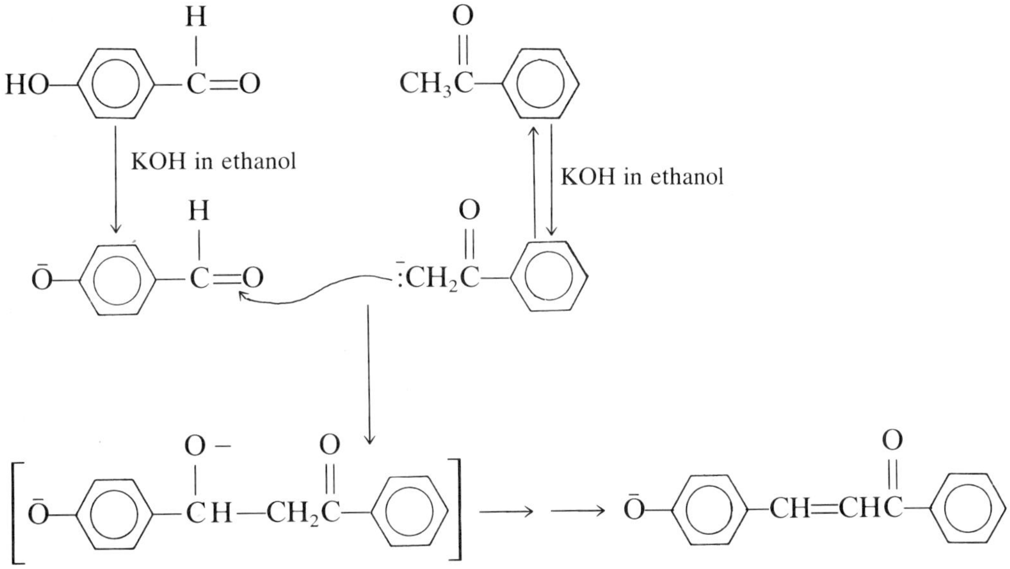

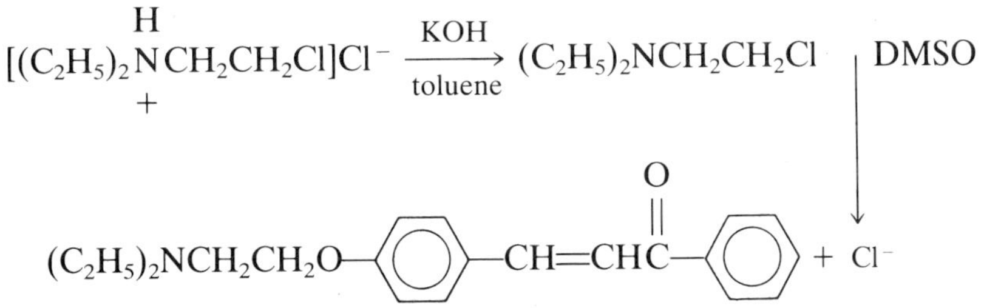

HO—C_6H_4—CH=O —(KOH in ethanol)→ $\bar{O}$—C_6H_4—CH=O

$CH_3C(=O)$—C_6H_5 ⇌(KOH in ethanol) $^-$:$CH_2C(=O)$—C_6H_5

$$\left[\bar{O}\text{—}C_6H_4\text{—}CH(O^-)\text{—}CH_2C(=O)\text{—}C_6H_5\right] \longrightarrow \longrightarrow \bar{O}\text{—}C_6H_4\text{—}CH{=}CHC(=O)\text{—}C_6H_5$$

$$[(C_2H_5)_2\overset{+}{N}HCH_2CH_2Cl]Cl^- \xrightarrow[\text{toluene}]{\text{KOH}} (C_2H_5)_2NCH_2CH_2Cl$$

DMSO

$$(C_2H_5)_2NCH_2CH_2O\text{—}C_6H_4\text{—}CH{=}CHC(=O)\text{—}C_6H_5 + Cl^-$$

A

The halide used for alkylation of the phenoxide is a difunctional compound, and a competing reaction is cyclization of the amino halide to a quaternary aziridinium salt (Question 2). To prevent the occurrence of this reaction, the halide is stored as the hydrochloride salt. Cyclization of the amine is favored by hydroxylic solvents such as water, which stabilizes the transition state leading to an ionic product. The bimolecular reaction with the phenoxide salt is favored by polar, aprotic solvents in which charge separation can occur, but the reactivity of the phenoxide as a nucleophile is not diminished by hydrogen bonding. Examples of such solvents are $(CH_3)_2S{=}O$ (dimethylsulfoxide, DMSO), $(CH_3)_2NCH{=}O$ (N,N-dimethylformamide), and $[(CH_3)_2N]_3P{=}O$ (hexamethylphosphoramide).

The product (Formula A) from the alkylation step is basic, because of the diethylamino group, and advantage is taken of this property in the isolation. The crude material is first extracted from aqueous DMSO solution with ether and is then removed from ether solution by extraction with aqueous acid, leaving neutral impurities behind (see Questions 6a and 6b). In the final operation, the aqueous acid solution is made basic and the product is again extracted. The final product is obtained as an oil (see Question 6c); the compound has not been described in crystalline form.

FORMATION AND DETERMINATION OF CAROTENES IN CITRUS FRUITS

Carotene pigments accumulate in both the pulp and peel of citrus fruits, and the level of carotene synthesis and metabolism in the two types of tissue can vary considerably, as seen in "white" and "pink" grapefruit. With mature fruit, the application of bioregulators by simply dipping or spraying the fruit affects carotenes in the peel only and leaves the pulp unchanged. In the second part of this experiment, the pigment formed in a citrus peel in response to a regulator will be extracted and the amount determined spectrophotometrically.

The long polyene chains in lycopene and other carotene pigments cause absorption of light in the blue end of the visible spectrum; removal of these wavelengths leads to yellow or red colors. The extent of absorption at a given wavelength is proportional to the number of molecules in the light path and thus depends on the concentration of the solution, c, and the thickness of the sample, l. If the initial intensity of the light beam entering the sample is I_0 and the intensity of the emergent beam is I, the absorbance of the solution, A, is given by the Beer–Lambert Law, as stated in Equation 34.1.

$$A = \log \frac{I_0}{I} = \epsilon c l \qquad \textbf{[34.1]}$$

In this equation, c is the concentration in moles/L and l is the path or cell length in cm. The constant ϵ, called the **extinction coefficient**, has a characteristic value for a given compound at a given wavelength. If ϵ for the pure compound is known, the concentration can be calculated directly from the absorbance, which is readily measured with a spectrophotometer. For compounds such as carotene pigments with large ϵ values in the visible region of the spectrum, a simple spectrometer or colorimeter is adequate for the measurement.

Experiments

A. 1-PHENYL-3-(*p*-DIETHYLAMINOETHOXYPHENYL)-2-PROPEN-1-ONE (Macroscale)

SAFETY NOTE THE 60% KOH SOLUTION USED IN THE FIRST STEP CAN CAUSE TISSUE DAMAGE; IF IT COMES IN CONTACT WITH THE SKIN, WASH THOROUGHLY WITH COLD WATER.

Procedure

In a 25 × 100-mm test tube, place 1.2 g of *p*-hydroxybenzaldehyde and 1.2 g (1.2 mL) of acetophenone. Add 3 mL of ethanol and warm the mixture to dissolve the solid. To the brown solution, add 5 mL of 60% aqueous KOH (contains 5 g of KOH in 3 mL of water).

Warm the tube to approximately 60°C on the steam bath, and stir the mixture until a thick red precipitate appears. Continue warming for several minutes more, and then chill the mixture in an ice bath. Filter the dark red phenoxide on a 2- or 3-inch Büchner funnel. (The filter paper in the funnel should be as large a diameter as possible without creasing, since the strong KOH solution dehydrates the paper and causes it to shrink.) Do not use any solvent to rinse or wash the solid. When the liquid has been sucked through, press the solid firmly on the paper to get it as dry as possible. Disconnect the aspirator, loosen the filter paper from the funnel with the tip of a spatula, peel the phenoxide away from the paper, place it in an 18 × 150-mm test tube, and add 6 mL of DMSO.

In another test tube, place 0.86 g of diethylaminoethyl chloride hydrochloride. Dissolve the salt in a minimum volume of water (approximately 2 mL) and add 5 mL of toluene. To this mixture, add 0.5 mL of 60% KOH (or 5 pellets of solid KOH). Stir the mixture so that the diethylaminoethyl chloride is extracted into the organic layer. When the layers have separated almost completely (some emulsion will persist), draw off the toluene layer using a pipet and bulb and transfer it through a small cotton plug in a glass funnel to the DMSO solution of the phenoxide.

Stir the deep red mixture of phenoxide and alkyl chloride solutions and heat it on the steam bath. Continue warming, with intermittent stirring, for 20 minutes, while a semisolid white layer separates on the wall of the test tube (see Question 5).

Pour the red mixture into 50 mL of cold water in a separatory funnel. Add 10 mL of solvent-grade ether, swirl but do not shake vigorously, and separate the layers. Place the upper layer in another flask, and extract the dark red aqueous lower layer with 15 mL of ether. Separate, combine the organic layers, extract the aqueous solution again with 10 mL of ether, and discard the aqueous layer (Question 6a). Place the combined organic layers in the funnel, wash twice with 10-mL portions of water, and discard the aqueous wash.

Shake the ether solution in the separatory funnel with 10 mL of 1 *M* HCl solution and note any changes in the colors of the phases. Separate the aqueous acid solution, drain off the ether layer (Question 6b), and return the aqueous solution to the funnel. Extract the aqueous solution twice with 2- or 3-mL portions of methylene chloride and discard the lower organic layers.

To the aqueous solution in the funnel, add 1 mL of 60% KOH solution and note any changes in appearance. Extract the mixture with two 5-mL portions of methylene chloride, dry the combined methylene chloride solutions with Na_2SO_4, and filter through cotton into a tared test tube. Evaporate most of the methylene chloride on a steam bath and remove the remainder under reduced pressure (see Fig. 4.5). Weigh the product and calculate the yield of the crude amine (Question 6c).

B. CAROTOGENESIS

Procedure

Wipe a lemon or grapefruit thoroughly with a paper towel soaked with hexane or petroleum ether. (This step removes a protective wax that is applied to prolong the freshness of the fruit.) Dissolve 100 mg of your amine in 10 mL of ethanol or isopropyl alcohol and pour the solution into an evaporating dish. Immerse the fruit and rotate it so that about half of the surface becomes wet with solution; try to keep a fairly sharp boundary between treated and untreated areas. Remove the fruit and permit the excess solution to drip off. Let the fruit stand at room temperature for the next few days; observe it occasionally and note changes in appearance.

After a well-defined color change has occurred, remove a thin outer layer of skin from the fruit with a vegetable peeler, keeping the treated and untreated peel separate. Remove just the outer surface, leaving the thick white fibrous tissue.

To measure the increase in lycopene caused by the bioregulator, the carotene content of both treated and untreated peel must be determined. The extraction procedure and analysis should be carried out in the same way for both samples.

Chop the samples of peel into small pieces with scissors and weigh out a 1- to 2-g portion of each. In order to extract the pigment, the oil cells in the peel must be ruptured. Place the peel in a mortar with approximately 1 teaspoon of sea sand, moisten the mixture with a few mL of hexane, and grind with a pestle for several minutes until the mixture has a mealy consistency. Transfer the mass to a small beaker, rinse the mortar with 2 to 3 mL of hexane, and pour the hexane into the beaker.

The pigment can now be extracted simply by stirring the ground peel with several portions of hexane or petroleum ether. Draw up the hexane solution with a pipet and bulb and transfer it to a dry graduated cylinder; if fine particles of peel cannot be avoided, filter the solution through a cotton pellet. Most of the pigment can be removed with three or four small portions of solvent. Adjust the final volume of hexane to 25 mL, record the volume, and proceed as directed by your instructor to determine the absorbance of the solution at 500 nm.

Subtract the absorbance of the control (untreated) sample from the value for the treated sample. From the difference and from Equation 34.1, calculate the increase in lycopene concentration. The value of ϵ for lycopene at 500 nm is $1.7 \times 10^5\ M^{-1}\ cm^{-1}$; the path length will depend on the type of cell used. From the concentration and volume of the solution, calculate the weight percentage of lycopene produced by the bioregulator in the treated peel. (The molecular weight of lycopene is 537.)

QUESTIONS

1. 1-Phenyl-3-*p*-hydroxyphenyl-2-propen-1-one has a lower pK_a value than does phenol. Account for this fact and the color of the phenoxide isolated in the experiment in terms of resonance structures for the anion.
2. Write the structure of the aziridinium salt formed by cyclization of diethylaminoethyl chloride. When a solution of the free base is allowed to stand, further reaction of the amine with the aziridinium salt gives a product with the formula $(C_{12}H_{28}N_2)^{2+}2Cl^-$. Suggest the structure of the dication.
3. The aldol condensation product, 1-phenyl-3-*p*-hydroxyphenylpropenone, has a melting point of 185°C. Outline the steps by which this compound could be isolated and purified, if desired.
4. Calculate the molar amounts of *p*-hydroxybenzaldehyde, acetophenone, and diethylaminoethyl chloride hydrochloride used in the procedure. Assuming that the yield of the phenoxide (which was not weighed) from the aldol condensation is 60% and the yield of diethylaminoethyl chloride

from the hydrochloride is 90%, which is the limiting reagent in the alkylation step?

5. What compounds are present in the semisolid white precipitate formed during the alkylation?
6. Since the phenoxide is not purified and retains some starting materials, and since the alkylation step does not proceed to completion, the red DMSO–toluene solution contains significant amounts of the following compounds: acetophenone, *p*-hydroxybenzaldehyde, 1-phenyl-3-*p*-hydroxyphenylpropenone, diethylaminoethyl chloride and the desired product, Formula A, 1-phenyl-3-diethylaminoethoxyphenylpropenone. (Some of these compounds are present in ionic form.)
   a. What compounds are present in significant amount in the aqueous layer remaining after the initial extraction with ether?
   b. What compounds are present in significant amount in the ether layer after extraction with acid?
   c. What compounds are likely to be present as significant impurities in the final product?
7. Extraction of lycopene from 50 g of tomato paste gave 500 mL of a final red extract. One mL of this solution was diluted to 100 mL and the absorbance of the diluted sample in a 1-cm cell was 1.20. Calculate the weight of lycopene that was isolated.

REFERENCE

Thomas H. Maugh II, Bioregulators: Alteration of Gene Expression in Citrus Fruit. *Science* **1974,** *184*, 655.

Name ______________________ Section ______________ Date ________

34 Synthesis of a Citrus Bioregulator

PRELABORATORY QUESTIONS

1. What is meant by the term "chalcone"?

2. Write the mechanism for the aldol-like condensation of acetophenone and *p*-hydroxybenzaldehyde.

3. Draw the structures of three aprotic solvents.

4. Write the equation for the reaction of diethylaminoethyl chloride hydrochloride with KOH.

5. Why are aprotic solvents used in the bimolecular substitution reaction involving the phenoxide ion and the free base, diethylaminoethyl chloride? Give mechanisms to explain your answer.

35

Benzyne

Benzynes, or dehydrobenzenes, were first recognized in 1953 as reactive intermediates in certain nucleophilic substitution reactions of aryl halides. Since their discovery, a number of different approaches have been devised for generating these intermediates. Several of the methods that have been used to prepare benzyne are shown in the following diagram:

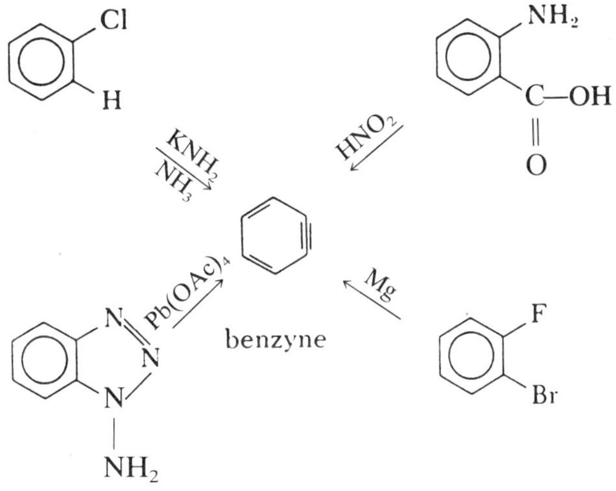

The main synthetic utility of benzynes arises from their great reactivity as electrophiles and dienophiles. As examples of the former property, when generated in water or liquid ammonia, benzyne rapidly reacts to form phenol or aniline, respectively.

The Diels–Alder reactivity of benzyne is illustrated in the experiment in this chapter, in which furan is used as the diene. The benzyne is generated in the presence of the furan and reacts in situ. The product is 1,4-dihydronaphthalene-1,4-endoxide, which can be isomerized very readily in the presence of acid to 1-naphthol.

Benzyne in this experiment is generated by the diazotization of anthranilic acid with a nitrite ester, followed by loss of carbon dioxide and nitrogen. To minimize side reactions of benzyne with the starting materials, the anthranilic acid and isoamyl nitrite are added at the same time so that neither is present in large excess.

$$\text{anthranilic acid} + C_5H_{11}ONO \text{ (isoamyl nitrite, d 0.87)} \longrightarrow \left[C_6H_4(N_2^+)(CO_2^-) \right] \xrightarrow{-N_2,\ -CO_2} \text{benzyne}$$

benzyne + furan (d 0.94) → 1,4-dihydronaphthalene-1,4-endoxide, mp 56°C $\xrightarrow{H^+}$ 1-naphthol, mp 96°C

Experiments

SAFETY NOTE ISOAMYL NITRITE IS TOXIC AND 1-NAPHTHOL IS A TOXIC IRRITANT. HANDLE BOTH COMPOUNDS WITH CARE AND IMMEDIATELY WASH OFF ANY THAT GET ON THE SKIN.

A. 1,4-DIHYDRONAPHTHALENE-1,4-ENDOXIDE (Microscale)

In a 25-mL round-bottom flask place 2 mL of furan, 2 mL of 1,2-dimethoxyethane (glyme), and a boiling stone, and attach the flask to a reflux condenser. In separate 10 × 13-mm test tubes, dissolve 1 mL of *iso*-amyl nitrite in 2 mL of glyme and 0.55 g of anthranilic acid in 2 mL of glyme. (If the anthranilic acid does not dissolve easily, heat gently on a steam bath). Heat the furan solution to reflux on a steam bath and at 3- to 4-minute intervals add 0.5 mL of each of the two solutions through the condenser. Use separate transfer pipets for each of the reagents, since they react with one another. One-mL graduated syringes are convenient for these transfers. After the additions are complete, continue refluxing for 3 minutes.

Prepare a solution of 0.1 g (1 pellet) of NaOH in 5 mL of water, and add this to the cooled reaction. Transfer the mixture to a separatory funnel with a Pasteur pipet and extract with 5 mL of pentane. Discard the aqueous layer and rinse the pentane solution with four 3-mL portions of water. Add charcoal to partially decolorize the pentane solution, filter through a pipet containing $MgSO_4$, and concentrate to approximately 1 mL on a steam bath. If an oil separates at this point, decant the warm solution into a clean test tube and rinse the oil with 1 to 2 mL of pentane. Let the solution cool and scratch to induce crystallization of the endoxide. If the solution does not crystallize, add pentane, place on a steam bath, reduce the volume, and attempt crystallization once again. Collect the product and recrystallize it from approximately 5 mL of pentane. Weigh, determine the melting point, and calculate the percentage yield of the 1,4-dihydronaphthalene-1,4-endoxide. If time permits, a purer product can be obtained by sublimation at 100°C under aspirator vacuum (see Chapter 18) rather than by recrystallization.

B. 1-NAPHTHOL (Microscale)

Place 0.1 g of 1,4-dihydronaphthalene-1,4-endoxide and 2 mL of ethanol in a distillation receiver or centrifuge tube, and add 1 mL of concentrated hydrochloric acid. Stir and let stand for 10 minutes. Add 4 mL of ether and 3 mL of water, stopper and shake the tube, and draw off the lower aqueous layer with a pipet. Wash the ether layer with two 1-mL portions of water and dry the ether layer by filtering it into a 15-mL Erlenmeyer flask through a column of Na_2SO_4 contained in a Pasteur pipet. Rinse the pipet with a small amount of solvent and evaporate the solution to an oil on a steam bath. Add 3 mL of hexane, heat to dissolve the product, and let cool to crystallize. Filter the slightly pink crystals and air-dry them in the funnel. Weigh and determine the percentage yield and the melting point of the 1-naphthol.

QUESTIONS

1. Write out complete equations for the four alternate preparations of benzyne given at the beginning of the chapter. Show intermediates and by-products of each reaction.
2. In the absence of added nucleophiles or dienes, benzyne forms a dimer and trimer. Give the structures and names of these compounds.
3. The 1,4-dihydronaphthalene-1,4-endoxide prepared in this experiment reacts with 2,3-dimethylbutadiene, followed by acid and then an oxidiz-

ing agent, to give 2,3-dimethylanthracene. Write equations with curved arrows (for electron shifts) showing how this product is formed.

4. Write equations involving benzyne intermediates for the following reactions:

a. *o*-bromoanisole $\xrightarrow[\text{liq } NH_3]{KNH_2}$ *m*-anisidine

b. *o*-fluorobromobenzene + Mg + anthracene $\longrightarrow$ triptycene

c. 3-(*m*-chlorophenyl)propionitrile $\xrightarrow{KNH_2}$ 1-cyanobenzocyclobutene

5. Careful analysis of the gas evolved in the reaction of anthranilic acid and isoamyl nitrite shows that nitrogen is lost first, followed by carbon dioxide. Suggest a possible uncharged intermediate that might be formed by the loss of nitrogen.

6. Suggest why four water rinses of the pentane layer are required in the isolation of the 1,4-dihydronaphthalene-1,4-endoxide.

7. Write the mechanism, showing intermediates, for the isomerization of 1,4-dihydronaphthalene-1,4-endoxide to 1-naphthol. Predict whether this reaction will be endothermic or exothermic, and explain why.

8. Discuss the bonding in benzyne in terms of molecular orbital theory.

REFERENCES

Fieser, L.F.; Haddadin, M.J. *Can. J. Chem.*, **1965**, *43*, 1599.

Gilchrist, T.L. In *The Chemistry of Functional Groups*, Supplement C; Patai, S.; Rappoport, Z., Eds. Wiley: New York, **1983**; Part 1, pp. 383–419.

Levin, R.H. *React. Intermed.* (Wiley) **1978** *1*, 1–26; **1981,** *2*, 1–14; **1985,** *3*, 1–18.

Name ______________________ Section ____________ Date ________

35 Benzyne

PRELABORATORY QUESTIONS

1. Draw the structure of benzyne.

2. When reacted with water, benzyne forms ____________________.

3. Draw the structure of furan.

4. Draw the structure of the product that forms when furan is reacted with benzyne.

5. In the experiment described in this chapter, benzyne is produced by the reaction of ____________________ with ____________________.

6. Calculate the molar equivalents of the reactants used (Question 5) in this experiment.

36

Synthesis of Phenytoin

Epilepsy is a neurological disorder that has been known since antiquity. Not until 1857, however, was the first effective anticonvulsant drug, potassium bromide, introduced for its treatment. Early in the twentieth century, the barbiturate phenobarbital began to be widely used, and further development of anticonvulsant drugs often involved barbiturates or compounds that are structurally similar to them. By 1940, phenytoin (5,5-diphenylhydantoin) had been developed and had become the leading drug for convulsive seizure disorders. It has been the drug of choice for grand mal epilepsy during the intervening years and is marketed under such trade names as Dilabid, Dilantin, and Divulsan.

In the United States, epilepsy is the second most frequently encountered neurological disorder, exceeded only by stroke. It has been estimated that there are over 1 million people who have recurrent seizures and that 2 million persons have had two or more seizures. In addition to phenytoin, which has been widely used in the treatment of epilepsy, other compounds such as phenobarbital, primidone, carbamazepine, and valproic acid have been found to be effective.

One preparation of phenytoin involves the reaction of benzil with urea under basic conditions. The course of the reaction sequence is illustrated on the following page.

Experiment

PREPARATION OF PHENYTOIN (Microscale)

SAFETY NOTE CONCENTRATED SODIUM HYDROXIDE SOLUTION IS QUITE CAUSTIC AND CAN CAUSE DAMAGE TO EYESIGHT WITHIN SECONDS. WEAR SAFETY GOGGLES AT ALL TIMES AND CLEAN UP ANY SPILLS IMMEDIATELY. IF ANY CONTACTS THE SKIN, WASH WITH COPIOUS AMOUNTS OF WATER. AVOID EXPOSURE TO PHENYTOIN; IT IS A SUSPECTED CARCINOGEN.

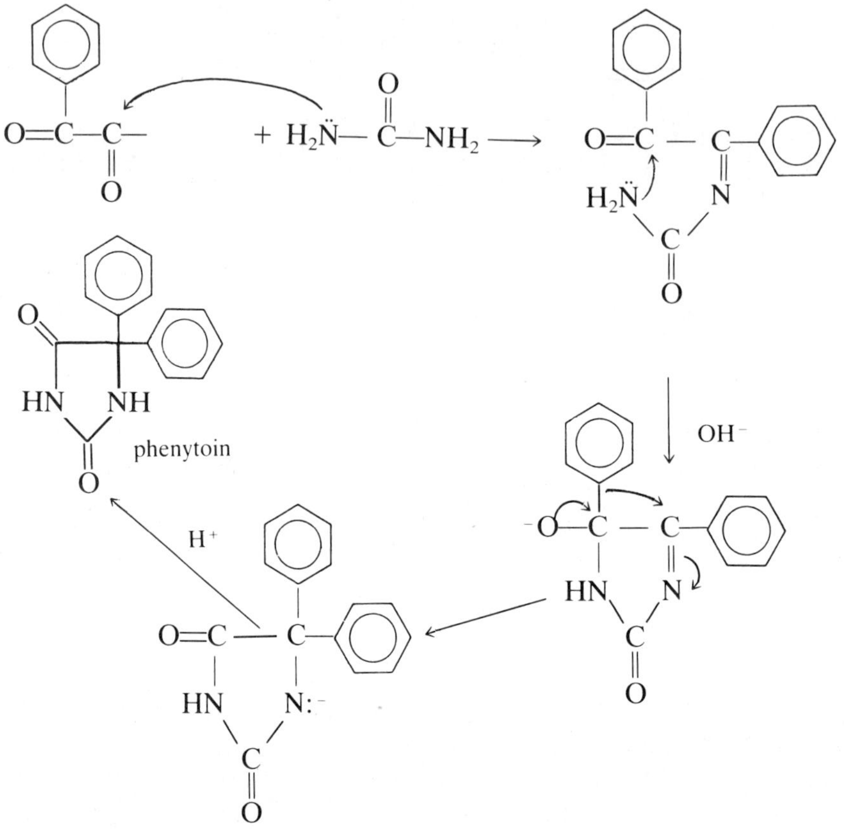

Procedure

In a 10-mL round-bottom flask dissolve 0.525 g of benzil (0.0025 mole) and 0.300 g of urea (0.0050 mole) in 6 mL of ethanol. Then add 1.2 mL of 6 *N* sodium hydroxide solution, add a boiling stone and reflux the mixture for 1.5 hours. (**CAUTION**: Ground joints may fuse when heated with caustic liquid; be sure to grease well the joint between flask and condenser, and disconnect the condenser after the reflux but before the apparatus cools.) Allow the mixture to cool, and filter it through a filter pipet to remove any solid material. Chill the filtrate in an ice-water bath and carefully neutralize it using 10% hydrochloric acid until it is acid to litmus paper. The 5,5-diphenylhydantoin that precipitates during the neutralization is collected by vacuum filtration and dried. The product is reasonably pure as isolated from the neutralization procedure, but may be recrystallized from 95% ethanol, if required. The melting point is 295 to 298°C (decomposition); the melting point should not be determined using an oil bath because of the high temperature.

Determine the weight and percentage yield of the product, record the IR and/or the NMR spectra, and submit the product to your instructor in an appropriately labeled container.

QUESTIONS

1. The infrared spectrum of phenytoin contains the following absorptions. Make assignments for these.

3275, 3205 cm^{-1}

3064 cm^{-1}

1740, 1719 cm^{-1}

1599, 1496 cm^{-1}

747, 690 cm^{-1}

2. The NMR spectrum of phenytoin shows peaks at 7.3, 9.2, and 10.9 ppm with relative areas 10:1:1. Make assignments for these peaks.

3. The pK_a of phenytoin is 8.3. Could the equivalent weight of this compound be determined by titration with standard NaOH in water or in water–acetone? Explain.

REFERENCES

Biltz, H. *Ber.* **1908,** *41*, 1391; **1911,** *44*, 411.

Pankaskie, M. C.; Small, L. *J. Chem. Educ.* **1986**, *63*, 650.

Philip, J., et al; In *Analytical Profiles of Drug Substances*; Florey, K. Ed.; American Pharmaceutical Association: Washington, DC, 1984; Vol. 13, p 417.

Name ______________________ Section ____________ Date ________

36 Synthesis of Phenytoin

PRELABORATORY QUESTIONS

1. List three substances that have been used as anticonvulsants.

2. Compare the chemical structure of a hydantoin to that of a barbiturate. What are some common features? Where are they different?

3. Draw the structure of benzil. Give the IUPAC name for it.

4. Draw the structure of the anion from which phenytoin is formed in the last neutralization step. What features contribute to its stability?

5. Suggest possible analogues of phenytoin that might have similar physiological properties. How would you alter the chemical structure?

37

Heterocyclic Syntheses

The most common heterocyclic compounds are those containing nitrogen and other heteroatoms in a fully unsaturated five- or six-membered ring. Such ring systems are isoelectronic with cyclopentadiene anion and benzene, respectively, and are aromatic compounds; typical examples are pyrrole, pyrazole, isoxazole, pyridine, and pyrimidine. These and other rings are present in important natural substances such as hemoglobin, vitamins, and nucleic acids, and they also form the basic structure of many synthetic drugs and dyes. Rings with a number of combinations of heteroatoms are readily prepared, and heterocyclic compounds constitute the bulk of all known organic molecules.

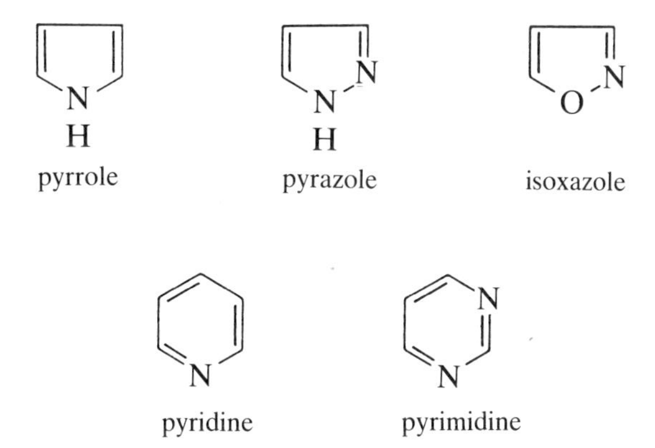

Synthesis of many heterocyclic compounds is based on simple reactions such as the condensation of carbonyl groups with amino and activated methylene groups in bifunctional molecules. The experiments in this chapter demonstrate how a number of the fundamental heterocyclic rings can be built from a single starting material, acetylacetone. The β-diketone system in acetylacetone is largely enolized, and the compound and its reactions should be represented with this tautomeric form. This enolic ketone contains two electrophilic centers and a highly reactive nucleophilic central carbon.

Condensations can occur between any two of these functional groups and two complementary groups in another molecule or molecules. This is indicated schematically below:

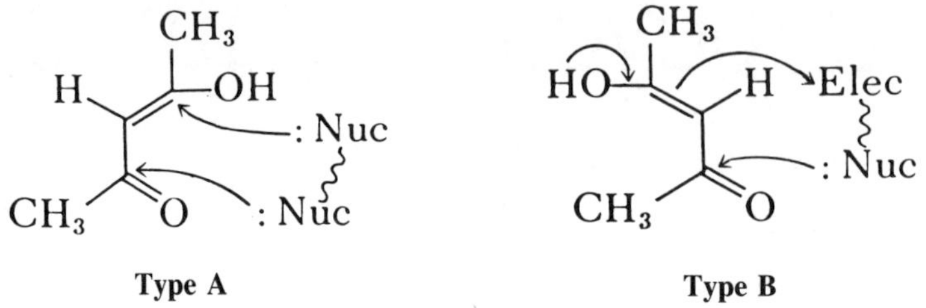

Type A **Type B**

TYPE A CONDENSATIONS

The first group of syntheses to be carried out are condensations of type A, in which two nucleophilic centers in the second component condense with the dicarbonyl system, eliminating two moles of water. Typical examples are the formation of the pyrazole (**1**) and the pyrimidine (**2**), illustrated with the enolic dibenzoylmethane (37.1).

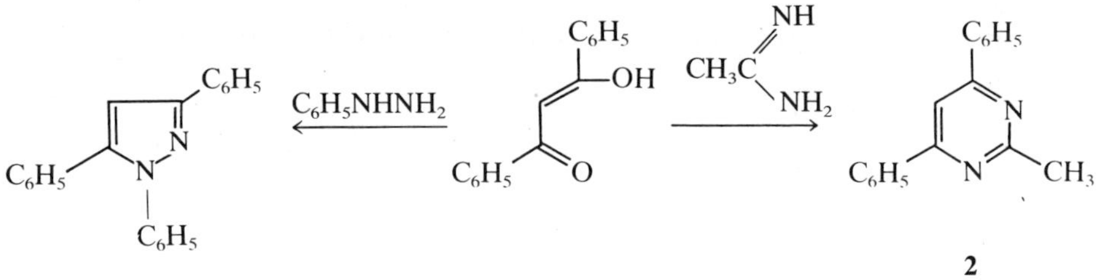

$C_6H_5NHNH_2$ $CH_3C(=NH)NH_2$

1 1,3,5-triphenylpyrazole

dibenzoylmethane

2 2-methyl-4,6-diphenylpyrimidine

[37.1]

Condensations of this type are very simple to carry out, and with the very reactive acetylacetone, these preparations involve little more than mixing the reagents in an appropriate medium. When the solubilities of starting materials and products permit, water is a very effective solvent.

Table 37.1 Starting Materials Used

STARTING MATERIAL	FORMULA
Cyanoacetamide	$NCCH_2CONH_2$
Guanidine	$(NH_2)_2C{=}NH$
Hydrazine	NH_2NH_2
Hydroxylamine	NH_2OH

Table 37.2 Products

2-Amino-4,6-dimethylpyrimidine, mp 156°C (from MeOH)
3-Cyano-4,6-dimethyl-2-pyridone, mp 290°C (from H_2O)
3,5-Dimethylisoxazole, bp 142°C
3,5-Dimethylpyrazole, mp 107°C (from H_2O)

The rate of these carbonyl condensations in aqueous solution is pH-dependent; a slightly basic solution is optimal for the reactions studied here.

Procedures are indicated in the following paragraphs for reactions of acetylacetone with the four substances listed alphabetically in Table 37.1. The names of the products are given, also in alphabetical order, in Table 37.2. Before proceeding with the experiment, determine the product that will be obtained from each starting material (see Question 2 at end of chapter). Then carry out one or more of the reactions as directed by your instructor. Only minimum directions are given; use our own judgment and experience in isolating the products. Submit products, labeled with yield, melting point, and structural formula, to your instructor.

Experiments

SAFETY NOTE IN THIS EXPERIMENT AND THE OTHER PARTS OF THIS CHAPTER, A NUMBER OF DIFFERENT REAGENTS ARE USED IN SMALL AMOUNTS. NONE OF THESE ARE PARTICULARLY TOXIC, BUT THE USUAL PRECAUTIONS SHOULD BE TAKEN TO AVOID INHALING ANY VAPORS OR CONTACT WITH THE SKIN. WASH YOUR HANDS THOROUGHLY WITH SOAP AND WATER IF ANY SPILLS OCCUR. HYDRAZINE IS CLASSIFIED AS A POSSIBLE CARCINOGEN.

A. REACTION WITH CYANOACETAMIDE

Dissolve 0.84 g of cyanoacetamide and 0.5 g of sodium carbonate in 15 mL of water, and add 0.01 mole (1 mL) of acetylacetone.

B. REACTION WITH GUANIDINE

Dissolve 0.9 g of guanidine carbonate and 0.5 g of sodium acetate in 2.5 mL of water, and add 1 mL of acetylacetone. After partial air-drying, recrystallize the product from a minimum volume of methanol; filter the hot solution to remove traces of inorganic solid.

C. REACTION WITH HYDRAZINE

Mix 0.5 mL of 85% hydrazine hydrate and 2.5 mL of water, and add 0.5 mL of acetylacetone.

D. REACTION WITH HYDROXYLAMINE

Dissolve 2.8 g of hydroxylamine hydrochloride and 3.3 g of anhydrous sodium acetate in 5 mL of water and add 4 mL of acetylacetone. Withdraw the oily product with a pipet, filter through granulated K_2CO_3 into a 10 mL side-arm flask and distil (see page 487).

TYPE A CONDENSATIONS

A variation of the type A condensation discussed earlier is illustrated in the preparation of 2,4,6-trimethylquinoline. The intermediate enamine from an aromatic amine and the β-diketone is prepared and isolated in a first step and then cyclized. The second step is a typical electrophilic aromatic substitution reaction involving the protonated carbonyl group of the enamine, as illustrated in 37.2.

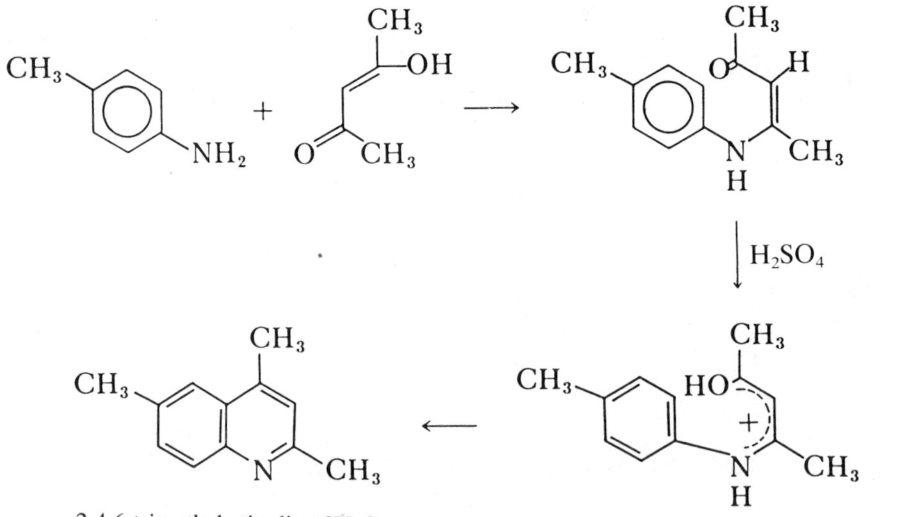

CH_3 NH_2 + CH_3 OH O CH_3 → CH_3 O CH_3 H N H CH_3

H_2SO_4

CH_3 HO CH_3 + N H CH_3 ← CH_3 CH_3 N CH_3

2,4,6-trimethylquinoline·$2H_2O$
(mp 62°C)

[37.2]

Experiments

A. 4-(*p*-TOLUIDINO)-3-PENTEN-2-ONE (Microscale)

In a 10-mL distilling flask, place 0.42 g of *p*-toluidine, 0.6 mL of acetylacetone, and a boiling stone. Insert a thermometer in the neck so that the bulb is immersed in the liquid as far as possible without touching the bottom. Use a 10 × 75-mm test tube as a receiver (see Fig. 46.1). Heat the flask over a wire gauze until the temperature reaches 140°C and boiling commences. Adjust the heating to maintain gentle boiling with a minimum of distillation into the side arm for 5 minutes, then increase the heating rate and collect water and excess acetylacetone until the temperature reaches 215°C. Cool the flask and pour the contents into another 10 × 75-mm test tube and rinse with 0.5 mL of pentane. Chill the pentane solution in an ice bath; after crystallization, collect the product and wash on the filter with a small volume of chilled pentane.

B. 2,4,6-TRIMETHYLQUINOLINE (Microscale)

In a 25-mL Erlenmeyer flask, place 2.5 mL of concentrated sulfuric acid and add 0.4 g of enamino ketone. (Reduce the amount of H_2SO_4 proportionately if less enamine is available.) Warm the solution for 20 to 25 minutes on the steam bath, then cool and add approximately 10 g of ice. Chill the solution in an ice bath; after the salt crystallizes, collect it on a Hirsch funnel and wash with a few mL of water. Suspend the moist salt in 2 mL of water and add concentrated ammonium hydroxide dropwise (see Question 6). Chill and crystallize the resulting oil, and collect the solid. Dissolve the moist solid in 1 mL of ethanol, add water until the solution is not quite turbid (approximately 1 mL of water), seed, stopper loosely, and allow the mixture to stand. The dihydrate of the quinoline crystallizes in long needles. Record the melting point and the yield (calculated on the basis of trimethylquinoline · $2H_2O$) and submit the product to your instructor.

TYPE B CONDENSATIONS

Type B condensations again involve the reaction of ammonia or an amine with the enolic diketone to give an enamino ketone, $-NH-\overset{|}{\underset{|}{C}}=\overset{|}{C}-\overset{|}{\underset{|}{C}}=O$. These compounds are stabilized by conjugation of the —NH and —CO groups, and are better regarded as "vinylogous" amides than amino ketones. The enamine system can provide one or two nucleophilic centers for further reactions.

One important application of this type of condensation is the **Knorr pyrrole synthesis**, shown in 37.3. An α-amino ketone is generated in the presence of the enol, and the intermediate enamine cyclizes to a pyrrole. In the following procedure, the amino ketone is formed by the reduction of isonitrosoacetophenone.

Experiment

3-ACETYL-2-METHYL-4-PHENYLPYRROLE (Macroscale)

Place 6 mL of acetic acid and 1.5 mL of water in a 50-mL Erlenmeyer flask and add 1.8 g of isonitrosoacetophenone and 1.3 mL of acetylacetone. To this solution add, in four portions, a total of 2.5 g of zinc dust. Swirl vigorously between each addition, cooling in an ice bath if the flask becomes too hot to hold in the hand. After all of the zinc has been added, heat the reaction mixture to gentle boiling for 2 to 3 minutes. Decant the hot solution from residual zinc into 50 g of ice. Add alkali until $Zn(OH)_2$ just begins to precipitate and then extract the mixture twice with ether. Wash the ether solution with dilute base and then water, dry over $MgSO_4$, and evaporate to about 5-mL volume. Cool, crystallize and collect the product, and recrystallize it from methanol. The pyrrole tends to crystallize in extremely fine needles that occlude the oily mother liquor, and thorough washing with cold solvent on the funnel (with the aspirator disconnected) is important. Determine the yield and melting point and submit the product, labeled with the compound name, structure, and melting point.

VARIATIONS OF TYPE B CONDENSATIONS

Two other condensations of type B are performed under very similar conditions, with an additional component present in one case. In these reactions, ammonia, conveniently added in the form of ammonium acetate to buffer the solution, gives the enamino ketone (Eq. 37.4). This intermediate can supply two nucleophilic centers for condensation with another molecule of acetylacetone (or a second molecule of enamine) (Eq. 37.5). If a more reactive carbonyl group is available, however, the reaction of two molecules of enamine can occur at this center. In the reaction with acetaldehyde described below, the acetaldehyde can very conveniently be added in the form of its ammonia addition product, called "acetaldehyde ammonia" (Eq. 37.6).

[37.4]

[37.5]

bp 116°C/20 mm

[37.6]

mp 156°C

Experiments

A. 3-ACETYL-2,4,6-TRIMETHYLPYRIDINE (Macroscale)

In a 125-mL Erlenmeyer flask mix 8 g of ammonium acetate and 10 mL of acetylacetone. Warm on the steam bath for 30 minutes, cool, and add saturated sodium carbonate solution in portions until CO_2 evolution stops. Extract the mixture with 25 mL of ether and then with 15 mL of ether. Dry the combined ether solutions (do not backwash with water) with $MgSO_4$ and evaporate to an oil. Transfer the oil with a pipet to a 50-mL round-bottom flask and set up for vacuum distillation using only a distilling head with thermometer, a vacuum take-off adapter, and a tared 50-mL round-bottom flask as a receiver; a condenser is unnecessary. Add several boiling stones, connect to aspirator suction, and distill with a heating mantle or a low flame. Some splashing will occur, but a nearly colorless distillate can be obtained. Report the yield and boiling range of the product.

B. 3,5-DIACETYL-2,4,6-TRIMETHYL-1,4-DIHYDROPYRIDINE (Macroscale)

Dissolve 0.6 g of acetaldehyde ammonia and 0.8 g of ammonium acetate in 10 mL of water and add 2 mL of acetylacetone. Warm for approximately 2 minutes at 50°C until all liquid droplets dissolve; allow to cool. Remove a few drops of the solution and chill to obtain a seed; then seed the main solution and allow it to crystallize for several days at room temperature if time permits. Collect the product and report the yield and melting point. (This preparation illustrates the point of view, held by some individuals, that organic chemistry is one of the more rewarding art forms.)

QUESTIONS

1. Explain how pyrrole qualifies as an aromatic compound according to the $(4n + 2)\pi$ electron rule. From these considerations, predict whether pyrrole or pyridine would be the more strongly basic compound.
2. Write equations with structural formulas for reactions of the four starting materials in Table 37.2 with acetylacetone.
3. A β-keto ester, $RCOCH_2CO_2R'$, is often used instead of a β-diketone in condensations of type A; the product then contains a carbonyl group adjacent to the heteroatom. Write the structural formulas for the products that would be obtained in the reaction of hydrazine and hydroxylamine with ethyl acetoacetate.
4. 4-Amino-3-penten-2-one, the intermediate in Equations 37.5 and 37.6, can be isolated as a low-melting solid if desired. Predict the product that would be formed if this compound were heated in an aqueous solution containing (a) ammonium acetate and (b) hydrazine acetate (NH_2NH_3OAc).
5. In the Knorr condensation (Eq. 37.3), no pyrrole is obtained if the acetylacetone is added after the addition of the zinc. A product with the formula $C_{16}H_{14}N_2$ can be isolated instead. Suggest the structure of this product.
6. Heterocyclic bases, and amines in general, can form two types of salts with sulfuric acid: sulfate $(R_3NH^+)_2SO_4^=$ and bisulfate $(R_3NH)^+HSO_4^-$. Suggest the formula of the salt that precipitates in the quinoline preparation. Describe the appearance of the reaction mixture as ammonia is added, and account for the observations in terms of the ionic species present.
7. Whereas the condensation of acetylacetone with cyanoacetamide leads to a single compound in quantitative yield, reaction with cyanoacethy-

drazide ($NCCH_2CONHNH_2$) can lead to several condensation products, depending on the conditions. In acidic solution a compound A ($C_8H_9N_3O$) is obtained; mild hydrolysis of A gives $C_5H_8N_2$, mp 107°C. In basic solution, acetylacetone plus cyanoacethydrazide give a compound B, isomeric with compound A. Reaction of compound B with HNO_2 gives N_2O and a compound, mp 290°C ($C_8H_8N_2O$). In buffered solution, compound C ($C_8H_{11}N_3O_2$) is obtained; treatment of compound C with acid gives compound A. Treatment of compound C with base gives still another product D, isomeric with A and B. Compound D contains no CN group. Write structures for compounds A to C (*Hint:* See Table 37.1), and suggest a possible structure for D. (See *Angew. Chem.*, **1958**, *70*, 344.)

REFERENCES

Texts

Acheson, R. M. *An Introduction to the Chemistry of Heterocyclic Compounds*, 3rd ed.; Interscience: New York, 1976.

Gilchrist, T. L. *Heterocyclic Chemistry*; Pitman: Marshfield, MA, 1985.

Katritzky, A. R. *Handbook of Heterocyclic Chemistry*; Pergamon: Elmsford, New York, 1985.

Newkome, G. R.; Paudler, W. W. *Contemporary Heterocyclic Chemistry*; Wiley-Interscience: New York, 1982.

Experimental Procedures

Fitton, A. O.; Smalley, R. K. *Practical Heterocyclic Chemistry*; Academic: New York, 1968.

Name ______________________ Section ______________ Date ________

37 Heterocyclic Syntheses

PRELABORATORY QUESTIONS

1. Name the following ring systems.

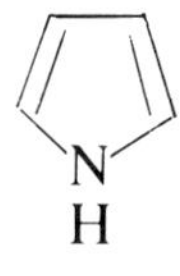

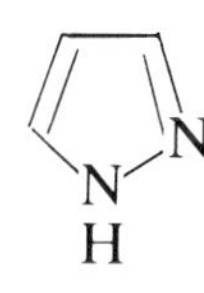

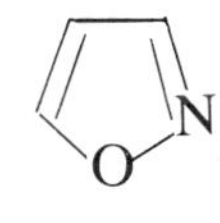

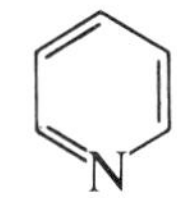

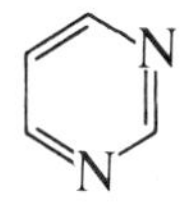

__________ __________ __________ __________ __________

2. The common starting material utilized in all of the syntheses described in this Chapter is ______________________.

3. In the synthesis of heterocyclic systems, the common starting material mentioned in Question 2 can be attacked by **a.** nucleophiles only, **b.** electrophiles only, **c.** free radicals only, **d.** both free radicals and nucleophiles, **e.** both electrophiles and nucleophiles.

4. The structure of an enamine is **a.** $-\overset{|}{\underset{|}{C}}-\underset{|}{C}=NH$, **b.** $-\underset{|}{C}=\underset{|}{C}-NH_2$,

c. $-\underset{|}{C}=\underset{|}{C}-\overset{|}{\underset{|}{C}}-NH_2$, **d.** $-\overset{|}{\underset{|}{C}}-\overset{|}{\underset{|}{C}}-N=\underset{|}{C}-\overset{|}{\underset{|}{C}}-$, **e.** $-\underset{|}{C}=C=NH_2$.

38

Synthesis of Sulfanilamide

Sulfa drugs, like some other medicinal agents, comprise a group of compounds having a key structural feature that imparts a specific pharmacological property. The common feature of sulfa drugs is the *p*-aminobenzenesulfonamido group, and the compounds are useful as antibacterial agents. In fact, the first discovered sulfa drug, Prontosil, is not a *p*-aminobenzenesulfonamide but is converted to the pharmacologically active *p*-aminobenzenesulfonamide in the body. When this fact was discovered, the latter compound, known as sulfanilamide, became the preferred agent for treatment. Attempts to increase its effectiveness toward specific infections led to the synthesis and testing of numerous derivatives. Sulfapyridine and later sulfadiazine and sulfamethazine became the agents of choice for streptococcal (throat) and pneumococcal (lung) infections such as pneumonia. Sulfathiazole is the drug of choice for staphylococcal (skin) infections, and sulfaguanidine for intestinal infections.

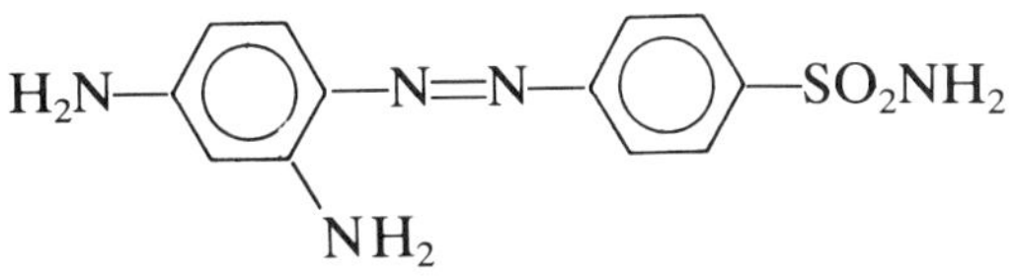

H_2N—$C_6H_3(NH_2)$—N=N—C_6H_4—SO_2NH_2

Prontosil
2′,4′-diaminoazobenzene-4-sulfonamide

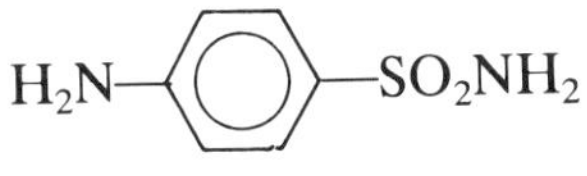

H_2N—C_6H_4—SO_2NH_2

sulfanilamide
p-aminobenzenesulfonamide

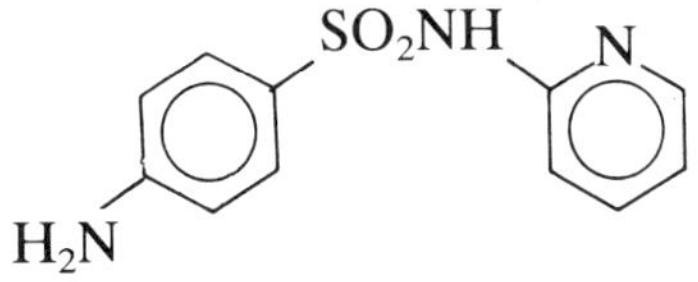

H_2N—C_6H_4—SO_2NH—(2-pyridyl)

sulfapyridine

H_2N—C_6H_4—SO_2NH—(2-pyrimidinyl)

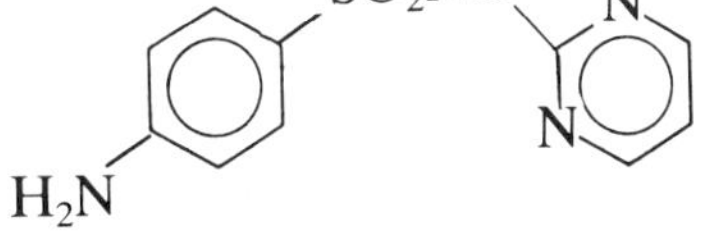

sulfadiazine

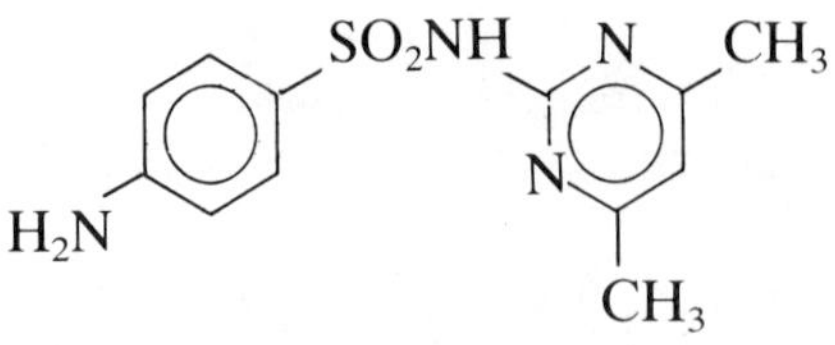

sulfamethazine

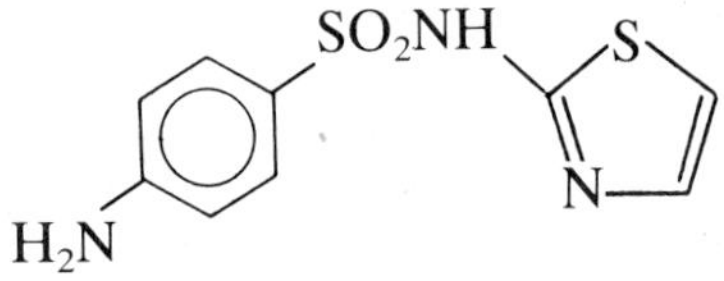

sulfathiazole

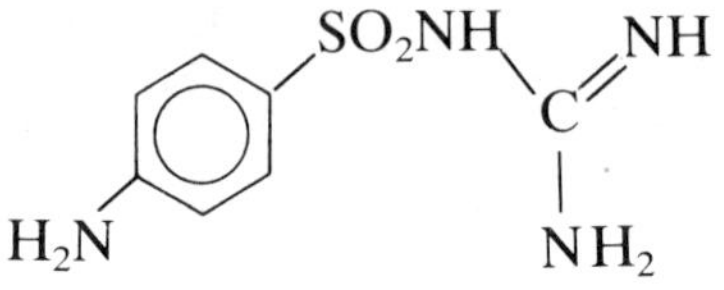

sulfaguanidine

The compounds act by competitively inhibiting the incorporation of *p*-aminobenzoic acid, an essential component for cell growth of the microorganisms. The structurally comparable sulfonamides, with a pK_a similar to that of the carboxylic acid, block the metabolic pathway. Thousands of analogs of sulfanilamide have been prepared and tested for antibacterial properties. A number of these compounds are used clinically, although they have been replaced to a considerable extent by naturally occurring antibiotics, such as penicillin and streptomycin.

The preparation of the parent compound, sulfanilamide, is a simple multistep process, involving the introduction and removal of an acetyl blocking group to control the course of the synthesis (see Chapter 25). The three steps, starting with acetanilide (see Chapter 3), can be carried out without purification of intermediates and can be completed in a single 3-hour laboratory period.

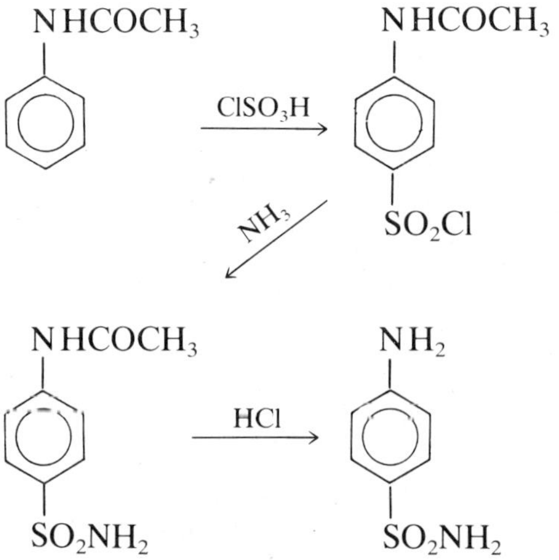

$C_6H_5NHCOCH_3 \xrightarrow{ClSO_3H} p\text{-}CH_3CONHC_6H_4SO_2Cl \xrightarrow{NH_3} p\text{-}CH_3CONHC_6H_4SO_2NH_2 \xrightarrow{HCl} p\text{-}H_2NC_6H_4SO_2NH_2$

Experiments

SAFETY NOTE CHLOROSULFONIC ACID IS A VERY HAZARDOUS REAGENT. *IT REACTS VIOLENTLY WITH WATER TO PRODUCE HCl GAS, AND CAUSES SEVERE BURNS* EVEN ON BRIEF CONTACT WITH THE SKIN. DO NOT POUR EXCESS REAGENT INTO THE SINK; DISPOSE OF IT CAREFULLY IN THE FUME HOOD. BE CERTAIN ANY GLASSWARE COMING IN CONTACT WITH CHLOROSULFONIC ACID IS ABSOLUTELY DRY. CONCENTRATED AMMONIUM HYDROXIDE FUMES ARE IRRITATING TO THE NOSE AND EYES. AVOID BREATHING VAPORS.

INSTRUCTOR'S NOTE The chlorosulfonic acid (use a fresh bottle) and the ammonia may be dispensed by an automatic pipet or buret. This results in better safety practice and less waste.

A. *p*-ACETAMIDOBENZENESULFONYL CHLORIDE (Microscale)

Carefully weigh 0.40 g of acetanilide into a centrifuge tube. Heat over a small flame and rotate the tube to form a thin layer of melted acetanilide on the bottom half of the tube. This provides a large surface area to insure better contact with the chlorosulfonic acid. Allow the tube to cool and stopper it with a cork.

Fit the centrifuge tube with a one-holed rubber stopper containing a short piece of glass tubing connected by rubber tubing to an aspirator or to a fume trap (see Fig. 24.1). If using the funnel system, be sure the system cannot "back up" and flood the reaction vessel by keeping the funnel just below the surface of the water in the beaker.

Using an automatic pipet or a buret, add rapidly 1.20 mL of chlorosulfonic acid to the cooled acetanilide, immediately replace the rubber stopper, and cool the tube in an ice bath. Remove the centrifuge tube and allow it to warm slowly to room temperature. If the reaction becomes too vigorous, replace the tube in the ice bath. When most of the solid has dissolved, warm the tube on a steam bath and continue heating it for 10 minutes after the solid has dissolved. Cool the tube, and in small increments, add 2 g of ice (with care). After each addition, replace the trap and shake the tube gently. Allow the ice to melt, disconnect the trap, and remove the supernatant liquid using a Pasteur pipet containing a cotton plug. Wash the precipitate with two 2-mL portions of cold water, each time removing the liquid with a pipet. Continue with Part B immediately since the product is sensitive to moisture and slowly decomposes on exposure to air.

B. SULFANILAMIDE (Microscale)

Rinse the stopper and trap to remove any residual HCl generated in *Part A*. Add 3.5 mL of concentrated ammonium hydroxide to the solid in the tube. Turn on the aspirator or replace the trap liquid in the beaker with dilute HCl to remove ammonia fumes and shake the entire apparatus to ensure that the reaction is proceeding. When no lumps of the sulfonyl chloride remain, warm the tube in a hot water bath for a few minutes and then cool it in ice. Remove the supernatant liquid using a Pasteur pipet, wash the solid with two 2-mL portions of cold water, and remove as much of the water as possible.

Add 0.6 mL of 6 *M* HCl to the solid in the tube and reassemble the trap to catch the acid fumes. Heat to boiling (use either a sand bath, aluminum heating block or a burner) until all of the solid has dissolved, and then boil for another 5 minutes. Cool the solution, add decolorizing carbon (if necessary), and filter through a cotton plug into a small Erlenmeyer flask. Add solid sodium bicarbonate in small portions to the filtrate (**CAUTION:** foaming results) until the pH is just neutral. Return the sample to the centrifuge tube and remove the supernatant liquid using a pipet fitted with a cotton plug. Wash the solid with two 1-mL portions of cold water, filter through a Hirsch funnel, and allow the product to air dry. If needed, the solid can be recrystallized from hot water. Determine the melting point and percent yield; submit the product for grading.

QUESTIONS

1. What product would be expected if aniline rather than acetanilide were treated with chlorosulfonic acid?
2. What is the compound in solution after boiling the acetamidosulfonamide in hydrochloric acid?
3. What would be the result if excess sodium hydroxide solution were used for the final neutralization after acid hydrolysis? (Hint: see page 318.)
4. Write a balanced equation for the overall reaction of acetanilide and chlorosulfonic acid and indicate the mechanism of the reaction(s).
5. Which enzyme system of a bacterium is affected by sulfanilamide (i.e., what is the mechanism of its biological activity)?
6. Prontosil is actually an orange-red dye which was found to be active in the treatment of patients with infections. What structural features of Prontosil impart its color?

Name ________________________ Section ______________ Date ________

38 Synthesis of Sulfanilamide

PRELABORATORY QUESTIONS

1. What is meant by the term "sulfa drug"?

2. Write the structure of the reactants and products in the synthesis of *p*-acetamidobenzenesulfonyl chloride from chlorosulfonic acid and acetanilide.

3. Write the mechanism for the hydrolysis of the carboxamide portion of the *p*-acetamidobenzenesulfonamide to give sulfanilamide.

4. What is the purpose of adding sodium bicarbonate after the hydrolysis reaction?

39

Resolution of α-Phenylethylamine

Chiral molecules exist as two mirror-image isomers, or **enantiomers**; that is, they exhibit handedness. Enantiomers are identical with respect to most physical and chemical properties such as melting and boiling points, solubilities, and reactivities with symmetrical reagents. An important exception is the behavior of enantiomers when placed in a beam of polarized light. When passing through samples of the individual enantiomers, the plane of polarization is rotated in equal but opposite directions. This phenomenon is termed **optical activity**, and enantiomers are sometimes called **optical isomers**.

The compound α-phenylethylamine contains one asymmetric carbon atom and is chiral. The commercial material, obtained by reductive amination of acetophenone, is optically inactive. The rotation is zero because the synthetic product is a **racemic mixture**; that is, both enantiomers are present in exactly equal amounts. Optical activity can be observed, of course, only when one of the enantiomers is present in excess over the other. The separation of a racemic mixture into its individual enantiomers is called resolution.

Two general ways of obtaining pure enantiomers are resolution of a racemic mixture by separation of diastereomeric derivatives and by the selective formation or removal of one enantiomer using a chiral reagent. The second method includes enzymatic resolutions; however, much simpler chiral reactants also can be used. The formation and separation of diastereomers is the most generally useful method and will be used in this experiment to resolve α-phenylethylamine.

When a racemic amine is treated with one enantiomer of a chiral acid, the resulting salts are diastereomers. If the two enantiomers of the amine are designated (+) and (−), and the acid is designated (+), the salts are (+) (+) and (−) (+); this is seen in Equation 39.1.

$$\left.\begin{matrix}(+)RNH_2\\ \\(-)RNH_2\end{matrix}\right. + (+)R'CO_2H \longrightarrow \begin{matrix}(+)RNH_3^+(+)R'CO_2^-\\ \\(-)RNH_3^+(+)R'CO_2^-\end{matrix} \qquad \textbf{[39.1]}$$

racemic amine — diastereomeric salts

Since the compounds no longer have a mirror-image relationship, they have different physical properties. If a solvent can be found in which one salt is more soluble than the other, they can be separated by fractional crystallization.

In this experiment the diastereomeric salts are obtained using tartaric acid, a chiral compound whose (+) enantiomer is produced as a by-product in wine making. The hydrogen tartrate obtained from the (+) acid and (−) α-phenylethylamine is less soluble in methanol than the (+) (+) salt, and the (+)(−) salt crystallizes in nearly pure stereoisomeric form. After separation of the pure diastereomeric salt, the amine enantiomer is obtained by treating the salt with excess strong base (Eq. 39.2).

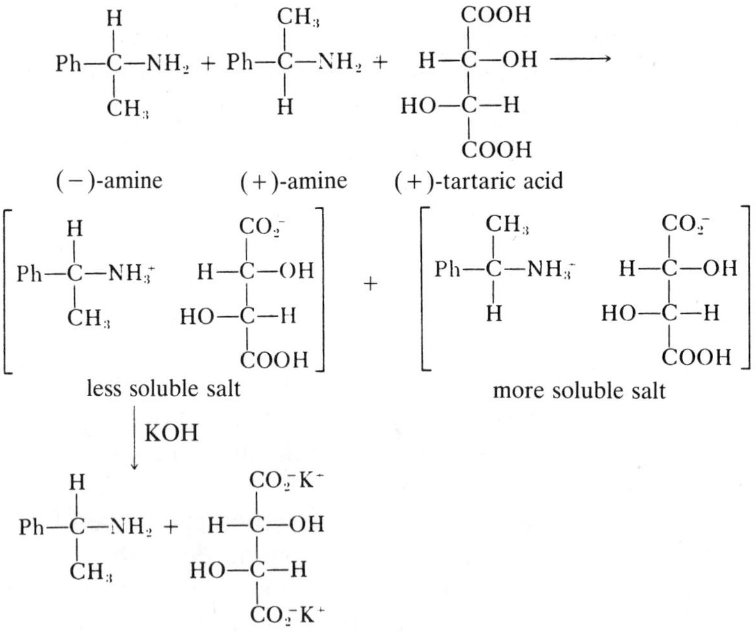

$$\underset{(-)\text{-amine}}{Ph{-}\overset{H}{\underset{CH_3}{C}}{-}NH_2} + \underset{(+)\text{-amine}}{Ph{-}\overset{CH_3}{\underset{H}{C}}{-}NH_2} + \underset{(+)\text{-tartaric acid}}{\begin{matrix}COOH\\ H{-}C{-}OH\\ HO{-}C{-}H\\ COOH\end{matrix}} \longrightarrow \qquad \textbf{[39.2]}$$

$$\underset{\text{less soluble salt}}{\left[Ph{-}\overset{H}{\underset{CH_3}{C}}{-}NH_3^+ \quad \begin{matrix}CO_2^-\\ H{-}C{-}OH\\ HO{-}C{-}H\\ COOH\end{matrix}\right]} + \underset{\text{more soluble salt}}{\left[Ph{-}\overset{CH_3}{\underset{H}{C}}{-}NH_3^+ \quad \begin{matrix}CO_2^-\\ H{-}C{-}OH\\ HO{-}C{-}H\\ COOH\end{matrix}\right]}$$

$$\downarrow \text{KOH}$$

$$Ph{-}\overset{H}{\underset{CH_3}{C}}{-}NH_2 + \begin{matrix}CO_2^-K^+\\ H{-}C{-}OH\\ HO{-}C{-}H\\ CO_2^-K^+\end{matrix}$$

THE MEASUREMENT OF OPTICAL ACTIVITY

The rotation of plane polarized light is measured with a relatively simple instrument called a polarimeter, shown schematically in Figure 39.1. Proceeding from left to right, the randomly oriented light from an incandescent

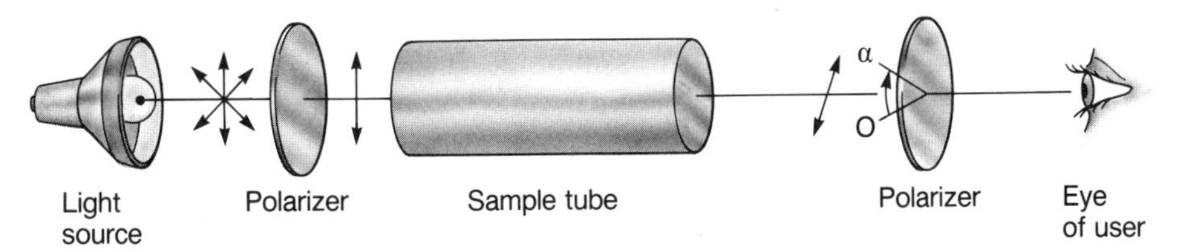

Figure 39.1 Schematic diagram of a polarimeter with a sample giving $\alpha \sim +30°$. Arrows indicate the direction of polarization of the light beam.

lamp is polarized by passing through a Nicol prism (calcite) or a sheet of Polaroid film. The light beam then passes through a sample tube containing a solution of the compound being analyzed, and if the compound is a liquid and a large enough sample is available, the neat liquid may be used. If the tube contains an optically active compound, the plane of the polarized light is rotated as it passes through the sample. The analyzer, another polarizer whose orientation is adjustable, is next in the path of the light. In using some polarimeters, the analyzer is rotated until it completely blocks the polarized light reaching it. A scale calibrated in degrees of arc is attached to the analyzer, and the amount of rotation is read from this scale. In a properly calibrated polarimeter, the reading will be 0° if there is no optically active sample in the tube. By convention, if it is necessary to rotate the analyzer clockwise (from the user's viewpoint) to cause extinction of the light, the rotation is designated positive, (+); counterclockwise rotation is designated (−).

The observed rotation of a substance depends on the structure of the compound and the number of molecules in the light path between the polarizers, the latter factor being a function of the path length and the concentration. A quantity termed **specific rotation**, [α], is defined as follows:

$$[\alpha] = \frac{\alpha}{lc}$$

where α is the observed rotation, l is the path length in decimeters, and c the concentration in g/mL. In addition, the temperature, solvent, and wavelength of the light used are variables that affect optical rotation and must be specified in reporting the value. Thus the specific rotation of (+)-tartaric acid has been reported as $[\alpha]_D^{20}$ + 12.0° (c = 20 g/100 mL in H_2O). The subscript D indicates that the wavelength of light used was that of the yellow D line from a sodium vapor lamp; that is, 589 nm, and the superscript 20 indicates that the measurement was made at 20°C. The concentration (in g/100 mL) is also reported, since the rotation is exactly proportional to concentration only in dilute solutions.

Commercial polarimeters come in a variety of models. The major differences are in the type of polarizers and the accuracy with which the positions of the analyzers can be measured. In models designed to be used with a light source that is not monochromatic, a filter or filters are incorporated in the light path. Provisions may also be made for controlling the temperature of the sample. Some polarimeters are available that automatically adjust the polarizers and provide a readout of the rotation on a meter or send data to a computer; a photocell replaces the user's eye in such an instrument. The accuracy obtained with the most expensive instruments is ±0.001 to 0.002°.

An inexpensive commercial polarimeter that is adequate for this experiment (accuracy ± 0.5°) is illustrated in Figure 39.2. Light from a strong flashlight or high-intensity desk lamp is reflected by the mirror up through a sheet of Polaroid films and through the sample, contained in a flat-bottomed tube. The polarized light then passes through a yellow filter to a second polarizer, which is rotated until the observed field reaches maximum darkness.

Experiment (Macroscale)

Procedure

Dissolve 11.9 g of (+)-tartaric acid in 165 mL of methanol in a 250-mL Erlenmeyer flask by heating the mixture on a steam bath. Slowly add 10.0 mL of racemic α-phenylethylamine to the warm solution while swirling the flask to mix well. Salt formation is exothermic and the solution will become fairly hot.

The desired salt crystallizes slowly as large clear prisms. The first crystals obtained are often fine white needles instead of prisms; these needles do not provide optically pure amine (see Question 2) and must be avoided. To ensure that the desired prisms form, seeds of the correct composition are needed. To obtain the seeds, transfer approximately 1 mL of the hot solution to a test tube and concentrate this solution to approximately half the original volume by heating on the steam bath. As soon as crystals form in the boiling solution, pour the suspension of crystals back into the flask containing the remainder of the solution. Cork and label the flask and set it aside until the next laboratory period.

After standing at least 24 hours, decant the solution from the crystals. (The solution can be processed for eventual recovery of the (+)-amine.) Add 10 mL of methanol to the flask containing the crystals and break the mass up into individual crystals with a stirring rod.

Using a small Büchner funnel and suction flask, collect the crystals. Wash any remaining crystals out of the flask with a little of the filtrate and rinse the crystals on the funnel with 10 mL of cold methanol. Spread the crystals on paper to dry, weigh the product, and calculate the percentage yield of (–) amine (+) hydrogen tartrate.

To recover the amine, place the crystalline salt in a 125-mL Erlenmeyer flask. Add 50 mL of 2 *M* (8%) aqueous NaOH solution (or use 40 mL of water plus 5 mL of 50% aqueous NaOH). Swirl the mixture until all of the crystals have dissolved. Transfer the mixture to a separatory funnel and rinse the flask with 10 mL of methylene chloride. Add the rinse to the separatory funnel, stopper the funnel, and shake well. Remove the methylene chloride layer into a clean, dry, 50-mL Erlenmeyer flask. Extract the aqueous phase twice more with 5-mL portions of CH_2Cl_2 and combine these extracts with the first. Dry the solution of the amine by adding approximately a gram of granular anhydrous K_2CO_3, swirling, and letting it stand for a few minutes. Remove the K_2CO_3 by filtering the solution through a cotton plug into a tared 25 × 100-mm test tube. Add a boiling stone, and carefully warm on a steam bath to evaporate most of the solvent. This operation should be carried out in a fume hood, or a tube can be connected to the aspirator vacuum and inserted in the neck of the tube to remove the vapors as they distill (Fig. 4.4). When nearly all of the methylene chloride has evaporated as indicated by cessation of boiling, remove the remainder of the solvent under reduced pressure (Fig. 4.5a) while heating the tube gently on the steam bath. Remove the boiling stone, weigh the test tube and its contents, and calculate the yield of amine.

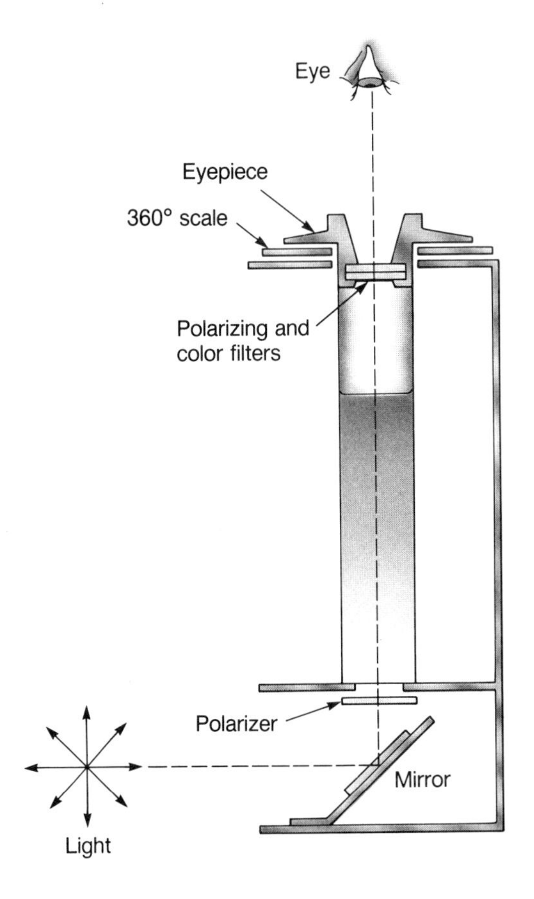

Figure 39.2 Diagram of a I^2R polarimeter.

To determine the optical purity of the amine, transfer it with a pipet to a polarimeter tube. Measure the length of the liquid column to the nearest millimeter (see Fig. 39.2), and measure its rotation, α. The rotation can also be measured in an appropriate solvent such as methylene chloride. The mechanics of filling the tube and measuring the rotation will depend upon the type of polarimeter available. Consult your instructor for specific directions. Calculate the specific rotation of the amine using the formula $[\alpha] = \alpha/cl$. Note that the path length is specified in dm and that the concentration of a neat liquid is equal to its density (0.97 g/mL in this case). The reported specific rotation of pure (−) α-phenylethylamine is $[\alpha]_D^{25} - 40.4 \pm 0.2°$ (neat). Calculate the optical purity of your sample (see Question 3).

QUESTIONS

1. If a pure (+) enantiomer has $[\alpha]_D + 100°$, what will be the rotation of a mixture containing 80% of the (+) enantiomer and 20% of the (−) isomer?
2. The α-phenylethylamine recovered from the hydrogen tartrate salt that crystallizes as needles was found to have a specific rotation of $-20 \pm 1°$. What is the percent composition of this mixture of (+) and (−) enantiomers?
3. Optical purity is defined as the ratio of the measured specific rotation to the specific rotation of the pure enantiomer. Calculate the optical purity of the amine described in Question 2.
4. Suggest how the procedure given in this experiment could be expanded to provide a product of higher optical purity.
5. Describe with equations how the (−)-α-phenylethylamine obtained in this experiment could be used to resolve racemic 2-phenylpropionic acid.

Name ______________________ Section ____________ Date ________

39 Resolution of α-Phenylethylamine

PRELABORATORY QUESTIONS

1. Draw Fischer projection formulas for R- and S-phenylethylamine.

2. Draw the structure of (–)-tartaric acid, that could be used to resolve the enantiomers of α-phenylethylamine.

3. Write an equation to illustrate the recovery of an amine from one of its salts.

4. What is meant by the term "plane-polarized light"?

5. How could an enzyme be used to separate a pair of enantiomers?

6. What are the units for the components, l and c, used to calculate specific rotation?

7. Describe how you could use the observed specific rotation of an aqueous sucrose solution to calculate the sucrose concentration.

40

Chemistry of Sugars

Sugars make up a specialized but important branch of organic chemistry. Because of their polyfunctional structures, sugars undergo a broad variety of reactions, including some complex rearrangements. In addition to transformations of sugars with acids, bases, and other simple reagents, enzyme-catalyzed reactions play a prominent role in carbohydrate chemistry, and sugars are good substrates for studying enzymatic processes.

Experimental work with sugars involves some special problems and techniques. Sugars are extremely soluble in water and only very slightly soluble in the usual organic solvents. They are complex and sensitive molecules and cannot be distilled or sublimed. Most sugars can be crystallized, although sometimes with great difficulty. One of the major problems in carbohydrate work is the characterization and differentiation of closely related sugars in aqueous solutions. To this end, a number of color reactions have been developed for detection and analysis of specific types of sugars.

SOME REACTIONS OF SUGARS

Monosaccharides (simple sugars) exhibit reactions of both hydroxyl and carbonyl groups plus others that depend upon interactions between these groups. Aldohexoses exist as equilibrium mixtures of cyclic hemiacetals with a negligible amount of the aldehyde tautomer present, but carbonyl reactions occur readily since the interconversion of these forms is rapid.

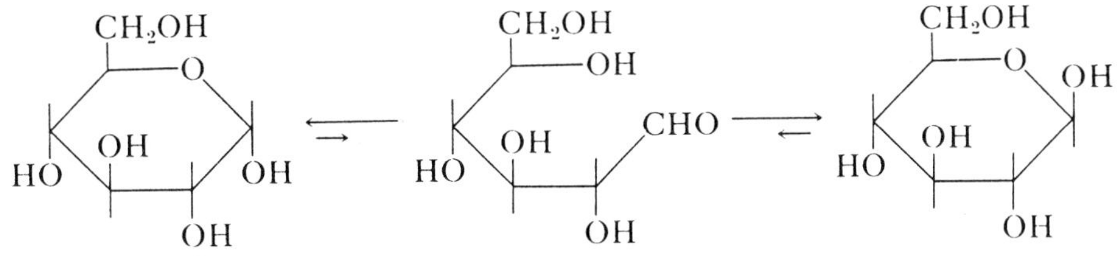

α-D-glucopyranose ⇌ aldehyde form of D-glucose ⇌ β-D-glucopyranose

Tests for reducing properties, condensations with amines, and certain TLC color reagents involve the reaction of the carbonyl group of aldose or ketose sugars. When the OH group of the hemiacetal is replaced by OR, as in glycosides, these carbonyl reactions do not occur unless the acetal is hydrolyzed under the conditions of the test. In most disaccharides, such as maltose or lactose, one hexose unit is tied up as an acetal or glycoside linkage but the second unit has an aldehyde $\rightleftharpoons$ hemiacetal group. An important exception is sucrose, in which the carbonyl groups of both hexose units are linked together as a "double acetal."

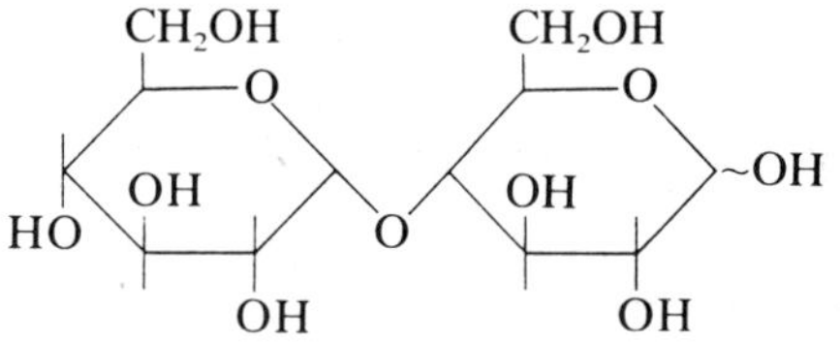

maltose

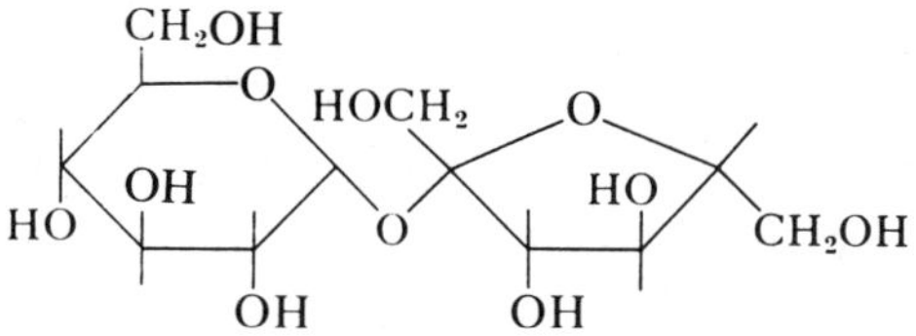

sucrose

Reducing Properties

A characteristic property of glucose and other sugars with a "free aldehyde" or hemiacetal group is oxidation by alkaline copper (II) ion. The blue cupric ion is reduced to red Cu_2O, and this reaction, with various modifications, is a general method for the detection and quantitative clinical analysis of reducing sugars. The standard reagent for qualitative work is Benedict's solution, which contains Na_2CO_3 and citrate ion to complex the cupric ion.

$$RCHO + 2Cu^{2+} + 5OH^- \rightarrow RCO_2^- + Cu_2O + 3H_2O$$

Reactions in Strong Acid

Disaccharides such as maltose and sucrose are hydrolyzed by warming with aqueous acid. Monosaccharides are relatively stable in acidic solution, but a series of dehydration reactions occurs when they are heated in strong acid, ultimately leading to furan derivatives. The process is carried out on an industrial scale to obtain furfural from pentose sugars found in grain chaff and corn cobs. These reactions occur most readily with sugars that exist in the furanose form, such as pentoses and 2-ketohexoses. Subsequent condensation of the furfural with a phenolic compound gives colored products that are useful for characterization. A typical example is the reaction of fructose with hydrochloric acid and resorcinol, called Selivanoff's reagent. A deep red color develops, which can be used to estimate the amount of fructose in a solution.

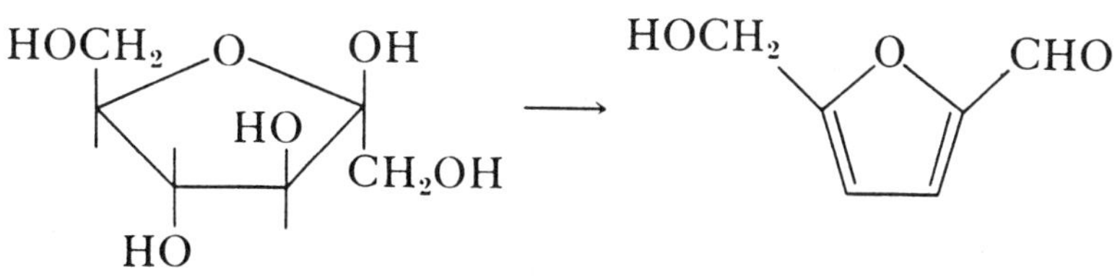

HOCH$_2$ O OH
HO
CH$_2$OH
HO
→
HOCH$_2$ O CHO

β-D-fructofuranose 5-hydroxymethylfurfural

Reactions in Base

In basic solution an equilibrium is established, by means of an enediol, between aldose and ketose sugars. When glucose is warmed briefly with dilute base, a mixture of glucose, fructose, and a small amount of mannose results. Further reactions then occur to give dark, extremely complex mixtures of other sugars together with saccharinic acids. Several types of acid have been isolated; these products arise from further enolization and rearrangement steps.

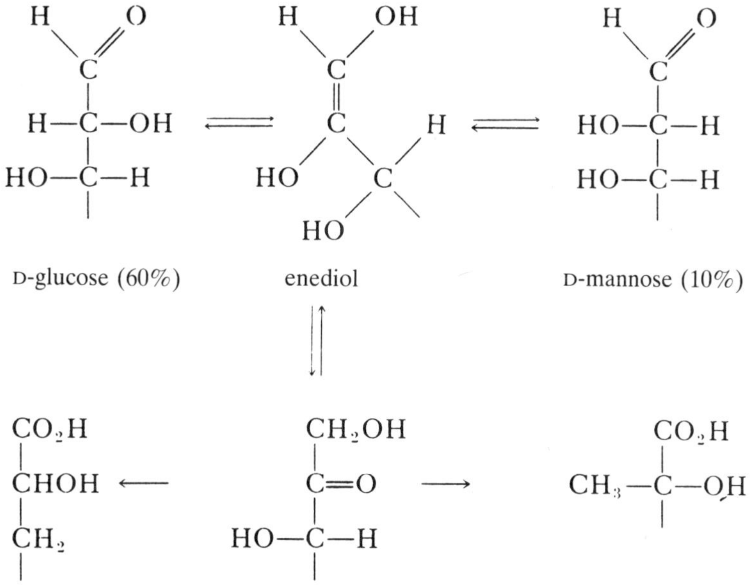

H O
C
H—C—OH
HO—C—H
⇌
H OH
C
C H
HO C
HO
⇌
H O
C
HO—C—H
HO—C—H

D-glucose (60%) enediol D-mannose (10%)

⇅

CO$_2$H
CHOH
CH$_2$
←
CH$_2$OH
C=O
HO—C—H
→
CO$_2$H
CH$_3$—C—OH

metasaccharinic acid D-fructose (30%) glucosaccharinic acid

Reactions with Amines

The reaction of sugars with ammonia or amines gives *N*-glycosyl compounds in which the hemiacetal —OH group is replaced by nitrogen. Derivatives of this type are intermediates in the biosynthesis of nucleosides and nucleic acids. Enolization of these *N*-glycosides occurs with a trace of acid, and a 1-amino-1-deoxyketose results; the overall reaction is called the **Amadori rearrangement**.

aldohexose ⇌ $\xrightarrow{RNH_2}$ N-glycoside $\xrightarrow{H^+}$ ... $\xrightarrow{-H^+}$ ... ⇌ 1-amino-1-deoxyketose

A traditional reagent for the characterization of sugars is phenylhydrazine. Excess reagent acts as a selective oxidizing agent for the adjacent OH group and the product is the yellow osazone, which is obtained from either of two epimeric aldoses or from the 2-keto sugar. Due to the presence of two aromatic rings, osazones of monosaccharides are only slightly soluble in water and provide derivatives that can be readily isolated from dilute aqueous solutions.

$\xrightarrow{3C_6H_5NHNH_2}$ $CH{=}N{-}NHC_6H_5$ / $C{=}N{-}NHC_6H_5$ (osazone) $+ C_6H_5NH_2 + NH_3 + 2H_2O$

Experiments

SAFETY NOTE SOME OF THE REAGENTS IN THIS EXPERIMENT ARE STRONGLY ACIDIC OR BASIC AND SHOULD BE HANDLED WITH CARE. PHENYLHYDRAZINE IS TOXIC AND A SUSPECTED CARCINOGEN; USE IT WITH PARTICULAR CAUTION. IF YOU GET ANY ON YOUR SKIN, WASH IT OFF IMMEDIATELY.

A. THIN LAYER CHROMATOGRAPHY OF SUGARS

Chromatographic methods are widely used in the separation and isolation of sugars. Qualitative examination of mixtures by chromatography on paper strips or TLC plates is an indispensable tool. Thin layer chromatography provides somewhat less resolving power than paper, but it is faster and more convenient. Since sugars are extremely polar compounds, mixtures of solvents such as butanol, acetic acid, and water are needed to obtain migration. Visualization of compounds on the developed chromatogram is accomplished by spraying with color-forming reagents. Combinations of aromatic amines and acids result in condensation products that have colors more or less characteristic of certain types of sugars. Nonreducing sugars such as sucrose usually yield much weaker color reactions and are more difficult to visualize.

Procedure

Commercial silica gel G strips on plastic backing should be used; plates prepared by the dipping procedure (Chapter 7) are not satisfactory for sugars. For applying samples, use the capillary tubes and the same general technique described in Chapter 7.

In 10 × 75-mm test tubes, obtain 10- to 20-mg samples of arabinose, fructose or glucose, maltose, and a commercial syrup such as Karo® or honey. Dissolve the sugars in 1 drop of water, add 0.2 mL of methanol, and shake or stir to mix the heavy aqueous syrup and solvent before spotting. Apply spots no larger than 1 mm in diameter on a 1-inch wide strip; four samples can be applied to each strip. Allow the strip to dry for 2 to 3 minutes after the samples have been applied. Develop the chromatogram in a solvent mixture of 1-butanol:acetic acid:ether:water in a volume ratio of 9:6:3:1. Approximately 20 minutes is required for development.

Dry the strips in air, and spray with a solution of *p*-anisidine and phthalic acid in ethanol, or a packaged aniline/phthalic acid aerosol spray. For spraying, an aspirator such as that shown in Figure 40.1 is used to provide a fine mist of the reagent solution. Attach the strip by one corner with Scotch tape to a large piece of cardboard and place it in the hood for spraying. Hold the aspirator tip about 6 inches from the strip and move it back and forth. Apply just enough pressure to the aspirator to produce a fine mist; the strip should not be soaked. Place the strip in an 80° oven or warm briefly over a hot plate to permit visualization of the spots.

B. CHARACTERIZATION REACTIONS

Procedures are given for three reactions that are useful in characterizing sugars. Arabinose, fructose, glucose, maltose and sucrose are representative sugars that can be used to illustrate these tests; samples of corn syrup

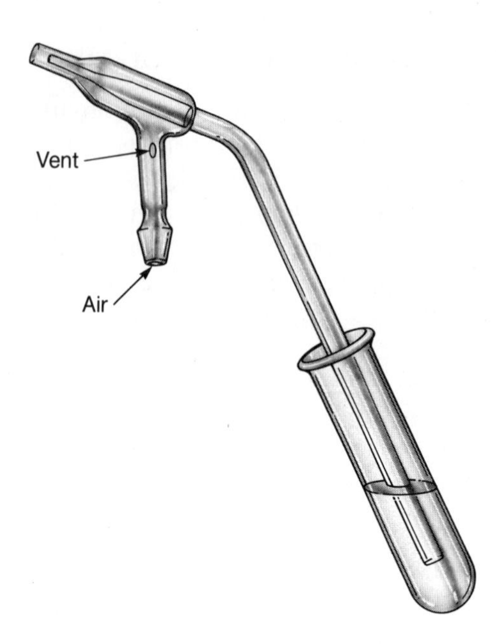

Figure 40.1 Aspirator for spraying TLC strips (Ace Glass Co.).

(Karo®) and honey can also be examined if desired. Obtain samples of a few sugars and/or unknowns as directed by your instructor, apply the tests, and account for your observations.

Benedict's Reagent for Reducing Properties

(The reagent solution contains 100 g of Na_2CO_3, 175 g of sodium citrate, and 17.3 g of $CuSO_4 \cdot 5\ H_2O$ per liter.) In a 13 × 100- or 18 × 150-mm test tube, dissolve approximately 10 mg of the sugar in a few drops of water (or use 5 drops of a 2% solution). For the corn syrup or honey, dilute 1 mL of the syrup to 50 mL and use 5 drops.

To each tube, add 3 mL of Benedict's solution and stir or shake to mix the solutions. Label the tubes with a marker or self-stick label that will not come off in hot water. Place all the tubes in a boiling water bath at the same time and remove them after exactly 2 minutes. The amount of reducing sugar in the various samples is best compared before the red cuprous oxide settles.

Selivanoff's Reagent (Furfural Formation)

(The solution contains 50 mg of resorcinol in 33 mL of hydrochloric acid diluted to 100 mL with water.) Empty the tubes from the Benedict test, rinse out the Cu_2O residue using a brush, and add samples of sugars as above. To each tube, add 2 mL of Selivanoff's reagent. Mix and heat in the boiling water bath for 2 minutes and compare the appearance of the tubes. The intensity of the red color is proportional to the fructose content.

Phenylhydrazine

(The reagent contains 60 g of $CH_3CO_2Na \cdot 3H_2O$ and 40 g of phenylhydrazine hydrochloride in 400 mL of water.) In 18 × 150-mm test tubes with hot water–resistant labels, dissolve 50 mg of the sugars to be tested in 2 mL of water (or use 2 mL of a 2% solution.) Add 5 mL of the phenylhydrazine solution and place the tubes in a boiling water bath. Keep the tubes in the bath for 15 minutes or until a precipitate appears. The osazone that precipitates can be collected and dried, and the melting point determined.

C. ENOLIZATION OF GLUCOSE

The objective of this experiment is to follow the reaction of glucose in base by use of the Selivanoff and Benedict reagents. In an experiment of this type, the course of reactions can be monitored by sampling the reaction mixture at intervals and observing changes in tests on successive samples. Relative amounts of substances present can be estimated by the intensity of a color reaction without determining absolute concentrations. In this approach, it is important that all samples are the same size and are handled in the same way.

Procedure

Arrange two series of test tubes (hot water–resistant) labeled 0, 5, 10, 20, and 30. To 10 mL of a 2% solution of glucose in water, add 1 mL of 2 *M* NaOH solution. Swirl to mix the solution, draw up a sample with a pipet and bulb, and place 5 drops of the solution in each of the tubes labeled 0. Place the glucose solution in a water bath at 70°C and, after 5-, 10-, 20-, and 30-minute intervals, transfer 5-drop samples to each tube as labeled. After taking the last samples, note the odor of the solution.

To one series of samples, add 2 mL of Benedict's reagent to each tube; to the second series add 2 mL of Selivanoff's reagent to each tube. Shake the tubes to mix the contents, and immerse all the tubes in each series at the same time in a boiling-water bath. After 2 minutes, remove the tubes and line them up according to reaction time. Record the intensities of the color or the amount of precipitate in the successive samples in each series and interpret the results. You should observe a rapid buildup and then a slower decrease in one series, but not in the other (Question 4).

D. *N*-*p*-TOLYL-D-GLUCOSYLAMINE AND *N*-*p*-TOLYL-1-AMINO-1-DEOXY-D-FRUCTOSE

The preparation of these derivatives illustrates the effect of minor changes in reaction conditions and provides an exercise in the art of crystallization. Both compounds form very readily; isolation is the challenge.

Procedure

Weigh out 2 g of D-glucose and 1.6 g of colorless *p*-toluidine, and mix the compounds in an 18 × 150-mm test tube. Add 0.2 mL of water and 0.2 mL of ethanol, and heat the mixture in a steam or hot-water bath. Shake the mixture intermittently while heating until a single liquid phase is present, and then transfer half of the solution to another tube.

To one tube, add 0.1 mL (2 to 3 drops) of 2 *M* (10%) acetic acid. Heat the tube in steam or boiling water for 20 minutes and add 10 mL of ethanol. To the other tube, add 10 mL of ethanol. Place a drop or two of each solution in a small test tube and cork the main solutions. Use the smaller samples to initiate crystallization. Evaporate the sample to a syrup, scratch with a rod, and add a drop of ethanol. Repetition of these steps and a few days' standing may be needed before crystals form. When crystals are obtained, seed the solution by rubbing a seed crystal against the wall at the surface of the solution. Collect the crystalline products, wash with a little ethanol, dry, and determine the melting points.

The melting point reported for *N-p*-tolyl-D-glucosamine is 109 to 113°C; the melting point reported for *N-p*-tolyl-1-amino-1-deoxy-D-fructose is 150 to 152°C.

QUESTIONS

1. Calculate the weight of glucose required to completely reduce 3 mL of Benedict's solution (17.3 g of $CuSO_4 \cdot 5H_2O$ per liter).
2. Raffinose (a trisaccharide occurring in cotton seeds), α,α-trehalose (occurring in yeast and various insects), and gentiobiose (present in several glycosides) have the structures shown below.
   - **a.** Which of these sugars would reduce Benedict's solution?
   - **b.** Which would give a positive test for fructose with Selivanoff's reagent?
   - **c.** How many different osazones would be formed after acid hydrolysis of each compound followed by reaction with phenylhydrazine?

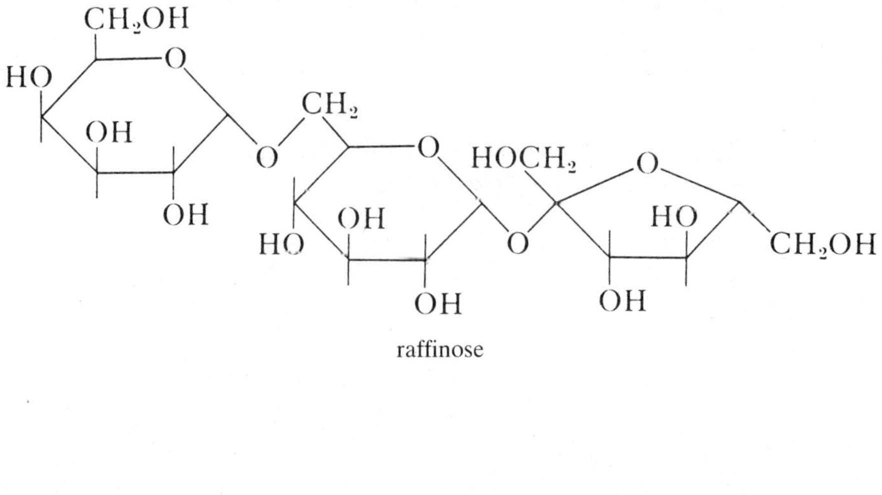

raffinose

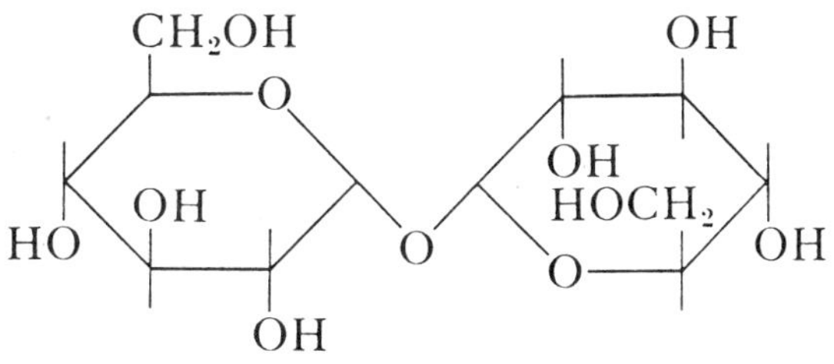

α,α-trehalose

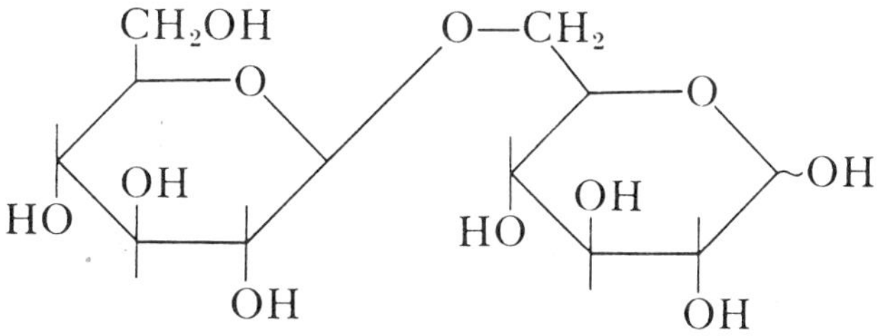

gentiobiose

3. Honey contains sucrose, fructose, and glucose. The Benedict test can be adapted for the quantitative analysis of reducing sugars. With suitable modifications and instruments, the Selivanoff test can be used for the determination of fructose plus sucrose in the presence of glucose. Outline a strategy by which the three sugars in honey could be determined using these two methods.
4. Account for the changes observed on treatment of D-glucose with base, and outline the reactions that occur.

Name ______________________ Section ______________ Date ________

40 Chemistry of Sugars

PRELABORATORY QUESTIONS

1. Write a balanced equation for the reaction of glucose with Benedict's solution (assume only the aldehyde group in glucose is oxidized, although the oxidation of α-hydroxyaldehydes in basic solution is more complex than this).

2. What is the color of Benedict's solution, that is, of the Cu(II) ion, complexed with citrate? What change of colors takes place as this reagent is reduced to Cu_2O?

3. a. Write a reaction depicting the formation of the osazone of D-glucose.

b. Write a reaction to show the formation of the osazone of D-fructose. How does it compare with the product in Question 3a?

41

Hydrolysis of Starch

Starch and cellulose are polymers formed by linking glucose units through $1 \rightarrow 4'$ glycoside bonds. These polysaccharides make up 60 to 80% of the dry weight of most plants. Because of the abundance of these polymers, their detailed structure and breakdown have been extensively studied.

The basic structural difference between starch and cellulose is the configuration of the glycoside bond, which is α in starch and β in cellulose. Cellulose has a regular structure and very high molecular weight. Naturally occurring cellulose is partially crystalline and forms strong fibers.

Starch, the main storage polysaccharide in grains and other food plants, is a mixture of two types of polymers. One fraction, amylose, usually approximately 25% of the total, is made up of unbranched chains of 250 to 300 D-glucose units linked by $1\alpha \rightarrow 4'$ glycoside bonds. The amylose chains are coiled into a helical structure. The cavity of the helix is the proper size to permit the entrapment of iodine molecules, and addition of iodine to starch gives an intense deep blue color. Amylopectin, the major component of most starch, consists of similar chains with side branches extending from the C—6 CH_2OH group at approximately every twenty-fifth glucose unit, giving a bush-like structure.

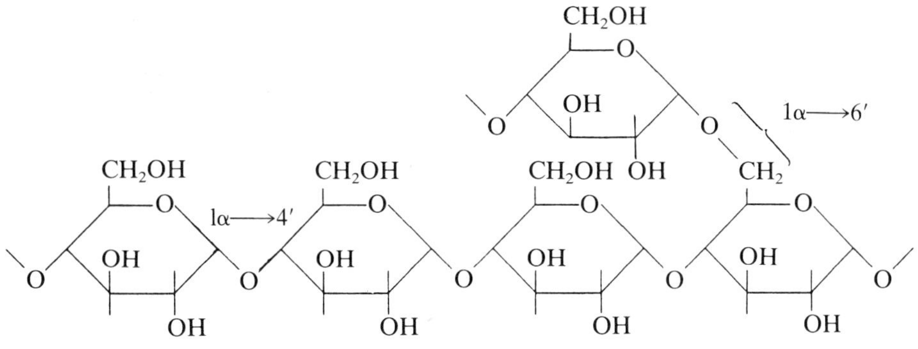

non-reducing end

Amylopectin (partial structure)

In digestion or in fermentation to produce alcohol, starch must be broken down to glucose units by hydrolysis of the α-glycoside bonds. This is accomplished by amylase enzymes present in saliva, digestive fluids, and yeast. There are two main types of amylase enzymes, designated α and β; both are specific for α-glycoside bonds, but they attack the starch chain in different ways. β-Amylase brings about cleavage of two-unit fragments, that is, molecules of the disaccharide maltose, starting from the nonreducing end of the chain. The action of β-amylase is interrupted by the presence of a branch in the chain.

α-Amylase cleaves the chain into fragments of six to seven glucose units very rapidly, causing loss of the iodine-complexing properties. Depending upon the type of α-amylase, further degradation to maltose and maltotriose (the three-glucose trisaccharide), and ultimately to glucose, occurs more slowly. α-Amylase of saliva has very little activity for cleavage of the disaccharide or trisaccharide fragments. Maltose and maltotriose therefore accumulate, but very little glucose is produced.

The breakdown of starch by acid-catalyzed hydrolysis occurs in a more random way, and the course of the reaction is quite different from that of the enzymatic hydrolysis. In this experiment the enzyme and acid-catalyzed reactions will be compared in three ways—iodine complexing, reducing power, and TLC.

Experiments

SAFETY NOTE ALTHOUGH NO PARTICULARLY HAZARDOUS CHEMICALS ARE USED IN THESE EXPERIMENTS, NORMAL CARE SHOULD BE OBSERVED. CAUTION IN THE USE OF BOILING WATER BATHS SHOULD BE TAKEN TO AVOID SPILLS AND/OR BURNS.

Place 0.6 g of soluble starch in a 125-mL Erlenmeyer flask, make a thin paste with a few mL of water, and then add 40 mL of boiling water. Boil the solution gently until all starch particles disappear, and allow the opalescent solution to cool.

A. ENZYMATIC HYDROLYSIS

Place half of the starch solution in a test tube. In each of six 10 × 75-mm test tubes, place 2 drops of 0.002 *M* iodine solution (0.05 g I_2 and 0.15 g KI in 100-mL water). To the first tube, add 0.5 mL of starch solution and note the color; this is the "zero time" sample.

Noting the time, add approximately 0.2 mL of saliva to the starch solution and mix the solution with a pipet. At 30-second intervals, remove

0.5-mL samples of the solution and add them to successive tubes of iodine solution. Compare the color of the successive samples against a white background. Note the time when the iodine solution becomes a pale lavender-brown color. At this point, remove 2 mL of the starch solution with a pipet, transfer it to a dry 18 × 150-mm test tube, and heat the sample to boiling to stop the reaction by denaturation of the enzyme. Remove further samples after 15, 30, and 60 minutes; boil each as it is removed and label with the time.

For TLC examination, use a 1 × 3-inch silica gel strip and apply spots from the samples taken after 15, 30, and 60 minutes in three lanes. In a fourth lane, spot a reference sample of a mixture of glucose and maltose (1% of each in water). The hydrolysis samples are quite dilute, and three or four applications should be made. Place a 1-mm spot of each solution on the strip, and then warm and blow gently over the spots until the white zone (due to water) has almost disappeared; then make another application exactly coinciding with the first. Develop and spray the strips as described in Chapter 40.

B. ACID HYDROLYSIS

After the enzymatic hydrolysis is under way, add 0.5 mL of concentrated hydrochloric acid to the other half of the starch solution. Arrange a series of empty 10 × 75-mm tubes, and then place the starch solution in a boiling-water or steam bath. At 5-minute intervals, place a 0.5-mL sample in one of the test tubes, cool to room temperature, and add 2 drops of iodine solution. (The samples must be cooled first, since the starch–iodine complex dissociates on heating.) Remove 2-mL samples for TLC after 15- and 30-minute heating and also 10 minutes after the iodine–starch test becomes light brown. Add a few drops of 2 *M* NaOH to these samples (until just neutral to indicator paper). Cool the remaining hydrolysis solution and run an osazone test (page 433).

Run a TLC on the neutralized samples, together with a glucose–maltose reference, as described for the enzymatic hydrolysis. (The final sample will be somewhat more concentrated because of heating, and only one or two applications are needed.)

After examining the TLC plates from the two hydrolysis reactions, run another TLC with selected samples from both reactions together on the same strip to make a direct comparison and to confirm conclusions.

To each of the six tubes (three from each hydrolysis) remaining from the TLC samples, add 5 mL of Benedict's solution. Heat the tubes in a boiling-water bath for 2 minutes and compare the amount of reducing sugar in the samples.

QUESTIONS

1. After partial acid hydrolysis and fractionation of the products by chromatography, cellulose gives one disaccharide, whereas the same treatment of starch gives two other disaccharides, one major and one minor. Write structured formulas for the three disaccharides.
2. From the information given in the discussion on the enzymatic hydrolysis of starch and your experimental findings in the two hydrolyses, compare the sequence of events with enzyme and with acid, and the products that are present at various stages in each case.

REFERENCES

Aspinall, G. O., *Polysaccharides*; Pergamon: Elmsford, New York, 1970.

Whistler, R. L. (Ed.), *Methods in Carbohydrate Chemistry*; Academic: New York, 1964; Vol. 4.

Name ______________________ Section ______________ Date ________

41 Hydrolysis of Starch

PRELABORATORY QUESTIONS

1. Cellulose and amylose are polymers of D-glucose. What structural feature makes them different?

2. Amylose and amylopectin, both polymers of D-glucose, are components of vegetable starch. How do they differ in chemical structure?

3. What is meant by the term "alpha glycoside"?

4. Distinguish between the terms "amylose" and "amylase."

5. What structural feature of starch causes the deep blue color when iodine is added?

6. What reagent is used to visualize the TLC strips of the hydrolysis of starch?

7. What types of compounds give a positive Benedict's test?

42

Acid-Catalyzed and Enzymatic Hydrolyses of Sucrose

The disaccharide sucrose, isolated from sugar cane or beets, is the most abundant sugar in nature, and pure sucrose is produced in larger amounts than any other single organic compound, natural or synthetic. Sucrose is a major food, and as with starch (Chapter 41), the first step in the digestion of sucrose is enzymatic hydrolysis to the constituent monosaccharides, glucose and fructose. The reaction is called **inversion** because the optical rotation of a sucrose solution changes from a positive to a negative value during the hydrolysis, due to the large negative rotation of fructose.

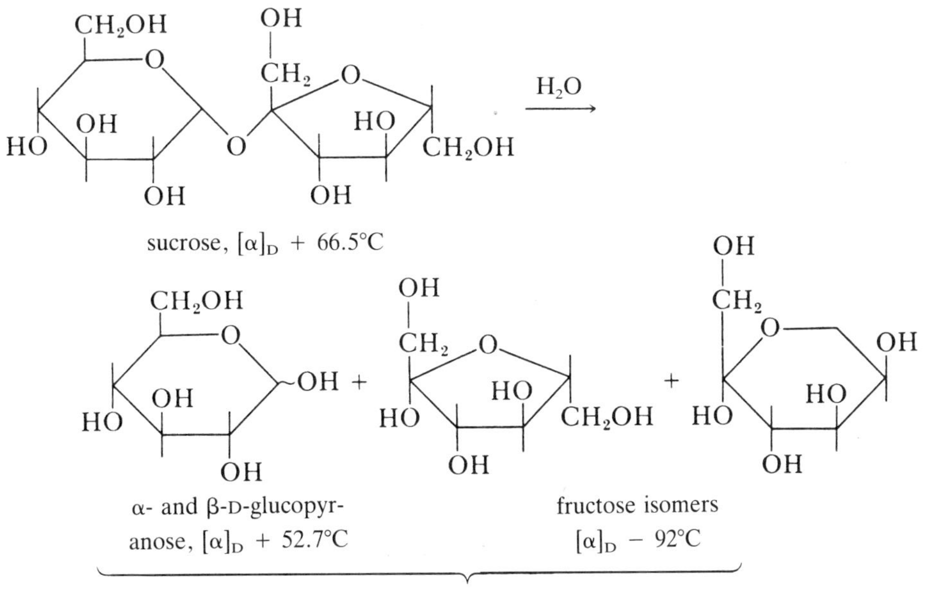

$\xrightarrow{H_2O}$

sucrose, $[\alpha]_D$ + 66.5°C

α- and β-D-glucopyranose, $[\alpha]_D$ + 52.7°C

fructose isomers $[\alpha]_D$ − 92°C

$[\alpha]_D$ − 20.6°C

The hydrolysis of sucrose can also be catalyzed by acid. Since the progress of the reaction can easily be followed from the large changes in optical rotation, sucrose hydrolysis provides a good means of comparing the rates of enzymatic and acid-catalyzed reactions directly. In this experiment the rates of hydrolysis of sucrose with the enzyme invertase and with hydrochloric acid will be measured, and the efficiencies and kinetic laws governing the two reactions will then be compared.

The concentration of sucrose at any time during the hydrolysis can be determined from the optical rotation of the solution, since the **observed rotation** α is directly proportional to the concentrations of the sugars present and their respective specific rotations. From the values $+66.5°$ for pure sucrose and $-20.6°$ for the final equilibrium mixture of glucose and fructose isomers, the rotation after complete hydrolysis is given by $\alpha_\infty = 0.31\alpha_0$. The concentration of sucrose can be expressed in terms of rotations, as shown in Equation 42.1.

$$\frac{C_{suc}}{C_{suc_0}} = \frac{\alpha - \alpha_\infty}{\alpha_0 - \alpha_\infty} = \frac{\alpha + 0.31\alpha_0}{1.31\alpha_0} \quad \textbf{[42.1]}$$

Experiments

A. ACID-CATALYZED HYDROLYSIS OF SUCROSE

The acid-catalyzed hydrolysis of sucrose is a second-order reaction; that is, the rate of the reaction is proportional to the concentrations of sucrose and H_3O^+ (Eq. 42.2). Since H_3O^+ is not consumed by the reaction (i.e., its concentration is constant), the hydrolysis is kinetically a pseudo-first-order reaction (Eq. 42.3). Integration of Equation 42.3 from time = 0 to time = t gives Equation 42.5, which states that the logarithm of the sucrose concentration decreases linearly with time. From the relationship between sucrose concentration and rotation, the rate can then be expressed in terms of rotation, as shown in Equation 42.6.

$$\text{Rate} = d[\text{sucrose}]/dt = -k_2[\text{sucrose}][H_3O^+] \quad \textbf{[42.2]}$$

$$d[\text{sucrose}]/dt = -k_1[\text{sucrose}] \quad \textbf{[42.3]}$$

$$k_1 = k_2[H_3O^+] \quad \textbf{[42.4]}$$

$$\ln\frac{[\text{sucrose}]}{[\text{sucrose}]_0} = -k_1 t \quad \textbf{[42.5]}$$

$$\ln\frac{\alpha + 0.31\alpha_0}{1.31\alpha_0} = -k_1 t$$

or

$$\log(\alpha + 0.31\alpha_0) = \log(1.31\alpha_0) - k_1 t/2.303 \quad \textbf{[42.6]}$$

To measure the pseudo-first-order rate constant, k_1, a plot is made of $\log(\alpha + 0.31\alpha_0)$ versus t. The slope of the plot when multiplied by -2.303 gives k_1. The second-order rate constant, k_2, can be calculated by dividing k_1 by $[H_3O^+]$ (Eq. 42.4).

Note on Procedures, Parts A and B The procedures given in these experiments are intended for use with the simple polarimeter shown in Figure 39.2, which requires a sample size of 20 to 30 mL for optimal readings. If a split-field polarimeter with a 2- to 5-mL capacity sample tube is used, the scale of the procedures can be reduced proportionately.

Procedure

Dissolve 20.0 g of sucrose in approximately 35 mL of water in a 50-mL volumetric flask. After all crystals have dissolved, add 10 mL of 6.0 *M* HCl (previously measured into a test tube from a pipet or buret), rinse the test tube with a few mL of water, and use this to fill the flask to the 50-mL mark. Record the time, and then mix the acid and sucrose by inverting the flask several times. Transfer the solution to a polarimeter tube, and as quickly as possible, measure the rotation of the solution and record the value and the time. At approximately 5-minute intervals for the next 30 to 40 minutes, measure the rotation of the solution and record the rotations and times in tabular form, as shown below:

TIME	t(min) = time − time$_0$	α	$\alpha + 0.31\alpha_0$	$\log(\alpha + 0.31\alpha_0) = y$

To obtain the values for the last two columns in the table, calculate the initial rotation α_0 from the specific rotation $[\alpha]$ of sucrose (+ 66°), the initial concentration c_0 (0.40 g/mL), and the path length of the solution in the tube, l (in dm):

$$\alpha_0 = [\alpha]c_0l$$

Plot the values of $\log(\alpha + 0.31\alpha_0)$ on the y-axis versus the time, t, on graph paper, and draw a straight line that best approximates the positions of the points. Using two points on the line, one near each end (t_1, y_1 and t_2, y_2), calculate the slope of the line, $m = (y_2 - y_1)/(t_2 - t_1)$, and from this determine the pseudo-first-order rate constant, $k_1 = -2.303\, m\ \text{min}^{-1}$. Since rate constants are conventionally expressed in units of seconds^{-1}, divide the $k_1(\text{min}^{-1})$ by 60 to get k_1 (sec^{-1}). Use the concentration of HCl in the solution to calculate k_2 ($M^{-1}\ \text{sec}^{-1}$).

B. ENZYMATIC HYDROLYSIS OF SUCROSE

The enzymatic hydrolysis of sucrose, such as that occuring in digestion, is also carried out by yeast in fermentation and by bees when producing honey. The enzyme in these biochemical processes is called invertase and is a **hydrolase** (catalyst for hydrolysis) specific for the β-glycoside bonds of fructose.

In enzyme-catalyzed reactions, the first step is the formation of a complex between the substrate, S, and the enzymé, E. The complex then breaks down to enzyme plus products (P):

$$E + S \underset{k_{-1}}{\overset{k_1}{\rightleftarrows}} ES \xrightarrow{k_2} E + P$$

The rate of sucrose disappearance depends upon the total enzyme concentration, [E] + [ES], and the three rate constants, k_1, k_{-1}, and k_2, that govern the buildup and breakdown of the complex, as shown in Equation 42.6. A term, $\boldsymbol{K_M}$, called the **Michaelis constant** for the reaction, is defined as $K_M = (k_{-1} + k_2)/k_1$.

$$\frac{d[S]}{dt} = \frac{-k_1k_2[E_{tot}][S]}{k_{-1} + k_2 + k_1[S]} = -\frac{k_2[E_{tot}][S]}{K_M + [S]} \qquad [42.6]$$

When the sucrose concentration is quite high ($[S] >> K_M$), Equation 42.6 reduces to Equation 42.7. In this situation, effectively all of the enzyme is in the complexed state. The rate of hydrolysis is therefore simply the rate of decomposition of the complex ES into products. In this experiment, $[S] >> K_M$ during most of the time that the reaction is occurring, and Equation 42.7 can be used to derive the rate constant k_2.

$$\frac{d[\text{sucrose}]}{dt} = -k_2[E_{tot}] \qquad [42.7]$$

Integration of Equation 42.7 results in Equation 42.8. Using the relationship between the optical rotation, α, and sucrose concentration in Equation 42.1, Equation 42.9 can be derived. This can be rearranged to Equation 42.10, so that a plot of the rotation, α, versus time should give a straight line with a slope $m = -1.31\alpha_0 k_2[E_{tot}]/[S]_0$. If the total enzyme concentration and the initial rotation and sucrose concentration are known, the rate constant, k_2, can be calculated from Equation 42.11.

$$[\text{sucrose}] = [\text{sucrose}]_0 - k_2[E_{tot}]t \qquad [42.8]$$

$$\frac{\alpha + 0.31\alpha_0}{1.31\alpha_0} = 1 - \frac{k_2[E_{tot}]}{[S]_0}t \qquad [42.9]$$

$$\alpha = \alpha_0 - (1.31\alpha_0 k_2[E_{tot}]/[S]_0)t = \alpha_0 + mt \qquad [42.10]$$

$$k_2 = -m[S]_0/(1.31\alpha_0[E_{tot}]) \qquad [42.11]$$

To calculate the molar concentration of enzyme, $[E_{tot}]$, from its weight concentration (mg/mL), the molecular weight of invertase is needed. Since purification of invertase is very difficult, an accurate molecular weight is not known, but it is thought to be approximately 135,000 g/mole.

Procedure

Place 20.0 g of sucrose in a 50-mL volumetric flask. Add water to dissolve, and bring the total volume to the 50-mL mark. Transfer the solution to a polarimeter tube and measure the optical rotation (α_0). Using a calibrated (Mohr) pipet, add 0.20 mL of the enzyme solution, which should contain approximately 5 mg invertase/mL. For 30 to 40 minutes, at approximately 5-minute intervals, measure the rotation of the solution and record the result with the measurement time in tabular form. Plot the rotation versus time (in minutes after adding the enzyme) on graph paper. With a straight-edge, draw a line that best approximates the points, and determine its slope, m, as previously described for the acid-catalyzed hydrolysis.

Use the initial concentration of sucrose and enzyme (molarity) and the initial rotation to calculate k_2, the rate constant for decomposition of the enzyme-substrate complex, from the slope and Equation 42.11. Divide the result by 60 to convert it from units of min^{-1} to sec^{-1}, which is the conventional form for expressing rate constants. The resulting k_2 is also known as the turnover number of the enzyme and is the number of molecules of sucrose hydrolyzed by each molecule of enzyme in one second.

QUESTIONS

1. Write the mechanism for the acid-catalyzed hydrolysis of sucrose showing all intermediates. Which is the slowest (rate-determining) step?
2. Compare the catalytic abilities of invertase and H_3O^+ in the hydrolysis of sucrose. Since the acid-catalyzed hydrolysis rate depends on the sucrose concentration and the enzyme catalyzed reaction does not, comparison must be made at a specific sucrose concentration, that is, the initial concentration used in these experiments. Thus, compare k_2 from the enzyme-catalyzed reaction with $k_2[\text{sucrose}]_0$ in the acid-catalyzed reaction. How many times more effective is invertase than H_3O^+ under these conditions?

Name ______________________ Section ______________ Date ________

42 Acid-Catalyzed and Enzymatic Hydrolyses of Sucrose

PRELABORATORY QUESTIONS

1. Draw the structure for sucrose.

2. What is meant by the term "inversion"?

3. What is the common name of the enzyme used in this experiment?

4. What effect on the rate of the acid-catalyzed hydrolysis would the doubling of the sucrose concentration have? Doubling the acid concentration?

5. What effect on the rate of the enzyme-catalyzed hydrolysis would the doubling of the sucrose concentration have? Doubling the enzyme concentration?

6. How do you account for your answers to Questions 4 and 5?

43

Analysis of Fats and Oils

The bulk of plant and animal tissue is composed of three main classes of compounds: carbohydrates, proteins, and lipids. The term **lipid** includes all substances in living tissues that are soluble in ether, methylene chloride, or similar organic solvents. The most abundant lipids are triesters of glycerol with long-chain "fatty" acids. The triesters, or **triglycerides**, from animal sources are usually low-melting solids and are called *fats*, while those from plants are viscous *oils* that solidify below 0°C. The general structures are the same, and the difference in properties arises from the fact that the vegetable oils contain a larger proportion of unsaturated acid groups, with one, two, or three double bonds in the chain. Triglycerides such as beef tallow and other depot fats of animals contain mostly acid chains with one or no double bonds. The degree of unsaturation is important in nutrition, since there is evidence that a high proportion of saturated fats in the diet leads to the deposition of cholesterol in blood vessels (atherosclerosis).

Fatty acids invariably have chains with an *even* number of carbon atoms, most commonly 16 or 18. A few of the more important fatty acids are listed below; these and several others can occur in any combination in a given triglyceride.

	TYPICAL FATTY ACIDS	TRIGLYCERIDE STRUCTURE
Lauric	$CH_3(CH_2)_{10}CO_2H$	$RCO_2—CH_2$
Myristic	$CH_3(CH_2)_{12}CO_2H$	|
Palmitic	$CH_3(CH_2)_{14}CO_2H$	$R'CO_2—CH$
Stearic	$CH_3(CH_2)_{16}CO_2H$	|
Palmitoleic	$CH_3(CH_2)_5CH{=}CH(CH_2)_7CO_2H$	$R''CO_2—CH_2$
Oleic	$CH_3(CH_2)_7CH{=}CH(CH_2)_7CO_2H$	
Linoleic	$CH_3(CH_2)_3(CH_2CH{=}CH)_2(CH_2)_7CO_2H$	
Linolenic	$CH_3(CH_2CH{=}CH)_3(CH_2)_7CO_2H$	

Table 43.1 Typical Fatty Acid Content of Selected Fats and Oils

	CONSTITUENT FATTY ACIDS (%)*									
No. Carbon Atoms:	*<12*	*12*	*14*	*16*	*16*	*18*	*18*	*18*	*18*	*>18*
No. Double Bonds:		*0*	*0*	*0*	*1*	*0*	*1*	*2*	*3*	
Human fat	—	—	3	24	5	8	47	10	—	2
Butterfat	4†	4	12	33	2	12	29	2	—	2
Lard	—	—	1	28	1	16	43	9	—	2
Coconut oil	15‡	46	18	10	—	4	6	—	—	—
Corn oil	—	—	—	10	1	2	25	62	—	—
Cottonseed oil	—	—	1	20	2	1	20	55	—	1
Linseed oil	—	—	—	10	—	4	23	57	6	—
Olive oil	—	—	—	9	—	3	78	10	—	—
Palm oil	—	—	1	45	—	4	40	10	—	—
Peanut oil	—	—	—	10	—	3	48	34	—	5**
Safflower oil	—	—	—	7	—	2	13	75	3	—
Soybean oil	—	—	—	9	—	4	43	40	4	—

*Blank spaces indicate that less than 1% of the corresponding acid is usually present.
†C_6 saturated (caproic acid) and C_{10} saturated (capric acid).
‡C_8 saturated (caprylic acid) and C_{10} saturated.
**C_{20} saturated (arachidic acid), C_{22} saturated (behenic acid), and C_{24} saturated (lignoceric acid).

Butterfat and coconut oil are atypical, since the mixtures of acids in these triglycerides contain a significant number of 8-, 10-, and 12-carbon chains. The compositions of a few fats and oils are given in Table 43.1. These values vary over a fairly wide range, depending on the source of the oil, and there is no one "correct" figure.

The classical procedure for obtaining the acids from a fat is hydrolysis (saponification) with alkali to give the salt (a soap), and then acidification. This hydrolysis is rather slow at room temperature, and at high temperatures, when a glycol is used as solvent, isomerization of the polyunsaturated acids occurs. For analytical purposes, where only a small amount of material is required, this difficulty is avoided by preparing the methyl esters directly from the triglyceride by transesterification (Eq. 43.1). By using a large excess of methanol, the equilibrium is shifted essentially completely to the right.

$$R{-}\overset{\displaystyle O}{\overset{\|}{C}}{-}OR' + MeOH \overset{acid}{\rightleftharpoons} R{-}\overset{\displaystyle O}{\overset{\|}{C}}{-}OMe + R'OH \qquad \textbf{[43.1]}$$

In this experiment, you will be given a sample of a commercially available fat or oil (e.g., a cooking oil) to determine its fatty acid composition by GC of the methyl esters. From this information you can then identify the oil from among a list of possibilities provided by your instructor, using the data in Table 43.1. Again, the percentages reported in the table are typical, but actual compositions may vary by ±5 to 10% among samples of the same oil from different sources.

Experiments

SAFETY NOTE CAUTION SHOULD BE OBSERVED IN THE USE OF CONCENTRATED ACIDS AND BASES. AVOID EXPOSURE TO SOLVENT FUMES BY USE OF A HOOD.

A. TRANSESTERIFICATION OF TRIGLYCERIDES (Microscale)

Weigh 0.10 g (approximately 8 drops) of the fat or oil into a 1-dram (4-mL) screw-cap vial. Add 2 mL of toluene to dissolve, and then add 1 mL of a solution of BF_3 in methanol (12 g BF_3/100 mL MeOH). Screw on the cap tightly, shake to mix, and place in a 50-mL beaker, half full of boiling water. Check to see that no vapor escapes; if hissing or bubbling is observed, cool and retighten the cap. After 30 minutes, remove the vial, cool, and add 1 mL of water. Shake well, and let the mixture stand to separate (10 to 15 minutes). The toluene (upper) layer contains the methyl esters.

B. GC ANALYSIS OF METHYL ESTERS

Inject 2 μL of the toluene solution onto a polar column (for example, a diethyleneglycol succinate liquid phase) in the GC unit at a column temperature in the range 185 to 200°C. By comparison with a GC of a standard mixture of fatty acid esters (provided), identify the components and calculate the approximate percent composition of the mixture. Relative areas may be measured by tracing the peaks onto bond paper using carbon paper, cutting out the tracings, and weighing them (±.01 g) on a balance (see Question 3). Compare the composition with those in Table 43.1 and identify the original oil.

C. ISOLATION OF TRIMYRISTIN (Microscale)

Place 0.5 g of crushed or ground nutmeg in a 5-mL round-bottom flask and add 2 mL of methylene chloride. Attach a condenser and reflux the mixture gently for 30 minutes. After allowing the mixture to cool somewhat, remove the solvent with a Pasteur pipet with cotton in the tip and place in a separate container. Repeat the extraction procedure using a fresh 2-mL portion of methylene chloride. Evaporate the combined extracts to dryness (in a hood!) to obtain the crude trimyristin. Weigh the crude material and calculate the percentage yield based on the weight of nutmeg used. Recrystallize 0.1 g of the product using ethanol, and reserve the remainder for hydrolysis in Part D of this experiment. Compare the recrystallized and crude samples using TLC (5% CH_3OH in CH_2Cl_2, silica gel).

D. HYDROLYSIS OF TRIMYRISTIN

$$\begin{array}{l} CH_2-O-\overset{\displaystyle O}{\overset{\|}{C}}-C_{13}H_{27} \\ | \\ CH-O-\overset{\displaystyle O}{\overset{\|}{C}}-C_{13}H_{27} \\ | \\ CH_2-O-\overset{\displaystyle O}{\overset{\|}{C}}-C_{13}H_{27} \end{array} + 3\ OH^- \longrightarrow \begin{array}{l} CH_2-OH \\ | \\ CH-OH \\ | \\ CH_2-OH \end{array} + 3\ C_{13}H_{27}\overset{\displaystyle O}{\overset{\|}{C}}-O^-$$

The crude trimyristin obtained is hydrolyzed by refluxing in 1 mL of 6 *M* NaOH and 2 mL of water for 1 hour. At the end of this time, the mixture should have largely clarified, indicating complete hydrolysis. Pour the basic solution with stirring into a mixture of 10 mL of water and 1 mL of 12 *M* hydrochloric acid. Collect the solid by vacuum filtration using a Hirsch funnel, wash with cold water, and allow it to dry. Determine the weight and melting point of the dry product, calculate the percentage yield, and submit the sample, in an appropriately labeled container, to your instructor.

QUESTIONS

1. What is the purpose of the BF_3 in the transesterification reaction? Give equations.
2. What happened to the glycerol from the triglyceride; that is, why was it not observed in the GC?
3. An alternative to cutting and weighing the peaks to determine their areas is the following procedure: Multiply the peak heights by their retention times, sum the products, and divide each product by the total. Suggest why this procedure works for this mixture of similar compounds (see pages 83 and 87), and if time permits, compare the results by both methods.

4. Assuming an essentially quantitative transesterification, estimate what amount (in grams) of ester corresponds to the smallest detectable peak in your GC tracing (i.e., the sensitivity of the detector).
5. Write an equation for the base-catalyzed hydrolysis of glyceryl trimyristate.
6. Why are compounds such as CCl_4, $C_2H_3Cl_3$, etc, used as "dry cleaning" solvents?
7. How could the equivalent weight of an acid obtained from the hydrolysis of a fat be determined?

REFERENCES

Gunstone, F.D. *An Introduction to the Chemistry of Fats and Fatty Acids*; Chapman & Hall: London, 1958.
Morrison, W.R.; Smith, L.M., *J. Lipid Res.* **1964**, *5*, 600.
Nichols, P.L., Herb, S.F.; Riemenschneider, R.W., *J. Am. Chem. Soc.*, **1951**, *73*, 247.

Name ______________________________ Section ______________ Date ________

43 Analysis of Fats and Oils

PRELABORATORY QUESTIONS

1. What is the chemical difference between a fat and an oil?

2. Draw the structure of a triacylglycerol (triglyceride) that could be optically active.

3. Draw the structural formulas for myristic, palmitic, and stearic acids.

4. Give an example of a *transesterification* reaction as applied to this experiment.

5. How do you explain the fact that trimyristin is a reasonably crystalline solid?

6. What is the structure of the polar, stationary phase used in the GC portion of this experiment?

44

Synthetic Organic Polymers

Polymers are large molecules that are built up from many small units called **monomers.** Several well-known polymers that occur naturally are cellulose, rubber, and proteins. In the past 50 years a wide variety of synthetic polymers have been developed for use as fibers, adhesives, protective coatings, and structural materials. Synthetic polymers are now produced in larger volume than any other chemical products.

Polymers can be prepared by two types of reactions: condensation and addition. **Condensation polymers** are obtained from monomers with two or more functional groups that can react to link the monomer units together by splitting out a small molecule such as water or alcohol. Typical examples of condensation polymers are polyesters, polyamides, and epoxy resins, which are one of the components of epoxy glues:

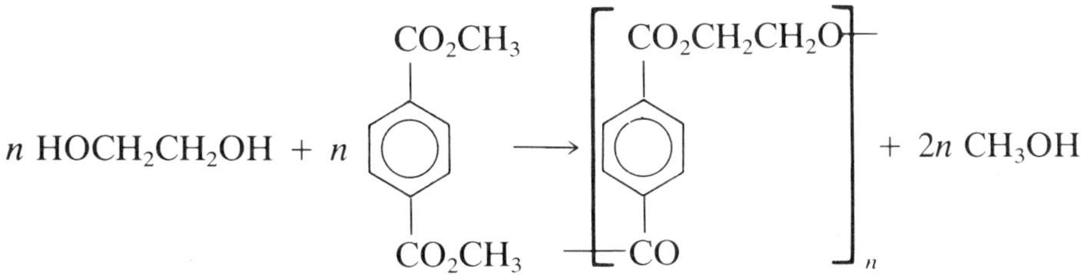

$$n\ HOCH_2CH_2OH + n\ CH_3O_2C{-}C_6H_4{-}CO_2CH_3 \longrightarrow \left[{-}CO{-}C_6H_4{-}CO_2CH_2CH_2O{-} \right]_n + 2n\ CH_3OH$$

ethylene glycol — diacid as diester — polyester (Dacron)

$$n\ NH_2(CH_2)_xNH_2 + n\ HO\overset{O}{\overset{\|}{C}}(CH_2)_y\overset{O}{\overset{\|}{C}}OH \longrightarrow$$

diamine — diacid

$$\left[NH(CH_2)_xNHCO(CH_2)_yCO \right]_n + 2n\ H_2O$$

polyamide (nylon)

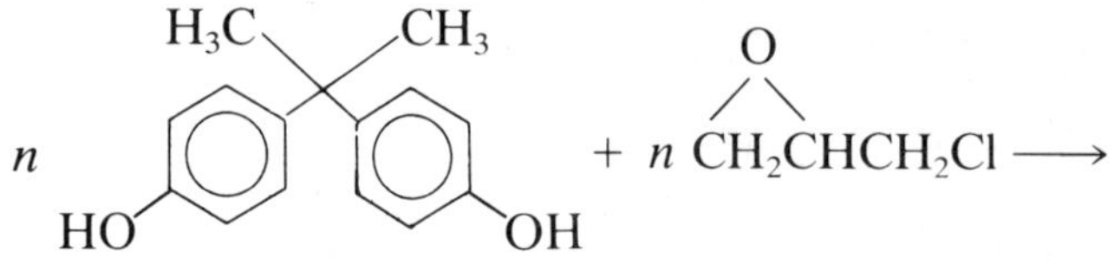

$$n\ \text{HO}-C_6H_4-C(CH_3)_2-C_6H_4-\text{OH} + n\ \overset{\text{O}}{\widehat{CH_2CH}}CH_2Cl \longrightarrow$$

bisphenol epichlorohydrin

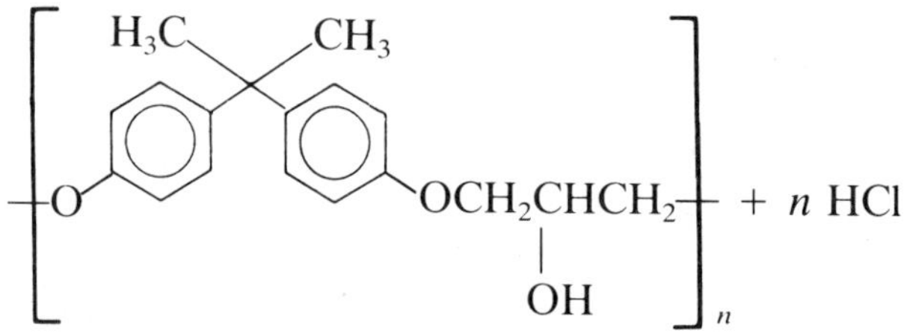

$$\left[-O-C_6H_4-C(CH_3)_2-C_6H_4-OCH_2\underset{\displaystyle OH}{\underset{|}{CH}}CH_2- \right]_n + n\ HCl$$

epoxy resin

Addition polymers are formed by end-to-end combinations of double bonds or rings, each monomer being added to the end of a growing chain. The most important examples are the simple vinyl polymers, including polyethylene, polyvinyl chloride (PVC), polyacrylonitrile (Orlon, Acrilan), polypropylene, polymethyl methacrylate (Plexiglass), and polystyrene. Depending on the monomer, the polymerization may involve electrophilic (cationic), nucleophilic (anionic), or free radical attack on the double bond. In each case, a small amount of an initiating reagent is used to start the polymerization by addition to the first monomer. These processes are illustrated below for several commercially important addition polymers.

Cationic

$$n\ CH_2{=}CHCH_3 \xrightarrow[TiCl_4]{Et_3Al} Et{+}CH_2\overset{\displaystyle CH_3}{\overset{|}{CH}}{+}_n$$

propylene polypropylene

Free Radical

$$n\ CH_2{=}CHC_6H_5 \xrightarrow{Z\cdot} Z{+}CH_2\overset{\displaystyle C_6H_5}{\overset{|}{CH}}{+}_n$$

styrene polystyrene

Anionic

$$n\ \overset{\text{O}}{\widehat{CH_2{-}CH_2}} \xrightarrow{RO^-} RO{+}CH_2CH_2O{+}_n$$

ethylene oxide poly(ethylene oxide)

The experimental procedures that follow describe the laboratory-scale syntheses of several polymers. On an industrial scale, where tons of polymer are produced daily, different, more economical methods are generally used. In addition, much greater care must be taken in controlling the reaction conditions so that a polymer with the desired properties is obtained reproducibly. Some of the properties looked for in specific polymers are plasticity versus rigidity, strength of a drawn fiber, clarity of a film, and stability with respect to elevated temperatures and chemical reagents.

Experiments

A. POLYSTYRENE (Microscale)

The polymerization of styrene can be effected using cationic, anionic, or free-radical initiators. Industrially, polystyrene is generally prepared by heating pure styrene slowly to about 180°C, taking advantage of the fact that above 80°C an added initiator is unnecessary. In the procedure that follows, styrene is polymerized in solution with benzoyl peroxide as the initiator. The O—O bond in benzoyl peroxide is weak and above 60°C breaks homolytically to form two benzoyloxy radicals, which rapidly undergo decarboxylation. The resulting phenyl radicals are the actual initiators that add to the double bond of styrene.

$$C_6H_5\overset{O}{\overset{\|}{C}}-O-O-\overset{O}{\overset{\|}{C}}C_6H_5 \xrightarrow{\Delta} 2\ C_6H_5\overset{O}{\overset{\|}{C}}-O\cdot \longrightarrow 2\ C_6H_5\cdot + 2\ CO_2$$

$$C_6H_5\cdot + CH_2{=}CHC_6H_5 \longrightarrow C_6H_5-CH_2-\underset{C_6H_5}{\underset{|}{CH}}\cdot$$

$$C_6H_5-CH_2-\underset{C_6H_5}{\underset{|}{CH}}\cdot + CH_2{=}CHC_6H_5 \longrightarrow$$

$$C_6H_5-CH_2-\underset{C_6H_5}{\underset{|}{CH}}-CH_2-\underset{C_6H_5}{\underset{|}{CH}}\cdot$$

$$\xrightarrow{n\ CH_2{=}CHC_6H_5} C_6H_5(CH_2-\underset{C_6H_5}{\underset{|}{CH}})\cdot_{n+2}$$

$$C_6H_5\text{-}(CH_2\text{—}\underset{\displaystyle C_6H_5}{CH})\cdot_{n+2} + C_6H_5\text{-}(CH_2\text{—}\underset{\displaystyle C_6H_5}{CH})\cdot_m \longrightarrow$$

$$C_6H_5\text{-}(CH_2\text{—}\underset{\displaystyle C_6H_5}{CH})_{\overline{n+2}}(\underset{\displaystyle C_6H_5}{CH}\text{—}CH_2)_m C_6H_5$$

polystyrene

SAFETY NOTE STYRENE IS SOMEWHAT TOXIC; AVOID BREATHING VAPOR OR CONTACT WITH THE SKIN. BENZOYL PEROXIDE IS UNSTABLE TO HEAT AND FRICTION AND MUST BE HANDLED WITH CARE. TRANSFER BY POURING FROM A PAPER CARTON ONTO A PIECE OF GLAZED PAPER, AND AVOID CONTACT WITH METAL (USE A PORCELAIN SPATULA). CLEAN UP ALL SPILLS WITH WATER, AND RINSE THE GLAZED PAPER WITH WATER BEFORE DISCARDING.

Procedure

Commercial styrene contains a phenol as a polymerization inhibitor, which must be removed before it can be used in this experiment. Extract 2 to 3 mL in a centrifuge tube with two 2-mL portions of 2 *M* (10%) NaOH solution using the microscale technique (Chapter 4). Dry the styrene over Na_2SO_4 and then pass it through a short column of 1 to 2 g of dry alumina contained in a pipet with a wad of cotton.

Place 0.5 mL of the purified styrene in a 13 × 100-mm test tube and dissolve it in 2.5 mL of toluene. Obtain approximately 25 mg of benzoyl peroxide (see Safety Note—estimate amount by volume, comparing with a weighed comparison sample). Add the benzoyl peroxide to the styrene solution, and swirl the mixture to dissolve the peroxide. Cork the tube loosely, and heat the solution in a boiling water bath placed on a hot plate. While the solution is being heated, replace water in the bath as it is needed and start one of the other experiments in this or another chapter as directed.

After heating the solution for 60 to 90 minutes, cool it to room temperature and pour it into 5 mL of methanol in an 18 × 150-mm test tube. Stir the mixture well to coagulate the precipitate, which will adhere to the walls. Decant the slightly cloudy liquid from the polymer, and pour it into a waste solvent container. Add an additional 2.5 mL of methanol to the test tube containing the polystyrene, stopper the tube, and shake it vigorously.

Collect the polystyrene in a Hirsch funnel, rinsing the test tube and the polymer in the funnel with another 2.5 mL of methanol. Air-dry the polymer, weigh it, and calculate the percentage yield.

A polystyrene film may be cast by transferring 50 to 100 mg of the polymer to a microscope slide, adding a few drops of toluene to dissolve the polymer (stir with a matchstick to mix well), and setting the slide aside until the toluene evaporates. Examine the film coating for clarity and hardness.

B. NYLON (Microscale)

A variety of nylons have been developed to fit different uses, and their production now exceeds 2.5 billion pounds per year. Their properties depend on the degree of flexibility or rigidity of the chain, molecular weight, and other properties. Extremely strong and tough fibers (aramids) are obtained by using aromatic dicarboxylic acids and aromatic diamines.

Nylons or polyamides were the first totally synthetic polymers to come into major use as textile fibers. Different nylons are named by designating the number of *carbon* atoms in each monomer unit. The original nylons, and still the most common type, are obtained by the condensation of diamines and acids. The amine salt obtained by mixing equimolar amounts of the two components is heated with a catalyst to split out water and form the amide bonds. The polymer is then spun from the melt into fibers:

$$\overset{+}{N}H_3(CH_2)_6\overset{+}{N}H_3 + \overset{-}{O}-\overset{\overset{O}{\|}}{C}(CH_2)_4\overset{\overset{O}{\|}}{C}-\overset{-}{O} \xrightarrow{\Delta} \left[NH(CH_2)_6NH-\overset{\overset{O}{\|}}{C}(CH_2)_4\overset{\overset{O}{\|}}{C}\right]_n + 2_n\ H_2O$$

nylon 66

A second type of nylon is obtained by the polymerization of monomer units containing both the amine and acid groups in the same molecule. This type is illustrated by the polymer of 6-aminohexanoic acid, which is actually prepared by the addition polymerization of the cyclic monomer ϵ-caprolactam.

$$\text{ϵ-caprolactam} \xrightarrow{\text{catalyst}} \left[NH(CH_2)_5\overset{\overset{O}{\|}}{C}\right]_n$$

ϵ-caprolactam nylon 6

Another method of condensation used for some of these nylons involves the use of the diacid chloride and the diamine. This process is called **interfacial polymerization** and is carried out by dissolving the amine in water and the acid chloride in an immiscible solvent. The two solutions mix only at the interface between the layers, and the resulting polymer can be removed as a film. This type of polymerization is illustrated below.

Preparation of Nylon 6,10

$$NH_2(CH_2)_6NH_2 + ClC(=O)(CH_2)_8C(=O)Cl \longrightarrow \text{-}[NHC(CH_2)_6NHC(=O)(CH_2)_8C(=O)]_n\text{-}$$

hexamethylene diamine — sebacoyl chloride, d 1.20 — nylon 6,10

SAFETY NOTE ACID CHLORIDES ARE IRRITATING SUBSTANCES; AVOID BREATHING VAPOR AND CONTACT WITH THE SKIN. AVOID CONTACT OF THE POLYMER STRAND WITH YOUR HANDS UNTIL THE STRAND HAS BEEN WASHED THOROUGHLY.

Procedure

Dissolve 0.1 g of hexamethylene diamine and 0.2 g of sodium carbonate in 2 mL of water in a 10-mL Erlenmeyer flask or 13 × 100-mm test tube. Mix separately in a 10-mL beaker, 0.1 mL of sebacoyl chloride and 2 mL of methylene chloride. Carefully and slowly pour the diamine solution down the side of the beaker so that it forms a layer above the methylene chloride solution. The polymer will form at the interface as a white film. Grasp the center of the film with a pair of forceps, and slowly pull the polymer out of the beaker. Note that as the polymer is being removed from the beaker, more forms at the interface, producing a continuous strand or "rope." Wrap the strand around a glass stirring rod and continue to pull the polymer from the beaker in this fashion, wrapping it around the rod until no more polymer forms. (**CAUTION**: The nylon may contain unreacted starting materials, and it should not be handled until the polymer is washed.) Wash the nylon under running water, then squeeze it in a paper towel, allow it to dry in air until the next class period, and record the weight.

When you have finished, **do not** discard the nylon **or** the remaining reaction mixture in the sink; place in designated waste receptacles.

C. POLYURETHANES (Microscale)

Polyurethanes are generally prepared from a diisocyanate and a dihydroxy compound. The name **polyurethane** is an exception to the standard nomenclature of polymers (e.g., polystyrene), in that polyurethanes are not polymers of urethane ($CH_3CH_2OCONH_2$, ethyl carbamate). Their importance arises from the large variety of R and R′ groups that may be used, with a resulting large variation in properties of the polymer.

$$O{=}C{=}N{-}R{-}N{=}C{=}O + HO{-}R'{-}OH \rightarrow \left(\overset{O}{\overset{\|}{C}}{-}NH{-}R{-}NH{-}\overset{O}{\overset{\|}{C}}{-}O{-}R'{-}O\right)_n$$

The diisocyanate may be aliphatic, such as hexamethylene diisocyanate, or aromatic, such as the tolylene diisocyanates (also called toluene diisocyanates). The diol may be as simple as ethylene glycol or 1,6-hexanediol, but generally larger molecular weight diols are used, such as polyethylene glycols, polypropylene glycols, and hydroxyl-terminated polyesters, such as poly(ethylene adipate). Branched (cross-linked) polyurethanes can be produced by using a triol instead of a diol.

Isocyanates

$OCN{-}(CH_2)_6{-}NCO$
hexamethylene diisocyanate

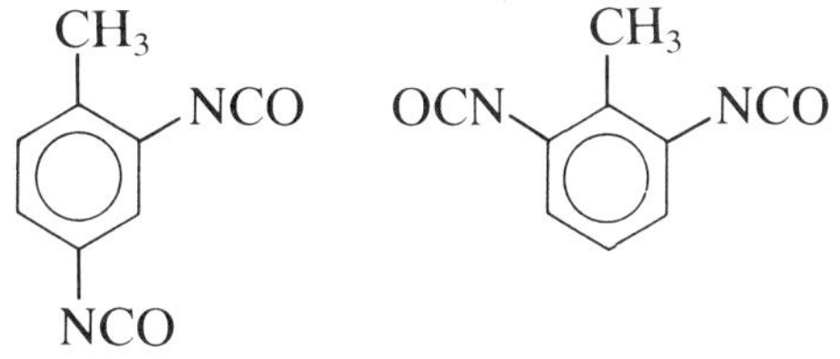

tolylene-2,4-diisocyanate (CH_3, NCO, NCO substituents on benzene ring)

tolylene-2,6-diisocyanate (OCN, CH_3, NCO substituents on benzene ring)

Diols

$HOCH_2CH_2OH$
ethylene glycol

$HO(CH_2)_6OH$
1,6-hexanediol

$HO{-}(CH_2CH_2O)_n{-}H$
poly(ethylene oxide)
"polyethylene glycol"

$HO{-}(\underset{CH_3}{\underset{|}{CH}}{-}CH_2{-}O)_n{-}H$
poly(propylene oxide)
"polypropylene glycol"

$HO{-}(CH_2CH_2O\overset{O}{\overset{\|}{C}}(CH_2)_4\overset{O}{\overset{\|}{C}}{-}O)_n{-}CH_2CH_2OH$
poly(ethylene adipate)

One of the most important applications of polyurethanes is in polyurethane foams. Other polymers can be produced as foams (e.g., styrofoam, which is foamed polystyrene) by dissolving volatile solvents in the polymer melt and allowing them to vaporize, but polyurethanes are usually foamed during polymerization by adding a small amount of water to the monomer mixture. The water competes with the diol for the isocyanate groups and forms an unstable carbamic acid, which decomposes, liberating the gas, carbon dioxide.

$$R{-}N{=}C{=}O + H_2O \longrightarrow R{-}NH{-}\overset{\overset{\displaystyle O}{\|}}{C}{-}OH \longrightarrow R{-}NH_2 + CO_2 \uparrow$$

In the experiment that follows, a variety of diols and triols can be used to provide foams with different properties. In determining the appropriate ratios of reactants, their equivalent weights must be considered. (Equivalent weight = molecular weight ÷ number of functional groups per molecule.) Several diols and triols that may be used with tolylene-2,4-diisocyanate (mol wt 174, equiv wt 87) and their equivalent weights are listed in Table 44.1. Castor oil is the glyceryl triester of a mixture of long-chain carboxylic acids, mostly ricinoleic acid [$CH_3(CH_2)_5CHOHCH_2CH{=}CH(CH_2)_7CO_2H$]. Since 88 to 90% of the acids is ricinoleic, castor oil can be considered as a mixture of 70% triol plus 28% diol and 2% monohydroxy triester, with an average of 2.7 hydroxyl groups per molecule.

In addition to the monomers and water, the procedures call for triethylamine and silicone oil. The tertiary amine functions as a base catalyst for the polymerization reaction and the silicone oil serves to lower the surface tension of the mixture and leads to smaller bubbles. Two specific procedures are given, and if time and materials are available, other combinations of reactants may be tested. In each case the ratio of hydroxyl to isocyanate groups should be kept approximately constant (Question 5).

SAFETY NOTE THE DIISOCYANATES USED IN THIS EXPERIMENT ARE TOXIC COMPOUNDS AND SHOULD BE DISPENSED IN THE HOOD. THE POLYMERIZATION MIXTURES SHOULD BE PREPARED AND KEPT IN THE HOOD UNTIL REACTION HAS OCCURRED. YOU MAY WANT TO USE DISPOSABLE GLOVES AND PIPETS WHEN HANDLING DIISOCYANATES.

Table 44.1 Diols and Triols for Copolymerization

	HYDROXYL COMPOUND	MOLECULAR WEIGHT (g/Mole)	EQUIVALENT WEIGHT (g/Eq)
Ethylene glycol	$HOCH_2CH_2OH$	62.1	31
Diethylene glycol	$(HOCH_2CH_2\!\!\rightharpoonup_2O$	106	53
Triethylene glycol	$(HOCH_2CH_2{-}O{-}CH_2\!\!\rightharpoonup_2$	150	75
Glycerol	$HOCH_2CHOHCH_2OH$	92	31
Castor oil		≈930	≈345
Water	H_2O	18	9

Procedure

1. Weigh into a 4-ounce (or smaller) paper cup 2.0 g of castor oil and 0.5 g of glycerol. Add 1 drop each of water, silicone oil, and triethylamine. With a glass stirring rod, mix these components well to form a creamy emulsion. Using a graduated pipet or buret in a fume hood, add 1.5 mL (1.83 g) of tolylene-2,4-diisocyanate (or a mixture of the 2,4- and 2,6-isomers). Stir the mixture vigorously until a smooth emulsion is formed and bubbles begin to form. Remove the stirring rod and set the cup aside in the hood for the polymerization to proceed. The foam will remain tacky for some time after the maximum volume is reached and should be left to cure for at least a day before attempting to remove it from the cup.
2. Weigh into a 4-ounce (or smaller) paper cup 0.6 g of glycerol and 0.5 g of triethylene glycol. Add 1 drop each of water, silicone oil, and triethylamine, and continue as described in the preceding procedure, again using 1.5 mL of diisocyanate. The final emulsion is more difficult to obtain owing to the more hydrophilic nature of the glycol mixture. Compare the foam obtained in this experiment with that from the first part of this procedure.

QUESTIONS

1. Other than increasing the rate of the polymerization, what would be the effect of using twice as much benzoyl peroxide initiator in the polystyrene preparation?
2. Calculate the percentage yield of the polymer obtained in the nylon 6,10 experiment.
3. What is the role of sodium carbonate in the nylon 6,10 experiment?
4. What other pair of functional groups might be used for the preparation of a polyurethane?
5. What is the ratio of hydroxyl to isocyanate groups in each of the procedures for polyurethane foam?

REFERENCES

Odian, G. Principles of Polymerization, 3rd ed.; Wiley: New York, 1990.

Nylon
Morgan, P.W.; Kwolek, S.L., *J. Chem. Ed.* **1959**, *36*, 182.

Name ______________________ Section ______________ Date ________

44 Synthetic Organic Polymers

PRELABORATORY QUESTIONS

1. The text mentions three well-known naturally occurring polymers. These are ______________, ______________ and ______________.

2. Polymerization can occur by one of three mechanistic processes. These are ______________, ______________, or ______________.

3. The catalyst used in the polystyrene experiment is ______________.

4. The repeating functional group found in nylon is the ______________ group.

5. Polyurethanes are prepared by reacting diisocyanates with ______________.

6. Calculate the molar ratios of the two reactants used in the nylon experiment.

45

Dyes and Dyeing

Dyestuffs include a diverse group of compounds that have in common the property of producing a permanent color on cloth, leather, or paper. All of the useful dyes are aromatic compounds with highly delocalized electron systems that absorb light at certain wavelengths. The most important structural types are azo compounds, triarylmethyl cations, and anthraquinones.

aniline yellow (azo compound) | malachite green (triarylmethyl cation) | alizarin (anthraquinone)

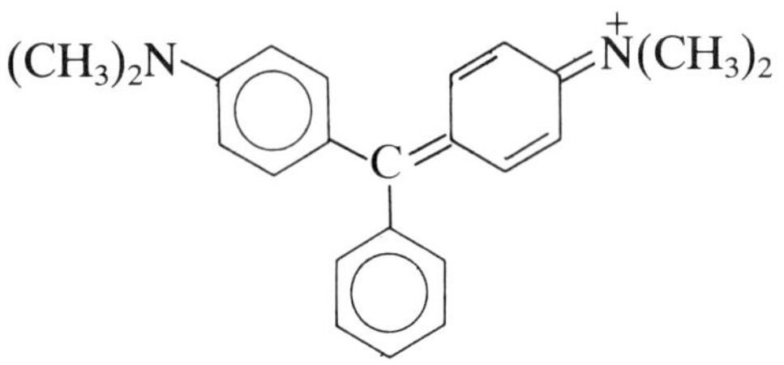

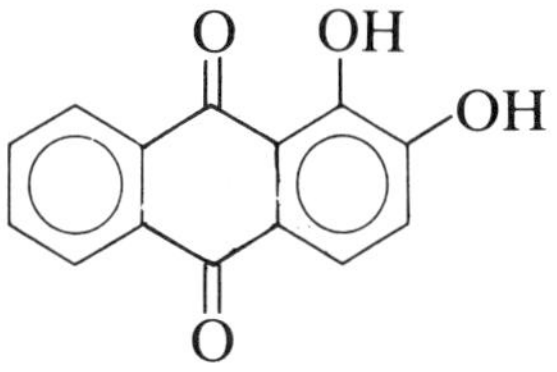

The chemistry of dyeing involves much more than the use of highly colored compounds. The dyeing process is an interaction between a dye and a fiber, and what counts is the final combination. A beautifully colored substance is of no use as a dye if it does not impart its color irreversibly to the fabric. The most important requirements from the standpoint of the user are pleasing shades and fastness or permanence to washing, air oxidation, perspiration, and exposure to strong light.

Another set of considerations comes in the manufacturing process. The dyer must be able to obtain a precisely reproducible shade and depth of color. For economic reasons the manufacturer must also achieve even, level color and at the same time utilize the dye as completely as possible. A major factor in textile dyeing is the large variety of chemical structures

Table 45.1 Fibers and Dye Types

NAME	TYPE	STRUCTURAL UNIT	TYPE OF DYE
Wool or silk	Protein	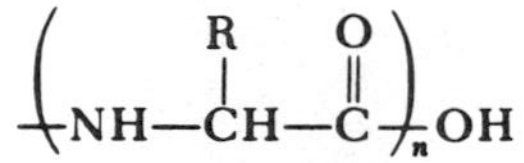 $-(NH-CH(R)-C(=O))_n-OH$ R contains CO_2H, NH_2, and OH groups	Cationic (Ar_3C^+) or anionic ($-SO_3^-$ groups)
Cotton or viscose	Cellulose	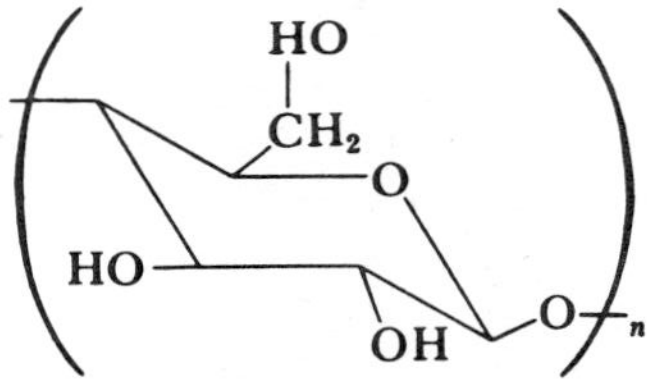 (glucose ring unit with HO, CH_2, HO, OH, $-O-)_n$	Substantive (azo) $-(N{=}N-C_6H_4-C_6H_4-N{=}N)-$; also vat and mordant
Acetate	Cellulose acetate	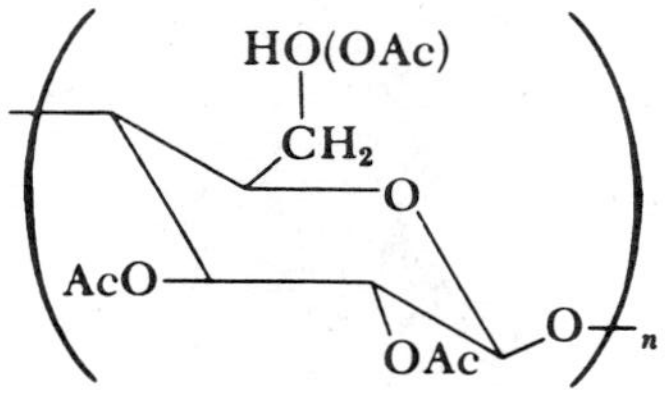 (glucose ring unit with HO(OAc), CH_2, AcO, OAc, $-O-)_n$	Disperse (anthraquinones and others)
Orlon	Acrylic	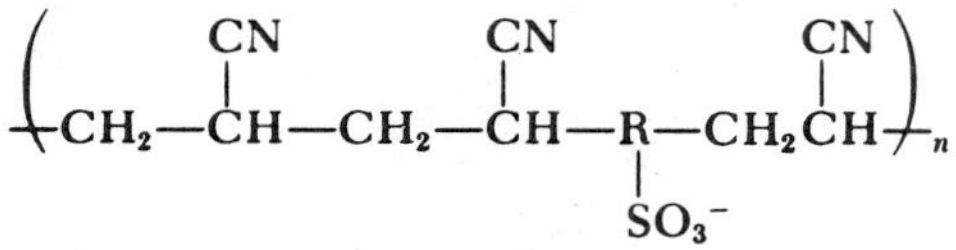 $-(CH_2-CH(CN)-CH_2-CH(CN)-R(SO_3^-)-CH_2CH(CN))_n-$	Cationic
Dacron	Polyester	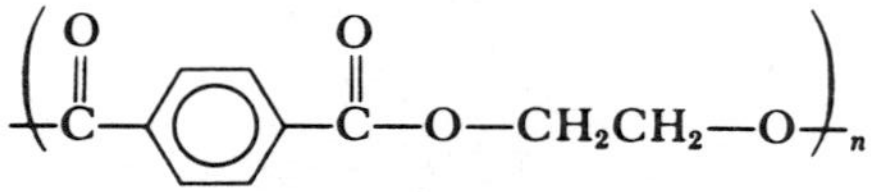 $-(C(=O)-C_6H_4-C(=O)-O-CH_2CH_2-O)_n-$	Disperse
Nylon	Polyamide	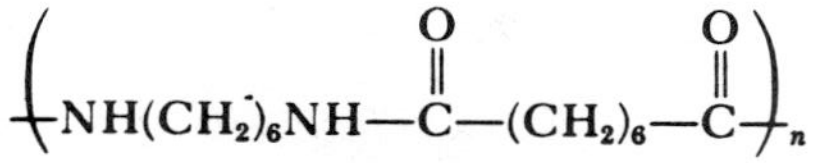 $-(NH(CH_2)_6NH-C(=O)-(CH_2)_6-C(=O))_n-$	Anionic, also disperse

present in modern fabrics. These include the natural fibers wool, silk, and cotton and several types of synthetic polymers (Table 45.1).

In dyeing and textile practice, dyes have traditionally been classified according to the *dyeing process* or the *fiber* that is dyed. For example, the terms "acid" and "basic" refer to dyes that are applied from an acidic or from an alkaline dye bath. **Direct dyes** are those that can be applied to cotton. **Disperse dyes** are applied from a *suspension* rather than from a solution of the dye. **Vat dyes** are applied by using a soluble, reduced form of the dye in the bath and then oxidizing it to an insoluble, colored form on the fabric. All of these classes of dyes can include various structural types. Thus, a disperse dye or an acid dye may have any type of chromophoric (color-producing) structure, provided that the dyeing process involves a suspension of the dye or an acid solution, respectively.

Acid and basic dyes are fixed to the fiber by ionic attraction, and the terms *anionic* and *cationic* are more descriptive and meaningful. Typical **"acid" dyes** contain $—SO_3^-Na^+$ groups and combine in acid solution with a fabric containing $—NH_2$ groups. The dyeing process is actually an ion exchange sequence in which the $—NH_3^+$ cation of the fiber replaces Na^+ in the dye. In **"basic" dyes**, the dye molecule has a *cationic* structure and combines with anionic centers in the fiber.

Silk and wool contain both cationic $—NH_3^+$ groups and anionic $—CO_2^-$ groups in the amino acid side chains and can be dyed with either type of ionic dye. Synthetic polyacrylic fibers contain $—SO_3^-$ groups incorporated in the polymer to provide dye sites. Nylon can be prepared with either $—NH_3^+$ end groups or $—CO_2^-$ end groups in excess; most nylon fabrics contain excess $—NH_3^+$ groups and can therefore be dyed with anionic (acid) dyes.

Acid dyeing:

$$\underset{\text{anionic dye}}{\text{dye—SO}_3^-\text{Na}^+} + \underset{\substack{\text{cationic fiber in}\\ \text{acid solution}}}{\overset{+}{\text{N}}\text{H}_3\text{—R}} \longrightarrow \underset{\text{dyed fiber}}{\text{dye—SO}_3^-\overset{+}{\text{N}}\text{H}_3\text{—R}}$$

Basic dyeing:

$$\underset{\text{cationic dye}}{\text{dye—}\overset{|}{\text{C}}\text{=}\overset{+}{\text{N}}\!<} + {}^-\text{O}_3\text{S—R} \longrightarrow \text{dye—}\overset{|}{\text{C}}\text{=}\overset{+}{\text{N}}\!< \ {}^-\text{O}_3\text{S—R}$$

or

$$\underset{\substack{\text{anionic fiber in}\\ \text{neutral or}\\ \text{basic solution}}}{{}^-\text{O—}\overset{\overset{\text{O}}{\|}}{\text{C}}\text{—R}} \quad \underset{\text{dyed fiber}}{\text{dye—}\overset{|}{\text{C}}\text{=}\overset{+}{\text{N}}\!< \ {}^-\text{O—}\overset{\overset{\text{O}}{\|}}{\text{C}}\text{—R}}$$

Cotton and modified cellulose (viscose, rayon) do not give good fastness (permanence) with simple anionic or cationic dyes. An important group of dyes that are direct or substantive for cotton are compounds in which azo groups occur at a distance that corresponds to the length of the repeating hexose unit in the cellulose chain. The dye is bound to the fiber by a series of hydrogen-bonding attractions. A single azo group is insufficient for dye fastness; "dis-azo" compounds, with two groups linked to a biphenyl system, provide excellent **substantive dyes.**

Polyester and cellulose acetate fibers have the least affinity for adsorption of dyes from solution, and these fibers are dyed by the disperse process. The fabric is heated in a suspension of an insoluble dye and a "carrier" that enhances penetration. The dye becomes dispersed in the hydrophobic fiber. Since there is no bonding to the fiber, **disperse dyes** are often less permanent than other types.

An important development in textile practice is the use of mixed and blended fibers. In dyeing these materials, the differential binding of dyes on the individual fibers provides textured colors. The blended fabric is treated with a mixture of dyes, and the *lack* of affinity of a fiber for certain dyes is just as important as its affinity for others. A blended fiber, or a fabric woven with different fibers, can be **cross-dyed** with a mixture of dye types. In this way, patterned colors can be achieved in a single dyeing operation.

An enormous range of dyes is available to cover the diverse requirements for colors and fabrics. The dyeing process is determined by the type of fabric and the dye, but a number of factors contribute to the final result. The concentration of dye, the presence of electrolytes, and the temperature are important variables in the dye bath. The degree of wetting and swelling of the fiber, the duration of the dyeing, and the addition of surface-active compounds are other factors that affect the overall outcome. Because of the large number of factors that can be varied and the complicated interactions between dye and fiber, textile dyeing is a highly empirical practice, requiring skill and long experience.

CHEMISTRY OF AZO DYES

The preparation of azo dyes involves two reactions—diazotization and coupling. Both reactions are very simple operations that are carried out in aqueous solution. In **diazotization**, an aromatic amine is converted to a diazonium ion with nitrous acid.

$$HONO + ArNH_2 \xrightarrow{H^+} \underset{\text{diazonium ion}}{ArN{=}N^+} + 2H_2O$$

The **coupling reaction** is an electrophilic substitution of a phenol, naphthol, or aromatic amine to give an azo compound. The electrophile is the $ArN{=}N^+$ ion, and substitution occurs at positions *ortho* or *para* to the —OH or —NH_2 group. To obtain soluble azo dyes, a naphthol containing —$SO_3^-Na^+$ groups is used for coupling. A typical reaction is the coupling with the salt of 2-naphthol-3,6-disulfonic acid ("R-salt"). The resulting azo compound can be used as an anionic or acid dye. If a naphthol without solubilizing —$SO_3^-Na^+$ groups is employed, the azo compound is insoluble and can be used as a disperse dye.

HO $SO_3^-Na^+$ HO $SO_3^-Na^+$

—N=N$^+$ ⟶ —N=N—

$SO_3^-Na^+$ $SO_3^-Na^+$

"R-salt" anionic azo dye

In certain dyeing applications, particularly textile printing, the azo coupling reaction is carried out *on the fabric*. The cloth is first immersed in an alkaline solution of the naphthol, or a paste of the naphthol is applied in a pattern. The fabric is then immersed in the diazonium solution, and the dye is formed on the surface of the fiber.

Experiments

Several aspects of dye chemistry and the dyeability of various fibers are illustrated in the experiments in this chapter. In a brief laboratory experiment it is not possible to obtain optimal dyeing conditions or to evaluate the permanence of a dye. However, the relative affinities of fibers for dye types can be observed simply by exposing samples of various fibers to different dyes under the same conditions. Multifiber cloth containing bands of several different fibers provides a simple means for evaluating and comparing dye affinities. One standard multifiber cloth, Number 10, has six bands in the following sequence: wool, Orlon, Dacron, nylon, cotton, and acetate.

The amines to be used in these experiments are shown in Figure 45.1. In Parts A and B, you will select one monoamine and one diamine from the list. The amines are available as 0.05 *N* solutions in 0.1 *M* HCl—each solution contains 0.05 mmole of monoamine or 0.025 mole of diamine and 0.1 meq of HCl per mL. These solutions are more dilute than those normally

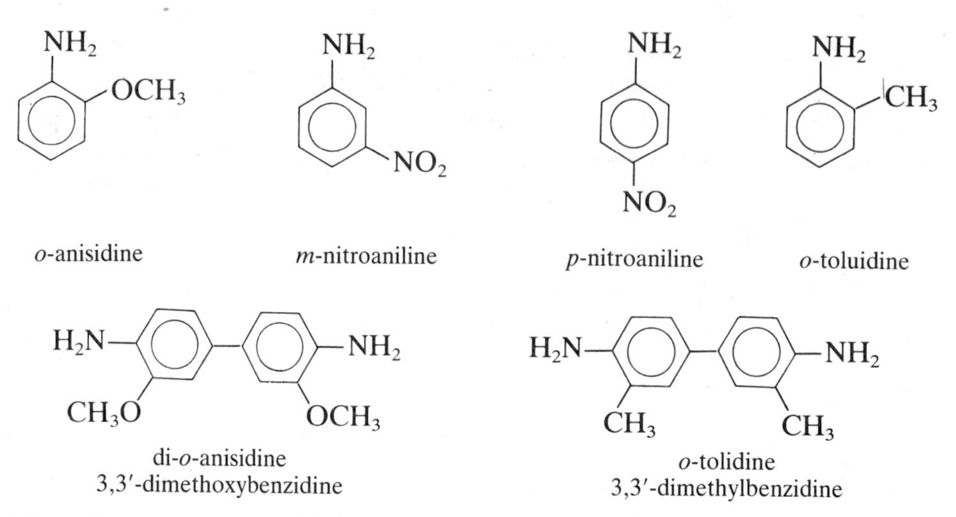

Figure 45.1 Amines to be used in dye syntheses.

used for diazotization, but the rather high dilution is advantageous for an experiment of this type and permits the reactions to be carried out rapidly by a standard procedure.

General Procedure for Diazotization

SAFETY NOTE THE AMINES AND DIAMINES USED IN THIS EXPERIMENT ARE NOT ON THE LIST OF SUSPECTED CARCINOGENS, BUT CERTAIN COMPOUNDS OF SIMILAR STRUCTURE ARE CARCINOGENIC. THE COMPOUNDS IN THIS EXPERIMENT ARE USED IN VERY SMALL AMOUNTS AND PRESENT NO SIGNIFICANT HAZARD. CARE SHOULD BE TAKEN TO AVOID CONTACT OF THE DIAZOTIZED SOLUTIONS, DYE SOLUTIONS, OR DYED CLOTH WITH THE HANDS OR CLOTHING, SINCE SKIN, FINGERNAILS, OR FABRIC CAN BE STAINED.

Place 5 mL (0.25 meq) of a solution of the monoamine or diamine in a graduated centrifuge tube (distillation receiver) and chill several minutes in an ice bath. Add 0.4 mL of 0.5 *M* $NaNO_2$ solution to the amine and note any change in appearance. The diazonium solutions must be kept at 0°C in an ice bath until they are used.

A. DYEING ON THE FIBER (Microscale)

Obtain one 2 × 4-cm swatch of cotton, lay it flat on a paper towel, and moisten each end with a 1.0 *M* solution of β-naphthol (the solution contains 0.144 g [1 mmol] of naphthol and 2 meq of NaOH per mL) by placing several drops of the solution on each end of the swatch. Blot with a paper towel and allow the swatch to dry.

To one end of the dried naphthol-treated swatch, add a single drop of a prepared monoamine diazonium solution and to the other end add a single drop of the diazonium solution prepared from a diamine. Rinse the patch thoroughly in running water and blot on a clean paper towel.

B. COMPARISON OF MONO- AND DIS-AZO DYES IN NEUTRAL AND ACID SOLUTION (Microscale)

Obtain in a graduated centrifuge tube 3 mL of a diazotized monoamine and diamine different from those used in Part A. If the monoamine used is one of the nitroanilines, add 0.4 mL of 1.0 *M* NaOH to the diazonium solution to neutralize the excess acid. To each diazonium solution add 0.3 mL of a 0.5 *M* solution of "R-salt" (this solution contains 0.174 g [0.5 mmole] of R-salt and 1 meq of NaOH per mL).

Place the dye solutions in a water bath at 70 to 80°C, and in each solution immerse a 1-cm-wide strip of number 10 multifiber cloth* and a 1 × 2-cm patch of cotton (this can be cut from a clean, cotton lab towel). Keep the solutions in the bath approximately 5 minutes, and then remove the tubes. Transfer the cloth samples with a glass rod to clean beakers and rinse the samples thoroughly with water and blot on a towel.

To the dye solutions, add 0.4 mL of 2 *M* HCl (or 0.2 mL of 4 *M* acid) and repeat the dyeing in each bath with a cotton patch and a strip of multifiber cloth. After 5 minutes, remove, rinse, and blot the samples. Save the dis-azo dye bath (from the diamine) and discard the other dye solution.

Compare the intensity of color of the cotton strips dyed with the mono- and the dis-azo dyes, and also compare the intensity of the color in the various multifiber bands for the two dyes in neutral and acid solutions.

C. DISPERSE AND CROSS-DYEING (Microscale)

Place 1 mL of the diazotized *m*-nitroaniline solution in a graduated centrifuge tube. Neutralize with 0.1 mL (approximately 1 drop) of 1 *M* NaOH and then add 0.1 mL (1 drop) of β-naphthol solution. Allow the solution to warm to room temperature and stir and rub, if necessary, to obtain a finely divided suspension of solid. Add approximately 0.5 mg of biphenyl and 1 drop of surfactant solution to the dye mixture and immerse a 1-cm-wide strip of the multifiber cloth. (Alternatively, if available, 1 mL of commercial yellow disperse dye suspension can be used instead of the dye prepared from *m*-nitroaniline.)

Heat the dye bath in a 90 to 100°C water bath for 10 to 15 minutes, and then cool and remove, rinse, and blot the sample. Record the relative intensities of the bands in the multifiber sample.

Combine the remaining dye suspension and the remaining acid solution

*Available from Testfabrics, Inc., P. O. Box O, Middlesex, NJ 08846.

of dis-azo dye from Part B, and repeat the dyeing at 90°C with another multifiber strip. If available, a sample of checked multifiber cloth (Number 7403) can be dyed also. Remove, rinse, blot, and record the appearance of the dyed sample.

D. CATIONIC DYES (Microscale)

Obtain 1 mL of a 0.1% solution of malachite green, and immerse a strip of multifiber cloth. Heat the solution on a steam bath for several minutes and then rinse, blot, and record the relative intensities of color in the bands.

QUESTIONS

1. Draw the structures of each of the dyes you prepared in this experiment and classify them as cationic, anionic, disperse, and/or direct dyes.
2. Which of the dyes tested had the greatest affinity for cotton? Account for this result.
3. Explain the differences observed from acid and neutral dyeing of the multifiber cloth in Part B.
4. Explain the results observed in the cross-dyeing experiment in Part C.

Name ______________________ Section ____________ Date ________

45 Dyes and Dyeing

PRELABORATORY QUESTIONS

1. Describe the process of
   a. vat dyeing

   b. disperse dyeing

2. How do anionic and cationic dyeing differ?

3. What is cross-dyeing?

4. Draw the structure of the so-called "R-salt." What role does it play in the dyeing process?

5. Write the mechanistic steps in the reaction of phenol with the diazonium salt derived from *p*-nitroaniline.

46

Identification of Unknowns

Along with synthesis and the examination of reaction mechanisms, an equally important part of organic chemistry has to do with the characterization and identification of compounds, which may be encountered in sources ranging from a laboratory reaction to exotic tropical plants. In any case, sufficient information must be accumulated to establish the identity of the compound in question with that of a previously described compound of known structure or else to determine, *ab initio*, the structure of the unknown.

In earlier days, organic chemists relied heavily on chemical behavior in the characterization of compounds and structure elucidation. Various reactions were applied to diagnose the presence of functional groups and structural units, and final evidence for a new structure usually involved systematic degradation to identifiable products. The development of spectroscopic methods has had a revolutionary effect on this area of organic chemistry. Today it is possible to characterize and arrive at the structure of a previously unknown compound entirely by physical and spectroscopic methods, without recourse to any "wet chemistry" at all. Very often, the structure stands revealed as soon as the ultraviolet, infrared, NMR, and mass spectra are in hand. With highly complex molecules, the entire structure can be determined by X-ray crystallography.

The approach to the problem of identifying or assigning the structure of an unknown organic substance will of course depend on the circumstances and the source of the sample. If the compound has been obtained as a component in a mixture of naturally occurring alkaloids or steroids, it will in all likelihood represent a variation of a known pattern, and the structural problem is relatively restricted, although subtle stereochemical differences, for example, may still present a challenging problem. Similarly, with an unknown arising as a by-product in a synthesis, one can generally assume some relationship to the starting materials, and after a few pieces of information are obtained, a probable structure may be inferred.

Another type of situation, sometimes mentioned in textbook problems but, hopefully, very rarely encountered in practice, is one in which the labels have come off all the bottles in the storeroom. In this case, the only premise that can be made is that most of the unknowns resulting from the disaster are to be found among the 25,000-odd entries in chemical suppliers' catalogs. Although artificial, this is essentially the context of this experiment, in which unknown samples selected from the entire range of simple organic compounds are to be identified.

In the classic approach to qualitative organic analysis, the main guidance that a student had in solving an unknown was a rather rigid classification scheme and a table of compounds for each of the principal functional groups, arranged according to increasing boiling point or melting point. This approach places a great deal of emphasis on these two physical properties and requires that most or all of the unknowns be included in relatively limited tables.

The present experiment is intended to be of a more open type, with infrared and NMR spectra as well as melting or boiling points providing orientation. With spectral data available, a broader range of unknowns is possible, and the approach in some cases assumes some of the character of a structural determination rather than simply narrowing a given list of compounds to one member.

The objective of the experiment should be to apply the methods available as efficiently as possible in arriving at a firm, well-documented identification. Spectral data can and should take the place of a number of the older chemical methods for detecting functional groups, but this identification experiment is not intended to be purely an exercise in spectral interpretation. It will often be necessary to seek information from hydrolytic or oxidative degradation of the unknown. In a particular situation, some special reaction or confirmatory test may be uniquely appropriate. Frequent reference to general textbooks and library sources will be essential. Although a thorough and complete job should be done, unnecessary or irrelevant steps should be avoided so that as many unknowns as possible can be done in the time allotted; some will require much more time and effort than others.

GENERAL APPROACH

The first step in the identification process will be to obtain physical constants, infrared and NMR spectra, and solubility properties of the unknown. These data will then be assessed, and further information as needed will be obtained to permit a tentative conclusion about structure. The identification is to be completed by locating the compounds or candidate compounds in handbook tables or other literature and by preparing derivatives for confirmation.

The unknowns provided in this experiment are of the purity normally encountered in commercial organic chemicals, which is usually in the range of 95 to 99%. Minor impurities will generally not interfere in the identification procedure, but preliminary recrystallization or distillation of a sample of the unknown may be desirable.

PHYSICAL PROPERTIES

The melting point of a solid or the boiling point of a liquid is traditionally one of the first physical constants cited in characterizing an organic compound; in early work these temperatures were among the few measurements that could be made. Several other properties, such as the refractive index and density of liquids and optical rotation of chiral compounds, are often recorded for additional characterization, although these have become less important since the advent of routine spectral data, and they are not particularly useful in locating a compound in the literature.

In using the observed melting point or boiling point of an unknown or a derivative for comparison with literature values, it is necessary to allow sufficient "leeway" and to consider a range of several degrees in literature values on either side of the observed temperature. Literature values as well as the observed ones are subject to inaccuracies; frequently, more than one value can be found because of differences among individual investigators.

The melting point of a compound conveys little structural information in itself, since it depends on such diverse factors as molecular size, symmetry, rigidity, and polarity of functional groups. Although there is a regular progression of melting points within a typical aliphatic homologous series, large disparities in the melting points of closely similar compounds can arise because of differences in molecular shape, as illustrated with the polyols erythritol and pentaerythritol.

$$HOCH_2{-}CH(OH){-}CH(OH){-}CH_2OH$$

erythritol, mp 121°C

$$HOCH_2{-}C(CH_2OH)_2{-}CH_2OH$$

pentaerythritol, mp 254°C

The boiling point of a liquid is much more directly related to the functional group and molecular size, since forces operating in the liquid state are less affected by symmetry and rigidity than those in a crystal. Within any straight-chain aliphatic homologous series, the boiling point increases in a regular way with increasing molecular weight, with increments between successive members becoming smaller, the longer the chain. Branching, particularly in the vicinity of a functional group, markedly lowers

Table 46.1 Boiling Point Ranges of Aliphatic Alcohols (°C)

TYPE	4-CARBON	6-CARBON	8-CARBON
Primary	108–116	148–156	183–194
Secondary	99	120–139	150–180
Tertiary	83	120–125	149–165

the boiling point within a group of isomers. The presence of an alicyclic or aromatic ring causes a significant increase in boiling point over that of an aliphatic compound having the same functional groups and number of carbon atoms. Table 46.1 contains representative data on the boiling points of a few simple monofunctional aliphatic alcohols. Within each group, the lowest boiling point is that of the most highly branched isomer, and the highest boiling point corresponds to the longest chains (*n*-alkanol).

It is obvious that a relatively low boiling point for an unknown, say, below 100°C, greatly limits the number of possible compounds and may even define the structure uniquely with little other data. A boiling point in the range of 100 to 160°C, together with information on functional groups, provides a rough indication of molecular size. On the other hand, the number of possible compounds increases enormously in this range, and the boiling point cannot be used to pinpoint one or two candidates if all known compounds are admitted as possibilities. If the boiling point is above 180°C, the value as a means of narrowing possibilities is practically nil. Moreover, boiling points above this temperature are likely to be very inaccurate. Your instructor will indicate whether the boiling point should be determined or not.

PROCEDURES

MELTING POINTS. For solid unknowns, the melting point is determined in the usual way, raising the temperature of an initial sample rapidly to get an approximate range, then repeating at a rate of 2 to 3°C per minute in this region. If a sample is recrystallized, the melting point should be checked before and after recrystallization.

BOILING POINTS. If sufficient sample is available (at least 3 to 5 mL) the boiling point can be observed by distillation in a simple 10-mL distilling flask (Fig. 46.1). The flask is mounted at an angle on wire gauze so that excessive heating of the glass is avoided. A test tube fitted over the side arm and chilled in ice serves as a condenser and receiver. Be sure to add a boiling stone; heat with a microburner. Care must be taken to distill slowly enough to permit thermal equilibrium to be reached. Walls of the

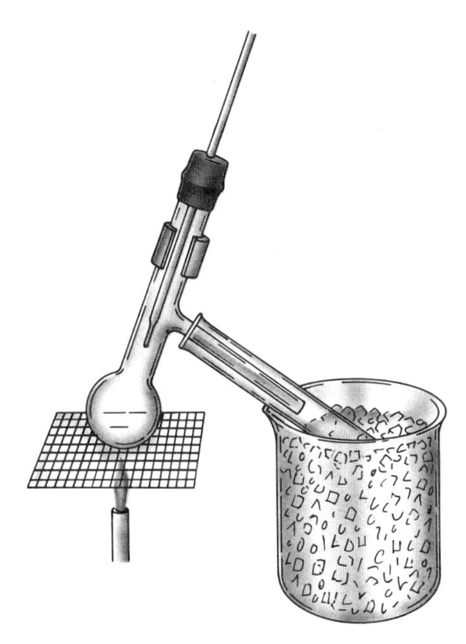

Figure 46.1 Distillation apparatus for unknowns.

flask above the thermometer should be wet with condensing vapor, and there should be a drop of condensate on the thermometer bulb. For strictly accurate results the barometric pressure should be taken into account.

An alternative procedure that is often more accurate and convenient is a **micro boiling point** determination. This method is described in Part F of the Experimental Section in Chapter 5. The determination should be repeated at least once. Remove the capillary tube and drain out the liquid before reheating.

SOLUBILITY CLASSIFICATION

A good deal can be learned about a compound from its solubility in a few media; the solubility classification complements spectral data and helps to determine the direction of further work. The solubility of the unknown is checked in water, dilute acid, dilute base, and concentrated sulfuric acid in that order, stopping when a positive result is obtained. If the compound is soluble in water, nothing is learned by testing in dilute acid or base; if it is soluble in any aqueous medium, it will also dissolve in or react with concentrated H_2SO_4.

A summary scheme of solubility classification is given in Chart 46.1.

WATER. Solubility of an organic compound in water reveals the presence of ionic or "polar" groups that can be solvated or can participate in hydrogen bonding. The extent of solubilization depends, of course, on the "ratio" of

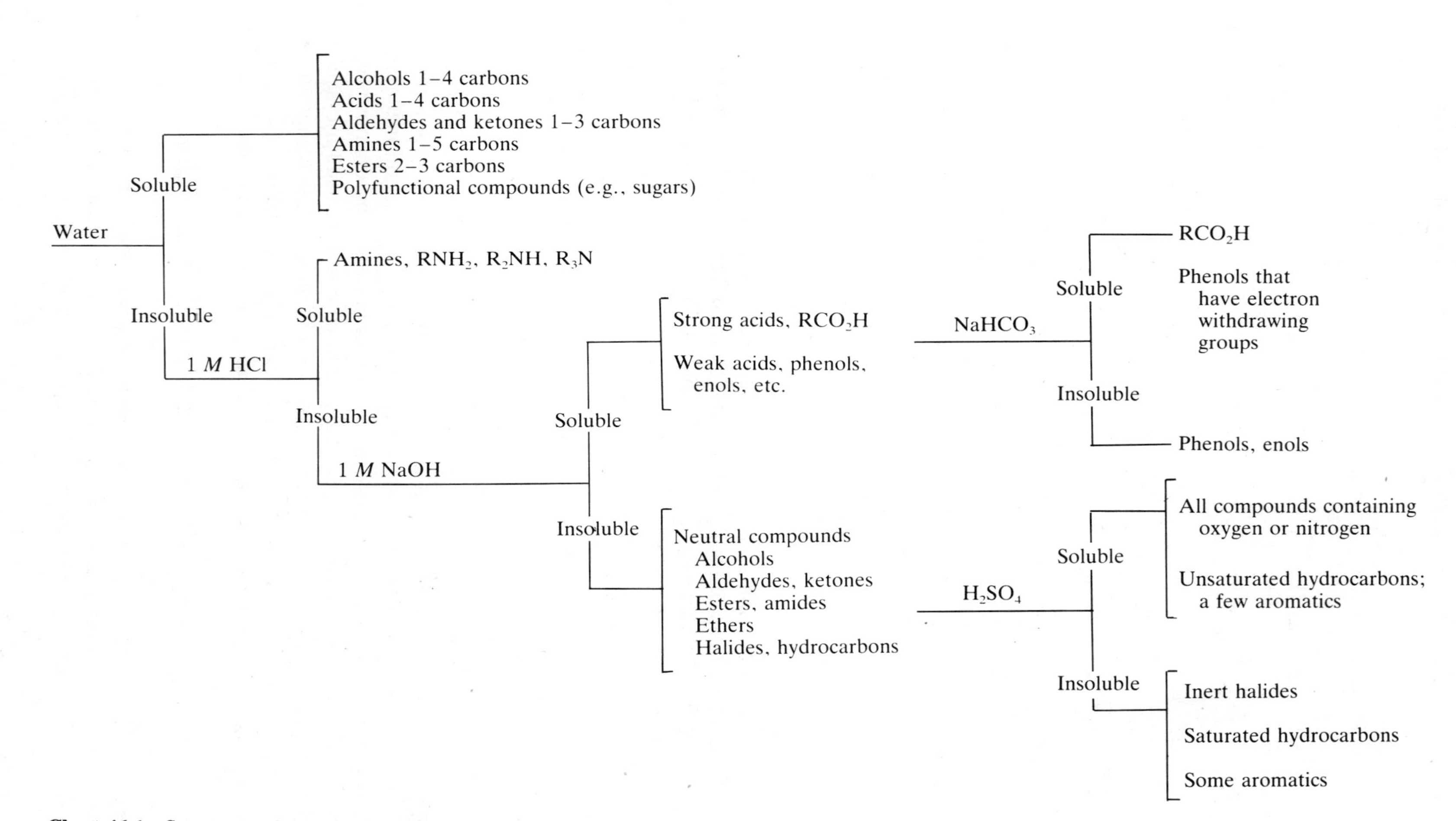

Chart 46.1 Summary scheme for solubility classification.

functional group to carbon skeleton. Liquids containing hydroxyl, carboxyl, amino, or amide groups and no more than four or five carbon atoms are miscible with water in all proportions (indicated by "∞" in solubility tables) or have appreciable solubility. With two or more of these groups, a much larger molecule will be soluble in water.

With solids, crystal structure has a major influence, and solubility in any solvent, water or organic, is related to melting point as well as to the polarity of the molecule. Thus oxamide, $NH_2COCONH_2$, which has the unusually high melting point of 410°C, is only very slightly soluble in water.

In testing for solubility in water, approximately 50 mg of solid or 0.1 mL (2 drops) of liquid is added to 1 mL of water in a 10 × 75-mm test tube. Large hard crystals of a solid may dissolve slowly and should be powdered and stirred well. If a clear solution is obtained, or a major amount of the compound dissolves with the amounts specified, the compound is considered "soluble" in water.

If the unknown is soluble in water, the pH should be estimated with indicator paper to detect the presence of acidic or basic groups in the molecule. The solubility in ether should also be checked if the unknown is readily soluble in water. A multiplicity of polar groups, as in a polyol, may render the compound insoluble in ether. With a solid, solubility in water and insolubility in ether may also suggest the possibility of an ionic salt, and this should be explored by treating the solution with acid or base, in which case the free acid or base may precipitate.

AQUEOUS ACID OR BASE. Compounds that can be converted to ionic species can be recognized by their solubility in water at certain pH values. Three major classes of compounds can be distinguished in this way: *strong acids*, *weak acids*, and *bases*. All compounds with $K_a > 10^{-12}$ ($pK_a < 12$) are converted to anions to a significant extent in 1 *M* NaOH (pH 14). Acids with pK_a in the range of 3 to 7, including all carboxylic acids and nitrophenols, are also soluble in aqueous sodium bicarbonate (pH 8). Solubility in NaOH, but not $NaHCO_3$, indicates a weak acid such as a phenol, enol, or aliphatic nitro compound. Organic bases with K_b 10^{-3} to 10^{-10} are soluble in 1 *M* HCl by virtue of the conversion to cations. The only compounds that can be protonated in dilute aqueous acid are amines.

$$RCO_2H + HCO_3^- \rightarrow RCO_2^- + CO_2 + H_2O$$
$$R_3N + H_3O^+ \rightarrow R_3NH^+ + H_2O$$

In detecting basic or acidic properties by these solubility criteria, the important point is whether the unknown is significantly more soluble in aqueous acid or base than in water. If a test is doubtful, neutralization of the test solution should cause reprecipitation of an amine or an acid that is dissolved because of salt formation. One possible complication that must be kept in mind is the relatively low solubility of certain salts, particularly in the presence of an excess of the common ion. For example, it is possible to mistake the formation of a slightly soluble hydrochloride for insolubility of a solid amine.

CONCENTRATED SULFURIC ACID. Virtually all compounds of moderate molecular size that contain a nitrogen or oxygen atom or a double or triple bond are protonated in 96% H_2SO_4 and therefore dissolve to some extent. This test should be deferred until after the infrared spectrum is obtained. If there is no clear indication of any functional group in the spectrum, it will then be useful to check the solubility in concentrated H_2SO_4 to detect the presence of ether oxygen, reactive double bonds, and so forth. The solution may become dark in color or polymer may separate; these reactions constitute a positive test, but slight darkening without actual solution may be due to trace impurities.

SPECTRA AND ANALYTICAL DATA

The two most widely useful types of spectra, infrared and NMR, will be obtained for each unknown. These spectra are not to be considered as extra data to be used if chemical analysis is unsuccessful; rather they should be the first data analyzed and should serve as the basis for deciding what additional steps are appropriate and necessary. Interpretation of both the NMR and infrared spectra should be included in your report of the analysis of an unknown. For general information on obtaining and interpreting these spectra, refer to the sections on spectral methods (Chapters 9 and 10).

In certain cases you may be given additional information about an unknown in the form of an elemental analysis (% C, H, N, and so forth) or a molecular weight. The latter, if determined by measuring accurately a mass spectrum, can fix the molecular formula. Approximate values ($\pm 10\%$), such as would be obtained by freezing-point determinations, are useful only in conjunction with other data.

In using an elemental analysis, it must be recognized that methods of analysis are such that the results are accurate to only $\pm 0.3\%$ (absolute) and that an empirical and not necessarily a molecular formula is obtained. As an example, consider a compound containing C, H, and possibly O, which analyzed for 74.02% C and 6.48% H. By difference it contains 19.5% O. The calculated molar proportions of C, H, and O are thus 5.06:5.27:1. Rounding to $C_5H_5O_1$, the theoretical analysis is 74.06% C and 6.21% H, in good agreement with that found. Since a compound containing only C, H, and O cannot have an odd number of hydrogen atoms, the molecular formula must be $C_{10}H_{10}O_2$ or some higher multiple.

On the following pages, three examples of typical unknowns are given, including data on physical properties and solubility and reproductions of infrared and NMR spectra (Figs. 46.2, 46.3, and 46.4). The discussions that follow the data illustrate the approach that should be followed in their interpretation and the conclusions that can be reached. The first and second examples are identified without considering their ^{13}C NMR spectra, since ^{13}C NMR data may not be available or necessary.

EXAMPLE 1

A liquid, bp 225 to 235°C, insoluble in water, aqueous acid and aqueous base, but soluble in concentrated H_2SO_4.

1. Infrared Spectrum

a. Broad strong band at 3360 cm^{-1} indicates bonded OH group typical of alcohols run in liquid film.
b. Bands at 3050 and at 2880 to 2950 suggest aryl or olefinic and aliphatic C—H, respectively.
c. Absence of strong bands in 1800 to 1650 region indicates no carbonyl groups; 1600 and 1500 cm^{-1} bands are typical for an aromatic ring.
d. Band at 1460 cm^{-1} is due to aliphatic C—H bending.
e. Strong absorption at 1030 to 1050 cm^{-1} is in the region of C—O stretching of alcohols and confirms an OH group.
f. Sharp bands at 700 and 740 cm^{-1} are very characteristic of a benzene ring with one substituent.

2. NMR Spectrum

Integration. The spectrum contains five groups of signals with relative areas, reading upfield (from left to right), of 7:2.9:1.5:2.8:2.8 (these are the vertical heights measured in cm on the original chart). Dividing by 1.4 leads to an integer ratio of 5:2:1:2:2 for the five peaks.

Chemical Shifts

a. The highest field multiplet at 1.83 ppm is within the range of either CH_3, CH_2, or CH groups in various environments, but the fact that this and the next two multiplets each represent two protons strongly suggests three $\mathbf{CH_2}$ groups with different combinations of substituents. Also, the complex splitting pattern rules out a CH_3 group; any CH_3 group would give either a singlet, a doublet, or a triplet, since it would have to be present in a grouping $\mathbf{CH_3}X$, $\mathbf{CH_3}CHX_2$, or $\mathbf{CH_3}CH_2X$.
b. The multiplet at 2.62 ppm requires a substituent such as a double bond or an aromatic ring; the latter is an obvious probability from other evidence.

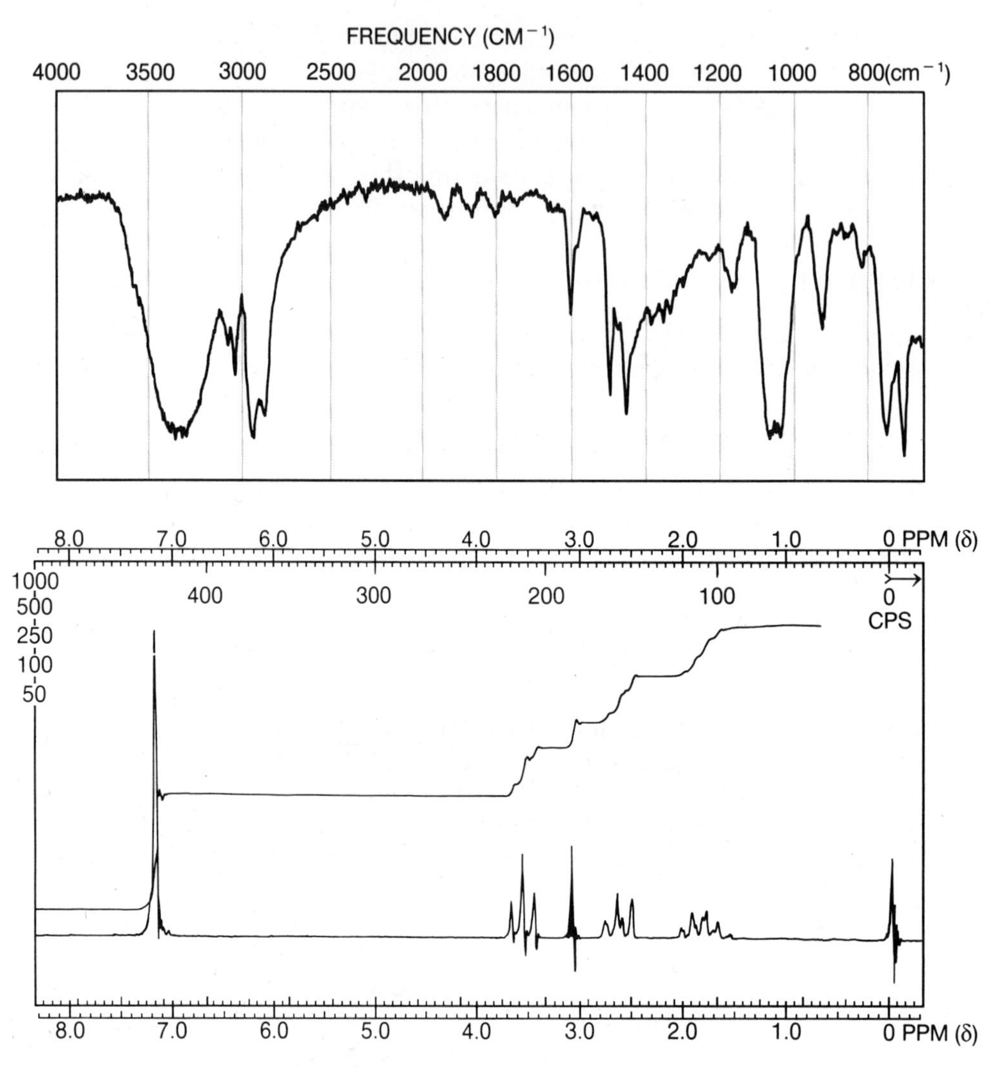

Figure 46.2 Spectra for Example 1.

c. The position of the most deshielded multiplet at 3.54 ppm requires an adjacent electronegative atom such as oxygen or halogen.
d. The filled-in singlet at 3.1 ppm indicates an **OH** proton. (The blackening of a signal in spectra provided with unknowns indicates an **OH** or **NH**, exchangeable with D_2O.)
e. The sharp 5-proton singlet at 7.12 ppm clearly indicates a C_6H_5 grouping.

ANALYSIS. Two major structural features that stand out in both spectra are C_6H_5 and OH groups. Lack of solubility in base confirms that the OH is not located on the ring. The rest of the molecule appears from the NMR spectrum to consist of three CH_2 groups, and the only construction that can be placed on this combination is a $CH_2CH_2CH_2$ chain, linking C_6H_5

and OH. The arrangement is exactly that inferred from the chemical shifts of the three CH_2 multiplets, i.e.,

$$\underbrace{}_{2.62}\ \underbrace{}_{1.83}\ \underbrace{}_{3.54}$$

$$C_6H_5—CH_2—CH_2—CH_2—OH$$

This structure is consistent with the splitting patterns of the CH_2 multiplets. The patterns are complex and cannot be analyzed exactly from this spectrum, but the highest field multiplet (1.83 ppm), which is assigned to the central CH_2 group on the basis of the chemical shift, has the largest number of neighboring protons and is the most highly coupled.

We therefore reach the tentative conclusion that the compound is 3-phenylpropanol, and the boiling point of this compound is recorded as 237°C, in the general range observed. This identification would be confirmed by the preparation of a derivative (p. 507). Of several possibilities reported, the 3,5-dinitrobenzoate ester (mp 92°C) is convenient and has a suitable melting point.

EXAMPLE 2

A liquid, bp 101 to 103°C, slightly soluble in H_2O, neutral toward 5% HCl and 5% NaOH.

1. Infrared Spectrum

a. Weak bands at 2960 and 3100 cm^{-1} suggest both aliphatic C—H and =C—H (either olefinic or aromatic), respectively. No OH or NH peaks are apparent in the 4000 to 3200 cm^{-1} region.

b. The strong band at 1740 cm^{-1} indicates a C=O group; the strong 1230 cm^{-1} band can be assigned to C—O—C stretching, which taken in conjunction with the 1740 cm^{-1} band suggests the possibility of an ester.

c. The weak band at 1650 cm^{-1} suggests C=C, probably not conjugated because of low absorbance.

2. NMR Spectrum

Integration. The spectrum consists of four separate groups of signals. The relative areas from the integration curve, from left to right, are 20:36:33:56 or 1:1.8:1.65:2.8. Dividing by 18 rather than 20 gives 1.10:1.97:1.82:3.09, which is within the 5 to 10 percent error of this measurement for 1:2:2:3.

Chemical Shifts

a. The singlet at 2.06 ppm (3 protons) clearly indicates a CH_3 group slightly deshielded by its substituent.

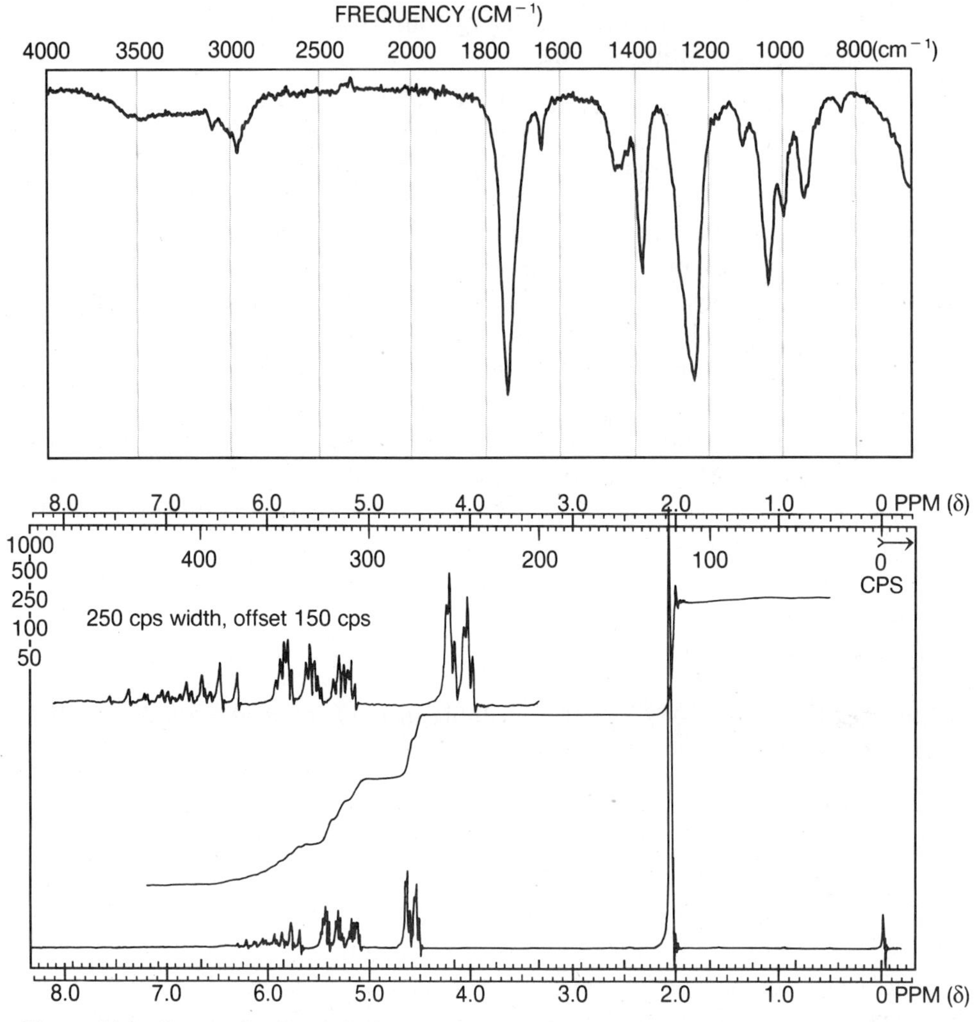

Figure 46.3 Spectra for Example 2.

b. The doublet with small additional splitting at 4.57 ppm (2 protons) suggests a CH_2 group flanked by two deshielding atoms or groups.

c. Multiplets in the range 5 to 6.5 ppm (3 protons total) can be assigned to olefinic protons (directly attached to C═C). The simplest possibility consistent with the extensive splitting, is a vinyl (—C═CH_2) group.

ANALYSIS. Taking the basic conclusions drawn (carbonyl, C—O—C, vinyl, methyl and methylene groups) without any of the qualifications given, numerous structures might be suggested, including ketone, ester, anhydride, and/or ether functional groups. Something of the nature of the carbonyl group can be learned by attempting to prepare a 2,4-dinitrophenylhydrazone of the compound (p. 503). The observation that none is formed indicates that the compound is not a ketone (or aldehyde). Since there is only one carbonyl band in the infrared spectrum, the compound cannot be an anhydride, and is therefore an ester.

The number of possible structures can be further narrowed by noting that the methyl group apears as a singlet in the NMR, indicating that it is not attached to the methylene or vinyl group. The deshielding of the CH_3 is not sufficiently large for it to be attached to an oxygen as in a methyl ester, but is in the correct range for a methyl adjacent to a carbonyl $CH_3CO—$.

From this analysis, allyl acetate, $CH_3CO_2CH_2CH{=}CH_2$, appears to be the most likely candidate. In agreement with this structure is the low field position (large δ value) and complex splitting of the allyl CH_2 signal. Data in Table 10.3 predict a chemical shift for such a methylene group of approximately 4.45 ppm, in good agreement with the 4.6 ppm shift observed. The splitting pattern shows a large (~6 Hz) coupling with the adjacent proton, and smaller coupling of the methylene protons of the vinyl group.

The reported boiling point of allyl acetate (104°C) is in good agreement with the range observed.

Confirmatory evidence for this structure could be obtained by hydrolysis, although isolation of the products, which would be volatile and water-soluble, could not be done easily on a small scale. Derivatives such as an amide of the acid portion could be obtained directly from the ester (p. 515). Another possibility in this case would be the synthesis of an authentic sample of the ester and comparison of their infrared spectra.

EXAMPLE 3

A solid, mp 84°C, insoluble in water, dilute acid, and aqueous bicarbonate, but appreciably soluble in aqueous NaOH.

1. Infrared Spectrum

a. The band at 3300 cm^{-1} is in the range of OH and NH stretching; the sharpness of this peak is inconsistent with OH and amine or amide NH_2 bands in solid-state spectra, and suggests the NH band in a monosubstituted amide, RCONHR.

b. Bands at 2950 and 3040 cm^{-1} (weak) indicate both alkyl and possibly aromatic or olefinic CH.

c. Strong sharp bands at 1705 and 1670 cm^{-1} reveal two carbonyl groups; the former is probably a ketone, and the latter possibly an amide.

d. The sharp 1595 cm^{-1} band is presumably due to an aromatic ring.

e. The strong band at 1530 cm^{-1} when taken together with NH and CO assignments can be attributed to the "amide II" combination and is a further indication of the NHCO grouping.

f. Of the remaining sharp peaks, the only one that can be assigned is the strong 755 cm^{-1} band, which indicates a benzene ring with four adjacent hydrogen atoms.

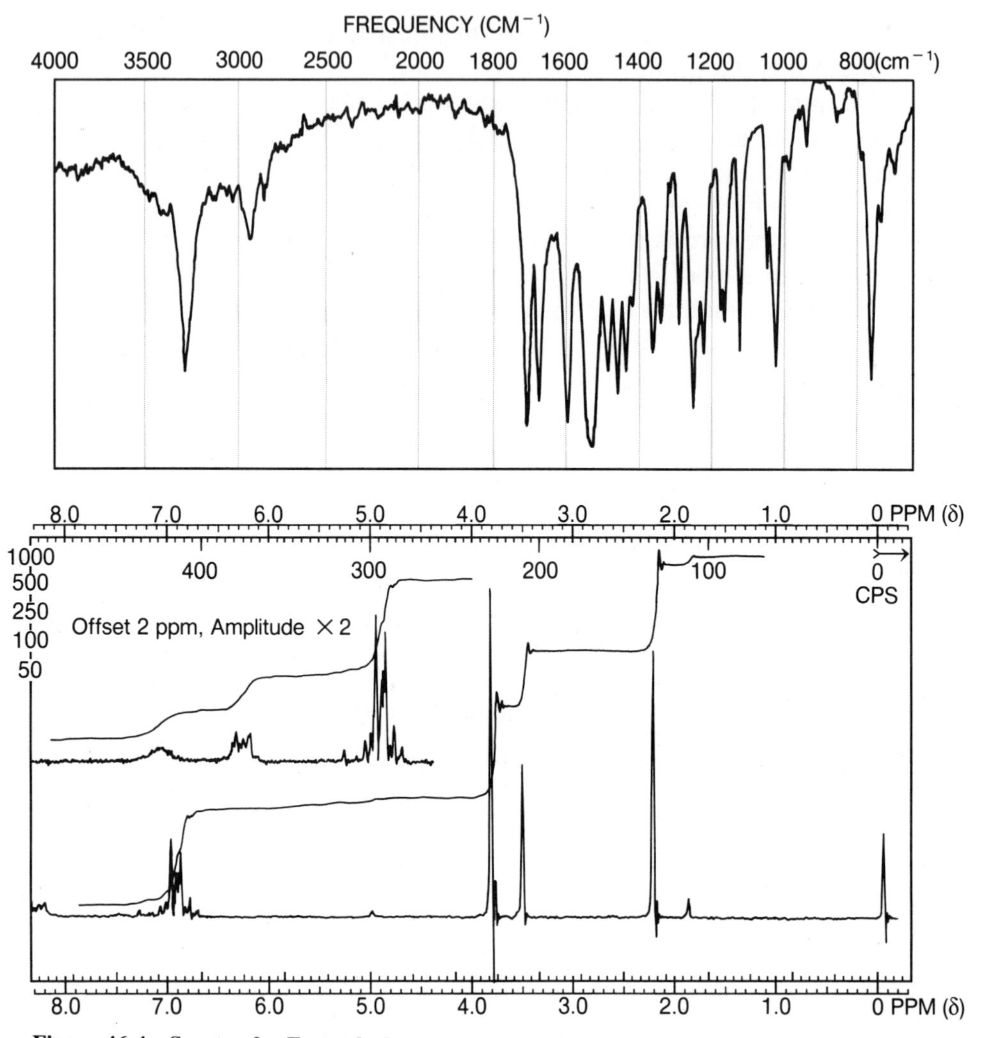

Figure 46.4 Spectra for Example 3.

2. NMR Spectrum

Integration. The spectrum contains six main peaks with an area ratio of 0.9:1.1:3.0:3.0:1.9:2.8, which reduces to 1:1:3:3:2:3, or a total of 13 protons. There are also two very small peaks (1.90 and 5.00 ppm), which are distinct from side bands and suggest the presence of a small amount of an impurity or a second compound.

Chemical Shifts

a. The 3-proton singlet at 2.28 ppm indicates $\mathbf{CH_3}$ adjacent to C=O.
b. The 2-proton singlet is evidently due to $\mathbf{CH_2}$ adjacent to at least one and possibly two deshielding substituents, such as C=O or aryl.

c. The 3-proton singlet at 3.87 must arise from a CH_3O group, and the position indicates that it is attached to an aromatic ring.
d. The 3-proton multiplet at 6.8 ppm suggests three protons with slightly different environments on a benzene ring.
e. The multiplet at 8.25 ppm (upper offset trace) appears to represent a fourth, highly deshielded aryl proton.
f. The broadened peak at 9.1 ppm is characteristic in both shape and position for an amide **NH**.

ANALYSIS. From the solubility data, the compound is a weak acid, probably phenolic or enolic; this inference could be confirmed independently by a color test with ferric chloride. Neither the infrared nor the NMR spectrum indicates an OH group, however.

Taking stock of the data, various lines of evidence establish the presence of the following components: (1) a benzene ring with four (adjacent) hydrogens, (2) *o*-$CH_3OC_6H_4$, (3) CH_3CO, (4) NHCO, accounting for 11 of the 13 protons in the NMR spectrum. The deshielded CH_2 group remains to be located, and the other groups must be connected in some way. There are four possibilities, as follows:

1. *o*-$CH_3OC_6H_4$—CH_2—CONH—$COCH_3$
2. *o*-$CH_3OC_6H_4$—CH_2—NHCO—$COCH_3$
3. *o*-$CH_3OC_6H_4$—CONH—CH_2—$COCH_3$
4. *o*-$CH_3OC_6H_4$—NHCO—CH_2—$COCH_3$

Of these alternatives, the diacylimide (**1**) and the β-ketoamide (**4**) both conceivably could have weakly acidic character. A distinction could be made readily by chemical means, since hydrolysis would lead to completely different types of compounds. The imide (**1**) on treatment with base would give *o*-methoxyphenylacetic acid, ammonia and acetic acid, whereas (**4**) would be hydrolyzed to *o*-anisidine.

1. *o*-$CH_3OC_6H_4CH_2CONHCOCH_3 \longrightarrow$ *o*-$CH_3OC_6H_4CH_2CO_2H$ $+ NH_3 + CH_3CO_2H$

4. *o*-$CH_3OC_6H_4NHCOCH_2COCH_3 \longrightarrow$ *o*-$CH_3OC_6H_4NH_2 + 2\,CH_3CO_2H$

In the event that refluxing the compound with aqueous NaOH gave an oily amine and not ammonia, the β-ketoamide (**4**) would be indicated. The melting point of *o*-acetoacetanisidide is found in tables to be 83 to 85°C, in agreement with that observed for the unknown. A derivative can be readily obtained by acetylation of the anisidine formed on hydrolysis.

Alternatively, the ^{13}C NMR spectrum could have been used to distinguish among the possibilities. The ^{13}C spectrum of the compound is listed below along with peak multiplicities from a partially decoupled spectrum (δ_c^{TMS}, $CDCl_3$ solvent):

204.0 (*s*)	120.2 (*d*)
163.5 (*s*)	110.2 (*d*)
148.4 (*s*)	55.8 (*q*)
127.5 (*s*)	50.9 (*t*)
124.1 (*d*)	30.7 (*q*)
120.9 (*d*)	

The signals at 204 and 163.5 ppm indicate two carbonyl carbons, probably a ketone and either an amide, imide, ester or anhydride, respectively (cf. Fig. 10.10 and Table 10.13). The six signals between 110 and 149 ppm suggest an unsymmetrically substituted benzene ring. The multiplicities indicate two substituted carbons (at 148.4 and 127.5 ppm) and four unsubstituted carbons. Since there are six signals, the ring must be *ortho* or *meta* disubstituted. (Why?) The methyl group at 55.8 ppm must be attached to an oxygen to be so deshielded (cf. ethane, 5.7 ppm, and Table 10.10); in fact, this must be the methoxy group on the ring indicated by the proton NMR, and it must be attached to the carbon at 148.4 ppm, since that is the only aromatic carbon that is sufficiently deshielded (cf. Table 10.11). The two remaining signals (at 50.9 and 30.7 ppm) are CH_2 and CH_3 carbons, respectively. The former is probably between two moderately deshielding substituents, but since α-substituent effects are not additive, nothing specific can be said about the flanking groups. A decision between structures **1** and **4**, which could not be made on the basis of the ^{1}H NMR spectrum, is possible from the ^{13}C shift of 30.7 for the CH_3 group. This value is consistent, within the expected error of ±5%, for the CH_3 in a ketone (δ 5.7 for CH_3CH_3 + 22 = 27.7 ppm; cf. Table 10.10), but is well outside the range for an amide. The ^{13}C NMR peak assignments for structure **4** are summarized in Figure 46.5.

A further point in the proton NMR spectrum is of significance in the light of the β-ketoamide structure. The two small peaks at 1.90 and 5.00 ppm are in the positions expected for the methyl and methine protons,

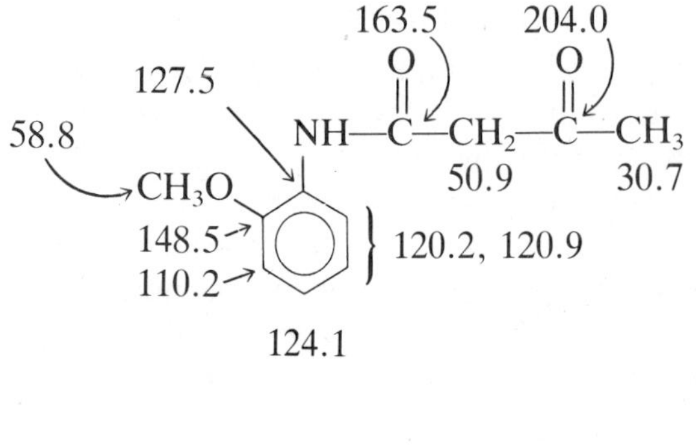

163.5 204.0
127.5
58.8 NH—C(=O)—CH_2—C(=O)—CH_3
CH_3O 50.9 30.7
148.5 120.2, 120.9
110.2
124.1

Figure 46.5 Assignments of ^{13}C NMR signals of Example 3.

respectively, of the enol form *o*—$CH_3OC_6H_4NHCOC\mathbf{H}{=}C(OH)C\mathbf{H}_3$. The integral of the 1.90 ppm peak matches the slight deficiency in the $\mathbf{CH}_3CO$ integral (2.8 vs. 3.0 for the other three-proton signals) and suggests about 5% of the enol. Such a small amount would not give recognizable **OH** peaks with the recording conditions used for either the infrared or NMR spectrum.

ADDITIONAL DATA ON UNKNOWNS

After assessing the information obtained from physical properties, solubility, and spectra, as illustrated in the preceding pages, it frequently will be possible to draw a tentative conclusion as to the identity of the unknown and proceed to selection and preparation of a derivative. In other cases, certain features will be established, but further data will be desirable to define the environment, or even the nature of the functional group, or to decide between two possible interpretations. Additional information can usually be obtained from one or more of the approaches discussed in the following paragraphs. More complete information on these and other procedures can be obtained from the references at the end of the chapter.

DETECTION OF OTHER ELEMENTS

One important item that may be needed is information on the presence of elements other than C, H, and O. Although C—Cl bonds can sometimes be seen in the infrared spectrum (800 to 600 cm^{-1}), these bands may be obscured; C—Br and C—I stretching bands occur in the far infrared (<600 cm^{-1}). A qualitative test for halogen may therefore be indicated, particularly if the NMR spectrum suggests an odd number of protons. The common functional groups containing nitrogen are usually revealed by the basicity (amines) or the infrared spectrum (amide CO, C≡N, NO_2), but a confirmatory test for nitrogen can readily be carried out.

Detection of N, Cl, Br, and I is accomplished by the total decomposition of the compound with hot metallic sodium, followed by detection of anions in the usual way. With rare exceptions, any compound containing a C—N bond will give cyanide ion under these conditions; excess carbon is usually converted to the amorphous element.

$$[C, H, O, N, X, S] \xrightarrow{Na} \xrightarrow{H_2O} C, OH^-, CN^-, X^-, S^=$$

In this and other tests that require observation of a positive or negative result, there is one all-important rule: *always run a control.* It is quite futile to attempt a conclusion from the reaction of an unknown when the observer does not know exactly the appearance of a positive and a negative result. Run a known compound first, on the scale that will be used for the unknown, to see the behavior and to insure that you are doing it properly. It may be equally important in some cases to run a *blank* in order to observe a negative result. If there is doubt about the result with a known, clear it up before turning to the unknown.

SAFETY NOTE METALLIC SODIUM IS EXTREMELY REACTIVE WITH WATER OR HYDROXYLIC SOLVENTS. HANDLE AND TRANSFER WITH A SPATULA; THE METAL IS SOFT, AND A SMALL PIECE CAN BE PICKED UP BY SPEARING IT WITH THE TIP OF A SPATULA. THE SODIUM FUSION SHOULD BE CARRIED OUT BEHIND A SHIELD OR HOOD WINDOW, AND ALWAYS WEAR SAFETY GOGGLES.

PROCEDURE

Obtain a small piece of sodium (a cube about 4 mm on edge) and place in a clean *dry* 10 × 75-mm test tube. Handle the test tube with a holder or clamp, not with your fingers. Have ready a sample of approximately 100 mg of a solid unknown or 0.1 mL (2 drops) of a liquid. Heat the sodium over a flame until it melts and begins to glow red. Add the sample to the tube, making sure that it falls directly on the molten sodium and not on the wall of the test tube where it may be volatilized. With a very volatile liquid, add a second 0.1-mL sample. Heat the tube briefly again and allow it to cool. Add 1 mL of ethanol dropwise and stir thoroughly with a glass rod to dissolve the excess sodium. After bubbling (hydrogen evolution) has stopped, cautiously add, dropwise, distilled water. Heat the mixture to boiling, filter through paper, and rinse with distilled water. If the solution is brown in color (indicating incomplete decomposition in the fusion), add a little charcoal and filter again.

TEST FOR N. This test, and the alternate one described below, are quite sensitive and care must be taken that a false positive is not obtained because of nitrogen containing impurities. If there is doubt, a standard should be run. In a 13 × 100-mm test tube, mix 10 drops of a 0.1 *M* solution of *p*-nitrobenzaldehyde in 2-methoxyethanol (methyl cellosolve), 10 drops of a 0.1 *M* solution of *o*-dinitrobenzene in 2-methoxyethanol, and 1 drop of 0.50 *M* aqueous sodium hydroxide, and then add 1 drop of the sodium fusion solution. The formation of a deep blue-purple color indicates the presence of nitrogen, while a yellow or tan color constitutes a negative test.

ALTERNATE TEST FOR N. To 1 mL of thc solution, add 2 drops of saturated ferrous ammonium sulfate solution and 2 drops of 30% potassium fluoride solution. Boil the mixture for 30 seconds and acidify by dropwise addition of 30% sulfuric acid until the iron hydroxide just dissolves. The appearance of a brilliant blue precipitate of Prussian blue indicates the presence of cyanide ion and nitrogen in the compound.

TEST FOR S. Acidify 1 mL of the solution with acetic acid and add several drops of 0.1 *M* lead(II) acetate solution. A black precipitate of lead sulfide indicates the presence of sulfur.

TEST FOR HALOGEN. Acidify a 2-mL portion of the alkaline fusion solution in a small beaker with 3 *M* nitric acid and boil briefly to expel any HCN that may be present. The appearance of a distinct white or yellow precipitate on addition of silver nitrate indicates the presence of halogen. A yellow color suggests bromide or iodide.

To differentiate the halogens in the event of a positive silver halide precipitate, acidify another 2- to 3-mL portion of the original solution and add approximately 0.3 mL of methylene chloride. Then add a few drops of fresh chlorine water, a few mg of calcium hypochlorite, or several drops of Clorox (be sure the pH remains acidic). A yellow-orange color in the CH_2Cl_2 layer indicates bromine; a violet color, iodine.

ALTERNATE TEST FOR HALOGEN. A rapid test for chlorine, bromine, or iodine can be carried out on the original sample using the **Beilstein procedure**, based on the formation of a volatile copper halide when an organic halide is strongly heated with copper oxide. The value of the test is that it is easily carried out and is highly sensitive. To perform the test, a small loop of copper wire is heated to redness in a Bunsen burner flame until the flame is no longer colored. The wire is cooled and the now oxide-coated loop is immersed into a small amount of the solid or liquid to be tested. On reheating in the Bunsen flame, a blue-green flame indicates the presence of chlorine, bromine, or iodine (copper fluoride is nonvolatile).

Although the test is rapid, there are potential problems. Occasionally, the blue-green flame is fleeting and one must look quickly so as not to miss it. Further, highly volatile substances may vaporize before the copper halide forms, and such compounds as urea, quinoline and pyridine derivatives can give a false positive because of the formation of copper cyanide. Thus, the test should always be confirmed by other methods such as those mentioned earlier.

EQUIVALENT WEIGHTS

There are many methods for the quantitative determination of various functional groups in organic compounds; these reveal the number of such groups in the molecule if the molecular formula is known. If the formula of the compound is not known, quantitative analysis of a particular group provides the *equivalent weight*, i.e., molecular weight ÷ the number of functional groups per molecule.

The simplest of all these quantitative methods is acid–base titration, which is used to determine the *neutralization equivalent* of acids. Practically any carboxylic acid can be titrated with standard base to a sharp end point with phenolphthalein, since the pH at complete neutralization for an acid of pK_a 3 to 6 is somewhat above 7. With care, the neutralization equivalent of a pure, dry acid can easily be determined with an accuracy of 1%, providing very useful information about an unknown acid. The neutralization equivalent of a monobasic acid is the molecular weight, whereas for dibasic or polybasic acids, it is some integral fraction of the molecular weight. The procedure given can also be used for the neutralization equivalent of amine salts, but it is generally difficult to obtain the latter sufficiently pure and dry for accurate results.

PROCEDURE

Weigh out a 150- to 200-mg sample of benzoic acid to ± 1 mg on an analytical balance. Place the acid in a 125-mL Erlenmeyer flask with 50 mL of water, add 2 to 3 drops of phenolphthalein solution, and titrate to a pink end point with ~0.1 *M* NaOH from a 25- or 50-mL buret. From the volume of titrant and weight of benzoic acid (eq wt 122) used, calculate the exact molarity of the NaOH solution. As a check of your technique and the result obtained, titrate a second sample of benzoic or another known acid.

Repeat the process using the standardized base with a similar accurately measured quantity of the unknown acid (*in duplicate*), and calculate its equivalent weight. Liquid acids should be freshly distilled and weighed in a capped vial with minimum exposure to air, since they are generally hygroscopic. Solid unknown acids should be recrystallized and thoroughly dried. Acids that are very insoluble in water can be titrated in aqueous alcohol. If this is necessary, the standardization should be carried out with the same amount of added alcohol.

Other quantitative methods, including the saponification equivalent of an ester or quantitative hydrogenation of an olefinic compound, may be desirable in certain cases. Procedures for these or other special methods should be obtained from the literature or the references at the end of the chapter, and arrangements should be made with the instructor for necessary reagents and equipment.

MISCELLANEOUS CHEMICAL TESTS

A large variety of qualitative chemical tests for organic functional groups have been developed. Their use in structural analysis has been largely replaced by spectral methods, but a few of the simpler tests may be found useful in resolving ambiguities in other data. *In each case the test should be run on known compounds similar to those suspected to be the unknown to insure that the reagent is correctly prepared and so as to be able to recognize the appearance of a positive test.*

PROCEDURES

CHROMIC ACID TEST FOR 1° AND 2° ALCOHOLS. This test, described in Chapter 15, is a reliable method for confirming the presence of a primary or secondary hydroxyl group (aliphatic aldehydes also react). A rapid color change to dark green constitutes a positive test. With amines or phenols, a brown color and a dark precipitate are usually seen.

FERRIC CHLORIDE TEST FOR PHENOLS. The presence of phenolic or enolic hydroxyl groups in a compound is usually indicated by formation of a red or violet color due to an Fe(III) complex when treated with ferric chloride solution. Some phenols give negligible or very weak tests because of interference by substituents in the ring.

The test is carried out by adding 1 drop or a few crystals of the unknown to 1 mL of freshly prepared 1% $FeCl_3$ solution.

IODOFORM TEST. Methyl ketones ($RCOCH_3$) and also methyl carbinols ($RCHOHCH_3$) react rapidly with iodine under basic conditions to give a carboxylic acid and iodoform. (Also see Chapter 20.)

$$RCHOHCH_3 \xrightarrow[OH^-]{I_2} RCOCH_3 \xrightarrow[OH^-]{3I_2} RCOCI_3 \xrightarrow{OH^-} RCO_2^- + CHI_3\downarrow$$

To carry out the test, dissolve approximately 20 mg or 1 drop of the unknown in 0.5 mL of water (if insoluble in water, dissolve in 0.5 mL of methanol) and add 0.5 mL of 10% aqueous NaOH. Add dropwise, with shaking, a solution of KI_3* until the dark iodine color persists. Warm the solution slightly, and if the color fades, add more KI_3 until the color remains for 1 to 2 minutes at 50°C. Then add a drop or two of NaOH solution to remove excess iodine, and dilute with water. Iodoform, if present, will separate as a dense, pale yellow solid, mp 119 to 121°C.

2,4-DINITROPHENYLHYDRAZINE TEST FOR ALDEHYDES AND KETONES. This reagent is commonly used for preparing dinitrophenylhydrazones of aldehydes and ketones and can also be used as a qualitative test for these functional groups to distinguish them from other carbonyl compounds, particularly esters, that do not react. The procedure used for preparing 2,4-dinitrophenylhydrazones (p. 513) is followed, and the formation of a red or orange precipitate constitutes a positive test. The test is also described in Chapter 20.

Since the reagent solution contains sulfuric acid, amines may give a heavy precipitate of the amine sulfate, which appears yellow in the solution and can be mistaken for a positive test.

---

*Prepare by dissolving 5 g KI and 2.5 g I_2 in 25 mL of water.

TOLLENS TEST FOR ALDEHYDES. Tollens reagent, a solution of silver ammonia complex $Ag(NH_3)_2{}^+$, provides a test (also described in Chapter 20) to distinguish between aldehydes and ketones. Aldehydes react by undergoing oxidation, and silver ion is reduced to form a metallic silver mirror.

$$RCHO + Ag(NH_3)_2^+ + 2OH^- \rightarrow RCO_2{}^-NH_4^+ + 2Ag\downarrow + H_2O + 3NH_3$$

In a clean test tube (water should drain out without leaving beads), place 1 mL of 5% $AgNO_3$ solution and add, dropwise, a dilute solution (1 to 2%) of aqueous ammonia, shaking after each drop, until the brown precipitate of silver oxide just dissolves (avoid excess ammonia). To this solution add 2 to 3 drops or a few crystals of the unknown. Formation of a silver mirror indicates a positive test.

SAFETY NOTE THIS REAGENT MAY FORM AN EXPLOSIVE PRECIPITATE ON STANDING; DO NOT MAKE UP A LARGER AMOUNT AND STORE FOR LATER USE. EXCESS TOLLENS REAGENT CAN BE DESTROYED BY THE ADDITION OF DILUTE NITRIC ACID.

HINSBERG TEST FOR AMINES. Primary and secondary amines react with benzenesulfonyl chloride in the presence of sodium hydroxide to give solid benzenesulfonamides. The sulfonamide from a simple primary amine is usually soluble in excess hydroxide solution because of the acidity of the —$NHSO_2Ar$ group. Tertiary amines do not react under mild conditions.

$$RNH_2 + C_6H_5SO_2Cl \xrightarrow{OH^-} RNHSO_2C_6H_5 \xrightarrow{OH^-} \underset{\text{soluble in aq NaOH}}{[R\ddot{N}SO_2C_6H_5]^-Na^+}$$

$$R_2NH + C_6H_5SO_2Cl \xrightarrow{OH^-} \underset{\text{insoluble in NaOH}}{R_2NSO_2C_6H_5}$$

The test can be ambiguous if the amine is insoluble in water or is a solid, since it is necessary to distinguish between unreacted amine in excess, unreacted benzenesulfonyl chloride, and an insoluble sulfonamide.

To carry out the test (also described in Chapter 27), mix approximately 50 mg or 1 drop of the amine and 0.2 g of benzenesulfonyl chloride in 4 mL of 10% aqueous NaOH solution, stopper the test tube and shake vigorously until the oily sulfonyl chloride has reacted (5 minutes or longer may be needed). If a clear or nearly clear solution is obtained, a primary amine is indicated; acidification should give a precipitate of $RNHSO_2C_6H_5$.

Formation of a significant amount of insoluble solid from a liquid amine indicates that the unknown is a secondary amine. To distinguish from the possibility of unreacted solid amine, acidify the mixture; an amine will be soluble.

DERIVATIVES

The final point in the identification is conversion of the unknown to a solid derivative whose melting point can be compared with a literature value. The derivative may be any compound that is formed in a reaction that is characteristic for the unknown. It is obviously desirable to choose, when possible, a derivative that can be obtained in good yield and readily isolated, and that has a melting point in the most convenient region, i.e., 80 to 180°C. In many cases, the derivative will be confirmatory evidence for a tentative conclusion, but occasionally it may be necessary to choose between two possible candidates; in this case, the derivative or reaction product from both compounds must be known.

In all cases the derivative should be recrystallized from a suitable solvent before measuring its melting point, since the melting point will be the basis for identification of the derivative. The product that precipitates or crystallizes from the reaction mixture may be quite impure, and an incorrect melting point is worse than none at all in this situation.

For acids, alcohols, amines, and carbonyl compounds, a variety of simple condensation products can serve as derivatives, and a number of these are given in the Tables of Derivatives at the end of this chapter. Procedures for the more useful of these standard derivatives are given below. Solid acids obtained by hydrolysis of esters, amides, or nitriles are usually satisfactory derivatives if the acid represents the major portion of the molecule. It may, however, be much better to characterize the entire molecule by a more specific reaction. For example, hydrolysis of acetanilide gives acetic acid and aniline, both liquids. Either the acid or amine or both could be isolated and converted to solid derivatives, but a much simpler and far more effective derivative is obtained by nitration to *p*-nitroacetanilide (see p. 300) or bromination to the *p*-bromo derivative.

A number of the unknowns may lend themselves to special derivatives that will be suggested by the tentative structure deduced from spectral and other data. Cyclization products can be readily obtained from many bifunctional compounds, and occasionally rearrangement or partial degradation will provide highly characteristic derivatives. In such cases, details for carrying out the reaction and isolation of the product should be obtained from the literature for the specific compound that is under consideration.

NOTE ON DERIVATIVES. In directions for preparing derivatives of unknowns, amounts of reagents are given for a compound of "average" molecular weight, and solvents suggested are for compounds with typical solubilities. For certain unknowns, these directions may not be optimum, and it may be desirable to modify the procedure according to the properties expected for the unknown in question. For experiments on a **microscale**, the quantities given can usually be carried out at one-half to one-tenth scale.

Merely reduce the amounts of sample, reagents, and solvents by the required amounts.

SAFETY NOTE IN MANY OF THESE PROCEDURES, HIGHLY REACTIVE REAGENTS, SUCH AS ACID CHLORIDES AND ISOCYANATES, OR STRONG ACIDS AND BASES, ARE USED. MOST SOLVENTS ARE FLAMMABLE. KEEP IN MIND THE SAFETY NOTES THAT YOU HAVE SEEN THROUGHOUT THIS BOOK AND THE STANDARD PRECAUTIONS. ALWAYS WEAR SAFETY GLASSES. PIPET ONLY WITH A BULB. WASH THOROUGHLY WITH WATER IF ANY CHEMICALS COME IN CONTACT WITH THE SKIN. IF IN DOUBT, WORK IN A HOOD.

ACIDS

The most satisfactory general derivatives of carboxylic acids are amides, particularly the anilides and *p*-toluidides, which are prepared by the general sequence:

$$RCO_2H + SOCl_2 \rightarrow RCOCl + SO_2 + HCl$$

$$RCOCl + 2\ R'NH_2 \rightarrow RCONHR' + R'NH_3^+Cl^-$$

The second reaction requires the use of some base to combine with HCl. In the derivatization of an acid, an excess of the amine is usually used, but if the amine is the important component, some other base, usually aqueous NaOH, is used as the acid acceptor.

It should be noted that in the reaction with thionyl chloride, polyfunctional acids will often give cyclization or condensation products rather than an acid chloride. Thus, dibasic acids may form anhydrides, and α-acylamino acids give oxazolones (azlactones).

$$HOOCCH_2CH_2COOH \xrightarrow{SOCl_2} \text{(cyclic } CH_2{-}CH_2{-}C(=O){-}O{-}C(=O)\text{)} \quad \text{(an anhydride)}$$

$$RCONHCH_2CO_2H \longrightarrow \text{(cyclic } N{-}CH_2{-}C(=O){-}O{-}C(R){=}N\text{)} \quad \text{(an oxazolone)}$$

PROCEDURES

ACID CHLORIDE. A very suitable procedure is that given for naphthalene-1-acetic acid in Chapter 32. Cool a mixture of 1 g or 1 mL of the acid plus 1 mL of thionyl chloride in a test tube and add 0.2 mL of dimethylformamide. Allow the mixture to warm and stand 30 minutes, and then proceed with the next step.

AMIDES. For conversion to the anilide or *p*-toluidide, dilute the acid chloride with 5 to 10 mL of methylene chloride, and add it to a solution of 1.5 g of aniline or *p*-toluidine in 10 mL of methylene chloride. After mixing, allow the reaction to stand for a few minutes, add 10 mL of water and transfer the mixture to a separatory funnel. Add more solvent if necessary to dissolve all of the amide. At this point it is convenient to add sufficient ether to make the organic layer lighter than water. Wash the organic phase with dilute HCl until the aqueous layer is acidic, then with bicarbonate solution and finally water. Dry the organic phase ($MgSO_4$) and evaporate the solvent on the steam bath and recrystallize and collect the derivative.

Conversion to the amide may be preferable if the acid is a high-melting solid. In this case, add the acid chloride, without dilution, *dropwise* to a mixture of 10 mL of concentrated ammonium hydroxide and an equal volume of ice.

ALCOHOLS

The most generally useful derivatives of alcohols are esters of substituted benzoic or carbamic acids; the carbamate esters are commonly called urethanes. The *p*-nitrobenzoates or 3,5-dinitrobenzoates are obtained by treatment of the alcohol with the acid chloride and pyridine, which serves as a catalyst and acid acceptor.

$$ArCOCl + ROH + C_5H_5N \rightarrow ArCO_2R + C_5H_5NH^+Cl^-$$

Urethanes are prepared from the alcohol and an aryl isocyanate;

$$ArN{=}C{=}O + ROH \rightarrow ArNHCO_2R$$

In any reactions with isocyanates, the following sequence leading to the diarylurea will occur if water is present.

$$ArN{=}C{=}O + H_2O \rightarrow [ArNHCO_2H] \rightarrow ArNH_2 + CO_2$$

$$ArN{=}C{=}O + ArNH_2 \rightarrow ArNHCONHAr$$

The urea is a high-melting insoluble compound (diphenylurea, mp 238°C) and can seriously interfere with the isolation of the desired derivative. Alcohols should be dry, and an excess of isocyanate must be avoided.

Chromic acid at room temperature converts primary alcohols to carboxylic acids and secondary alcohols to ketones. The latter transformation is often useful in dealing with an aliphatic alcohol, since derivatives of the ketone may be more reliable. The experimental procedure in Chapter 19 for the preparation of norcamphor can be followed, adjusting the scale to the amount of unknown available.

PROCEDURES

***p*-NITRO- AND 3,5-DINITROBENZOATES.** Dissolve 1 mL of the alcohol in 3 mL of pyridine and add 0.5 g of the acid chloride corresponding to the derivative desired. Warm the solution for a few minutes and pour into 10 mL of water. If a well-crystallized ester separates, this is collected; otherwise, extract the product with ether and wash the ether solution free of pyridine with dilute HCl, dry, and evaporate.

URETHANES. To approximately 0.1 mL or 0.1 g of the anhydrous alcohol, add about 0.1 mL of phenyl isocyanate (**CAUTION**: lachrymator) and warm the solution for a few minutes in a beaker of hot water (do not expose to steam bath vapors). Cool, add about 1 mL of high boiling petroleum ether and heat to dissolve the product. If a very slightly soluble precipitate remains, it is probably the urea. This must be removed by filtration, followed by cooling the filtrate in ice, and scratching to induce crystallization.

PHENOLS

Nitrobenzoates or urethanes of phenols can be obtained by the same procedures described for alcohols. For a few phenols, only the benzoate esters are described: these should be prepared by the procedure given for eugenol benzoate (Chapter 8).

For a few phenols, derivatives can be obtained by reaction with chloroacetic acid to give the aryloxyacetic acid.

$$\text{ArOH} + \text{ClCH}_2\text{CO}_2\text{H} \xrightarrow{\text{NaOH}} \text{ArOCH}_2\text{CO}_2\text{Na} \xrightarrow{\text{H}_3\text{O}^+} \text{ArOCH}_2\text{CO}_2\text{H}$$

PROCEDURE

ARYLOXYACETIC ACIDS. To 1 g or 1 mL of the phenol add 5 mL of 30% NaOH solution and 1.5 g of chloroacetic acid. Stir and warm the mixture, and if solid is present, add water dropwise to obtain a clear solution. Heat the solution on the steam bath for 30 to 40 minutes, cool, and dilute with 10 mL of water. Acidify the solution (test paper) with 6 *M* HCl and extract the acid with ether. Wash the ether solution with a little water and then extract with saturated bicarbonate solution. Cautiously acidify the bicarbonate solution; stir to prevent loss from foaming. Collect the precipitate of aryloxyacetic acid and recrystallize from hot water.

ETHERS

A variety of derivatives are possible for aryl ethers; most of these involve reactions of the aromatic ring rather than the ether linkage. Examples include bromination, chlorosulfonation followed by conversion to the sulfonamide, and nitration. The benzene ring of simple aryl ethers is also sufficiently electron-rich that stable charge transfer complexes with picric acid can be isolated and characterized by melting points. For highly substituted aryl ethers, a more appropriate derivative may be the corresponding phenol obtained by cleavage of the ether linkage.

Enol ethers may be hydrolyzed to a carbonyl compound as described on page 514. No generally useful derivatives exist for saturated aliphatic ethers.

PROCEDURES

SAFETY NOTE BROMINE CAUSES SEVERE BURNS ON CONTACT WITH THE SKIN. AS FIRST AID FOR BROMINE BURNS, APPLY A SOLUTION OF SODIUM THIOSULFATE TO THE AFFECTED AREA. IN SEVERE CASES, SEEK MEDICAL HELP. AVOID BREATHING THE VAPORS OF THE BROMINATING REAGENT AND CARRY OUT THE REACTION IN A HOOD IF POSSIBLE.

BROMINATION. Dissolve 0.2 g or 0.2 mL of the aryl ether in 2 mL of acetic acid, and add a solution of 10% bromine in acetic acid in 0.5-mL portions until the yellow color persists for at least a minute after mixing. Add 5 mL of water, mix well, and collect the precipitate by filtration. Rinse the solid with water, and recrystallize from aqueous ethanol or hexane.

NITRATION. Aryl ethers can be nitrated under the conditions used for acetanilide in Chapter 25, adjusting the procedure to use 0.2 to 0.5 g or less.

CHLOROSULFONATION. Aryl ethers (and halides) can be converted to the sulfonyl chlorides with chlorosulfonic acid, and further to the sulfonamides (Chapter 38). The latter are generally more useful derivatives because of their higher melting points. Dissolve 0.2 g of the unknown in 1 mL of chloroform (**CAUTION:** carcinogen) in a test tube and cool the solution in an ice bath. In the hood, add 1 mL of chlorosulfonic acid dropwise, and let the mixture warm to room temperature. If the ring is deactivated by one or more halogen substituents, warm the solution in a beaker of warm (50 to 70°C) water for 10 to 15 minutes. Add 5 mL of ice water and mix well by stirring or shaking, pipet off the aqueous layer, and pour the chloroform solution into 5 mL of concentrated ammonium hydroxide. Mix well and evaporate the chloroform by heating on a steam bath in the hood. Add 2 mL of 10% NaOH to dissolve the sulfonamide, filter out any undissolved impurities, and reacidify with dilute hydrochloric acid. Collect the sulfonamide and recrystallize it from ethanol or aqueous ethanol.

PICRIC ACID COMPLEXES. Dissolve 0.2 g of the aryl ether in a minimum volume of warm chloroform and add a solution of 0.25 g of picric acid in 1 mL of boiling chloroform (**CAUTION:** Chloroform is carcinogenic—carry out this procedure in a hood). Mix well, let the solution cool, and collect the crystals. Determine the melting point of the complex as soon as possible since some picrates decompose on standing.

CLEAVAGE OF ETHERS. Place 0.2 g or 0.2 mL of the alkyl aryl ether, 2 mL of acetic acid, and 2 mL of concentrated hydriodic acid in a test tube. Add a boiling stone and heat in a steam bath *in the hood* for 1 hour. Cool the solution, and pour into 25 mL of water. Add, in small portions, sufficient solid $NaHCO_3$ to neutralize the solution (pH ~8), mixing well between additions. Transfer the mixture to a separatory funnel and extract the phenol with two 5-mL portions of methylene chloride or ether. Wash with water, dry, and evaporate the solvent to obtain the phenol.

AMINES

A variety of amides and ureas are available as derivatives of 1° and 2° amines. Amides derived from secondary amines are often much more difficult to isolate because the R_2NCOAr structures are not hydrogen-bonded, and the compounds are therefore lower-melting and more soluble. The ureas are generally better derivatives. Picrate salts are useful derivatives for certain tertiary amines.

PROCEDURES

ACETAMIDES. Acetamides of aromatic amines are readily prepared by dissolving approximately 0.5 g of the amine in 1 to 2 mL of acetic anhydride and heating for a few minutes. Then add a few mL of water and warm the mixture until the excess acetic anhydride is destroyed (second liquid layer disappears). If the amide does not crystallize directly, extract with ether, wash the ether solution with bicarbonate solution, dry, and evaporate.

BENZAMIDES. Mix the amine (1 g) with 10 to 15 mL of 10% sodium hydroxide solution and add 2 mL of benzoyl chloride. Shake or stir the mixture for 10 minutes and isolate the amide by collecting the solid and washing with water. In some cases it may be desirable to add a solvent such as methylene chloride and recover the amide from solution.

SUBSTITUTED UREAS AND THIOUREAS. The reaction of an amine with an isocyanate or isothiocyanate leads to the corresponding urea; the precautions about water mentioned under alcohols must be kept in mind with isocyanates. Isothiocyanates are much less reactive and the reaction with amines occurs on warming, but hydrolysis with water is negligible and an excess of the reagent can be removed by recrystallization.

$$ArN{=}C{=}O + RNH_2 \rightarrow ArNHCONHR$$

$$ArN{=}C{=}S + RNH_2 \rightarrow ArNHCSNHR$$

To 1 g of the amine, add 0.5 mL or less of the isocyanate or isothiocyanate; a few mL of methylene chloride or toluene can be used as diluent. Warm the solution if an isothiocyanate was used. Excess amine should be removed by washing the solution with dilute acid before isolating the urea.

PICRATES. Dissolve 0.2 g of the tertiary amine in 5 mL of 95% ethanol and add the solution to 5 mL of a saturated solution of picric acid in ethanol. Heat the mixture to boiling; let it cool to room temperature. Collect the crystals by filtration and recrystallize the product from methanol or ethanol.

AMIDES AND NITRILES

Alkaline hydrolysis of these compounds leads to the acid salt and ammonia or the free amine.

$$RCONHR' + H_2O \xrightarrow{NaOH} RCO_2Na + R'NH_2$$

$$RC{\equiv}N + H_2O \xrightarrow{NaOH} RCO_2Na + NH_3$$

Amides or esters of carbamic acids give amines, alcohols, and CO_2.

$$CH_3NHCONH_2 + H_2O \xrightarrow{NaOH} CH_3NH_2 + NH_3 + CO_3^{=}$$

$$C_6H_5NHCO_2C_2H_5 + H_2O \xrightarrow{NaOH} C_6H_5NH_2 + C_2H_5OH + CO_3^{=}$$

With high-melting amides or nitriles of aromatic acids, the rate of hydrolysis and the solubility are low, and alcoholic alkali or hydrolysis in acid is recommended. Most nitrogen-containing compounds are quite soluble in 40 to 60% sulfuric acid and fairly high temperatures can be obtained by refluxing such solutions.

$$ArCONHR' + H_2O \xrightarrow{H_2SO_4} ArCO_2H + R'NH_3^+ \ HSO_4^-$$

An additional possibility for the characterization of *aromatic* nitriles is conversion to the amide. This reaction, which strictly speaking is not a hydrolysis but a hydration, is carried out by treatment of the nitrile with alkaline hydrogen peroxide; hydroperoxide ion (O_2H^-) is the specific reagent involved:

$$RC{\equiv}N + 2\,H_2O_2 \xrightarrow{OH^-} RCONH_2 + H_2O + O_2$$

The hydrolysis and hydration procedures given are generally useful only for amides or nitriles of aromatic acids or amides of aromatic amines. Low molecular weight aliphatic amides or nitriles cannot easily be converted to useful derivatives by hydrolysis, since isolation of the product is usually impractical. Reduction of these compounds to amines with sodium borohydride in the presence of cobalt(II) chloride is one alternative that can generally be used to obtain a derivative:

$$R'CONHR \xrightarrow[CoCl_2]{NaBH_4} R'CH_2NHR$$

$$RC{\equiv}N \xrightarrow[CoCl_2]{NaBH_4} RCH_2NH_2$$

PROCEDURES

ALKALINE HYDROLYSIS. Add 1 mL of a liquid or 1 g of a solid amide or nitrile to 30 mL of 10% NaOH or KOH solution and reflux the solution for 30 minutes. Test for the presence of a volatile amine or ammonia by holding a piece of moist indicator paper at the top of the condenser. Cool the solution to room temperature. If an insoluble amine is present, or if the presence of an aliphatic amine of intermediate solubility is suspected from the odor, extract the solution with several 10- to 15-mL portions of ether, wash the ether solution, dry with $MgSO_4$ or K_2CO_3, and evaporate the solvent to obtain the amine. A low molecular weight aliphatic amine will probably be lost in this procedure because of its solubility in water and its volatility during removal of ether.

After removing the amine, if any, acidify the aqueous solution and collect the acid if it is a solid. A low molecular weight aliphatic acid can be isolated by thorough extraction, but will usually be of little value as a derivative.

ACID HYDROLYSIS. Slowly add 5 mL of concentrated sulfuric acid to 10 mL of water. To this solution add 1 g of the amide or nitrile (this procedure will generally be used with high-melting solids). Reflux the solution for 30 to 60 minutes, then cool in ice and dilute with an equal volume of water. If a solid acid separates, this is collected, washed with water, and characterized.

The amine is recovered by making the aqueous solution basic by the addition of 10% NaOH. A heavy precipitate of inorganic salts usually separates, and more water must be added. The amine is then recovered as described in the section on alkaline hydrolysis.

CONVERSION OF NITRILE TO AMIDE. In a 100-mL round-bottom flask, place 1 mL or 1 g of the nitrile, 5 mL of 20% hydrogen peroxide, and 1 mL of 6 *M* (20%) NaOH solution. If the nitrile is insoluble, add 5 to 10 mL of ethanol. The reaction should occur exothermically, and cooling may be necessary at first. Keep the temperature at 40 to 50°C by cooling or gentle warming for 2 hours. Neutralize the solution and concentrate it by distillation to remove most of the alcohol. The amide will usually separate from the aqueous solution on chilling; if it does not, extract with methylene chloride, and isolate the amide by evaporating the solvent from the dried extract.

REDUCTION TO AMINES. In a 125-mL Erlenmeyer flask, dissolve 0.5 g of the nitrile or amide and 2 g of $CoCl_2 \cdot 6H_2O$ in 25 mL of methanol. In the hood, add 2 g of $NaBH_4$, in small portions, shaking vigorously after each addition. Hydrogen is evolved and a black precipitate usually separates. When the addition is completed, add 3 *M* HCl until the solution is acidic (5 to 10 mL) and heat the solution on the steam bath to evaporate the methanol. Make the residual aqueous solution basic with NaOH, extract the amine with ether, and prepare a derivative of the amine.

ALDEHYDES AND KETONES

Semicarbazones and 2,4-dinitrophenylhydrazones are the most generally satisfactory derivatives of simple aldehydes and ketones:

$$RCOR' + NH_2NHCONH_2 \rightarrow \underset{R}{\overset{R'}{}}\!\!>C{=}NNHCONH_2$$

$$RCOR' + NH_2NHC_6H_3(NO_2)_2 \rightarrow \underset{R}{\overset{R'}{}}\!\!>C{=}NNHC_6H_3(NO_2)_2$$

PROCEDURES

(These derivatives are also described in Chapter 20.)

SEMICARBAZONES. Prepare a solution of 0.5 g of semicarbazide hydrochloride and 1 g of sodium acetate in 2 mL of methanol; the large crystals are ground together with a glass rod, and finely divided sodium chloride separates. Add approximately 0.5 mL of the carbonyl compound and allow the solution to stand for 15 to 30 minutes. Add 1 mL of water to dissolve the NaCl, chill the solution, and collect and recrystallize the derivative.

2,4-DINITROPHENYLHYDRAZONES. Dissolve approximately 0.2 g of the solid or liquid ketone or aldehyde in 1 mL of ethanol and add dropwise 3 mL of the 2,4-dinitrophenylhydrazine solution (1 g of 2,4-dinitrophenylhydrazine, 5 mL of concentrated sulfuric acid, 8 mL of water, and 25 mL of ethanol). The dinitrophenylhydrazone should precipitate from the solution. If a product does not precipitate, warm the solution briefly on the steam bath; this is necessary for sterically hindered carbonyl compounds. Isolate the derivative by filtration and recrystallize it from ethanol or aqueous ethanol.

ACETALS, KETALS, AND ENOL ETHERS

Acetals, ketals, and related carbonyl derivatives such as enol ethers, imines, and enamines may be hydrolyzed in warm dilute aqueous or alcoholic acid. The aldehyde or ketone that is liberated can be isolated by extraction and converted to a derivative as described in the preceding section. If the 2,4-dinitrophenylhydrazone is desired, generally it is sufficient to warm the unknown directly with the acidic 2,4-dinitrophenylhydrazine reagent; the hydrolysis takes place and the hydrozone is formed and can be isolated by filtration.

ESTERS

Hydrolysis (saponification) of an ester to the acid and alcohol is a very general reaction, but the choice of a procedure depends on the type of ester and the product that is to be isolated. For esters of aromatic acids the acid is nearly always the most practical derivative, and hydrolysis should be carried out in aqueous alcohol by the first procedure given; this procedure, of course, does not permit characterization of the alcohol portion.

With an ester of an aliphatic acid, isolation of the acid is usually impractical. In order to isolate the alcohol, the hydrolysis can be carried out in the high-boiling sovent diethylene glycol ($HOCH_2CH_2OCH_2CH_2OH$, bp 244°C); the alcohol from the ester is isolated by distillation.

Another approach for the derivatization of an aliphatic ester is the direct conversion to a substituted amide. An aromatic amine is converted to the more reactive amide anion by treatment with ethylmagnesium bromide, and the ester is then refluxed with this reagent.

$$ArNH_2 + C_2H_5MgBr \rightarrow ArNH^- \ MgBr^+ + C_2H_6$$

$$ArNH^- \ MgBr^+ + RCO_2R' \rightarrow RCONHAr + R'OMgBr$$

PROCEDURES

HYDROLYSIS—ETHANOLIC BASE. In a 50-mL round-bottom flask, place 5 mL of ethanol, 5 mL of water, 1 mL or 1 g of the ester, and 1 g of KOH. Add a boiling stone and heat the solution under reflux for 30 to 40 minutes. Cool, add 5 mL additional water, and arrange the condenser for distillation. Distill the solution until 5 to 6 mL is collected to remove alcohol, cool the solution, and acidify with 2 *M* HCl. Collect the acid or, if it is a liquid, extract with ether, dry, evaporate, and convert it to a derivative.

HYDROLYSIS—DIETHYLENE GLYCOL. In a 10-mL distilling flask (Fig. 46.1), place 3 mL of diethylene glycol (bp 244°C), 0.5 g of KOH pellets, and 0.5 mL of water. Heat the mixture until the alkali has dissolved and then cool and add 1 to 2 mL of the ester. Heat again until the ester dissolves and then more strongly, distilling the alcohol into a cooled test tube. The potassium salt of the acid may separate as a solid during the reaction.

CONVERSION OF ESTER TO SUBSTITUTED AMIDE. In a 50-mL round-bottom flask, prepare a solution of ethylmagnesium bromide from 10 mL of anhydrous ether, 1 mL of ethyl bromide, and 0.2 g of magnesium (see Chapter 17). After the formation of the Grignard compound and brief refluxing to ensure completion of the reaction, add a solution of 1.5 mL of aniline or 1.5 g of *p*-toluidine in a small amount of dry ether. After reaction of the Grignard reagent is complete, add 1 mL of the ester and reflux the mixture for 10 minutes. Cool, add 10 mL of 2 *M* HCl and shake to extract unreacted amine, adding more ether if needed. Isolate the derivative by the usual procedure from the ether solution.

ALKYL HALIDES

Alkyl halides that contain no reactive functional groups can be derivatized by conversion to the Grignard compound and treatment of the latter with an aryl isocyanate to give a substituted amide of the homologous acid. This method is quite general for halides that can form Grignard compounds.

$$\text{R—X} \xrightarrow{\text{Mg}} \text{RMgX} \xrightarrow{\text{ArNCO}} \text{R—}\overset{\overset{\displaystyle\text{O}}{\|}}{\text{C}}\text{—NHAr}$$

PROCEDURE

Convert 1 to 2 mL of the halide (the sample must be anhydrous) to the Grignard compound in the usual way (see Chapter 17) with approximately 0.2 g of magnesium and 10 mL of anhydrous ether; a crystal of iodine can be added if needed. To this solution add, in small portions, a solution of 0.3 to 0.5 mL of the isocyanate (an excess must be avoided, **CAUTION:** lachrymator) in 5 mL of ether or methylene chloride. After hydrolysis with 2 *M* HCl, isolate the amide from the ether solution in the usual way.

NITRO COMPOUNDS

For aromatic nitro compounds, the oxidation of alkyl side chains or further nitration often provides satisfactory derivatives (see following section). Another general approach is the reduction to the amine, which can then be converted to an amide.

PROCEDURE

In a 100-mL round-bottom flask with reflux condenser, place 2 g of tin (granulated or mossy) and 1 g of the nitro compound. Then add 20 mL of 2 *M* HCl through the condenser, in small portions with constant shaking. After warmimg the solution on the steam bath for 10 minutes, allow it to cool and slowly add 40% NaOH solution to liberate the amine, which can then be steam distilled or extracted with ether and further characterized by a derivative.

ARYL HALIDES AND AROMATIC HYDROCARBONS

As noted in the discussions of phenol and aryl ether derivatives, the benzene ring can often serve as a satisfactory functional group for derivative preparation. For aryl halides and aromatic hydrocarbons, ring substitution may be the most practical reaction available. Nitration (Chapter 25) and chlorosulfonation (pages 415 and 509) are the two most generally useful reactions. In addition, aryl bromides can be converted to Grignard reagents and converted with CO_2 to benzoic acids (Chapter 17) or with phenyl isocyanate to the corresponding anilide as described for alkyl halides (page 515).

Alkylbenzenes can be degraded oxidatively to the corresponding benzoic acids. Other aliphatic side chains can also be oxidized, e.g., 2-phenylethyl chloride or phenylacetic acid to benzoic acid. The reaction is not generally useful for dialkylated or polyalkylated benzenes because of the inconveniently high melting points of the acids produced.

PROCEDURE

SIDE CHAIN OXIDATION. In a 50-mL round-bottom flask place 2 g of $KMnO_4$, 25 mL of water, 5 mL of 10% NaOH, 0.5 g or 0.5 mL of the unknown, and a boiling stone. Reflux gently over a flame for 1 hour or until the purple color has disappeared. Cool and acidify with dilute sulfuric acid. Reheat the mixture to boiling and add a few grains of sodium bisulfite to dissolve any tan manganese dioxide present. Cool the solution and collect the acid by filtration. Recrystallize from ethanol or aqueous ethanol.

REFERENCES

Pasto, D. J.; Johnson, C. R. *Laboratory Text for Organic Chemistry*. Prentice-Hall: Englewood Cliffs, N.J., 1979.

Pasto, D. J.; Johnson, C. R. *Organic Structure Determination*. Prentice-Hall: Englewood Cliffs, N.J., 1969.

Rappoport, Z. *Handbook of Tables for Organic Compound Identification*, 3rd ed. Chemical Rubber: Cleveland, 1967.

Shriner, R. L.; Fuson, R. C.; Curtin, D. Y.; Morrill, T. C. *The Systematic Identification of Organic Compounds*, 6th ed. Wiley: New York, 1980.

Name ______________________ Section ____________ Date ________

46 Identification of Unknowns—Part 1 (Physical Properties, Solubility Classification, Detection of Elements)

PRELABORATORY QUESTIONS

1. What effect (if any) does branching of the carbon chain have on the boiling point of a substance?

2. Traditionally, what is one of the first (and most important) physical constants cited for the characterization of a solid?

3. The solubility tests include water, aqueous $NaHCO_3$, dilute NaOH, dilute HCl, and concentrated H_2SO_4. If given in the order shown, where in the sequence would the following first be soluble?

a. cyclohexene ____________

b. ethyl iodide ____________

c. toluene ____________

d. *m*-bromobenzoic acid ____________

e. aniline ____________

f. acetamide ____________

g. *p*-chlorophenol ____________

h. 2,4-dinitrophenol ____________

4. The sodium fusion analysis allows one to test for what elements?

5. Lead acetate is one of the test reagents. For what element is it used? ____________ Write an equation showing the reactants and products for this test.

Name ______________________ Section ____________ Date ________

46 Identification of Unknowns—Part 2 (Equivalent Weights, Miscellaneous Chemical Tests, Derivatives)

PRELABORATORY QUESTIONS

1. For malonic acid (HOOC—CH_2—COOH), how does the equivalent weight compare with the molecular weight? Explain.

2. The following tests or procedures are used for what functional groups?

a. Chromic acid test ____________

b. Iodoform test ____________

c. Hinsberg test ____________

d. Tollens test ____________

e. 2,4-Dinitrophenylhydrazine ____________

f. Ferric chloride test ____________

g. Neutralization equivalent ____________

3. Give a generally useful derivative for each of the following.

a. acids ____________

b. alcohols ____________

c. ketones ____________

d. amines ____________

4. List the class or classes of compounds for which the following are useful derivatives (i.e., are products of the derivatization procedure).

a. acids ____________

b. phenols ____________

c. nitroaromatics ____________

d. Grignard followed by reaction with an aryl isocyanate

____

TABLES OF DERIVATIVES

Table 46.2 Acids

COMPOUND	Mp, °C	Bp, °C	DERIVATIVE, Mp, °C *Amide*	*Anilide*	p-*Toluidide*
Acetic		118	82	114	147
Acetylanthranilic	185		171	167	
o-Anisic (methoxybenzoic)	100		128	131	
p-Anisic	184		162	169	186
Benzoic	122		128	163	158
o-Bromobenzoic	150		155	141	
Butanoic		163	115	95	72
Chloroacetic	63		118	134	120
o-Chlorobenzoic	140		139	118	131
m-Chlorobenzoic	158		134	122	
p-Chlorobenzoic	242		179	194	
4-Chloro-3-nitrobenzoic	182		156	131	
Cinnamic	133		147	153	168
Dichloroacetic		189	98	118	153
2,4-Dichlorobenzoic	158				
3,4-Dichlorobenzoic	208		133		
Diglycolic (oxydiacetic)	148		135 (mono)	118 (mono)	
3,4-Dimethoxybenzoic	182		164	154	
2,2-Dimethylsuccinic	140				
Diphenylacetic	146		167	180	
p-Ethoxybenzoic	198		202	169	
Glutaric	97		174	224	
Hippuric (benzoylglycine)	187		183	208	
p-Hydroxybenzoic	215		162	202	204
Itaconic	165 (*d*)		192 (di)		
p-Methoxyphenylacetic	85		189		
Methylpropanoic (isobutyric)		155	129	105	109
Methylsuccinic	115		165 (mono)		
Naphthalene-1-acetic	133		181	155	
1-Naphthoic	162		205	161	
m-Nitrobenzoic	140		142	155	162
p-Nitrobenzoic	241		201	217	
Phenylacetic	76		154	117	
2-Phenylbutanoic	42		86		
3-Phenylpropanoic	48		82	92	
5-Phenylpentanoic	60		109	90	
Phthalic	208 (*d*)		149		
Propionic		140	81	106	126
Salicylic (*o*-hydroxybenzoic)	158		139	136	156
Succinic	188		242 (mono)		
o-Toluic	102		142	125	144
m-Toluic	111		97	126	118
p-Toluic	177		158	140	160
3,4,5-Trimethoxybenzoic	170		176		

Table 46.3 Alcohols

COMPOUND	Bp, °C	Mp, °C	DERIVATIVE, Mp, °C *3,5-Dinitrobenzoate*	*Phenylurethane*
Benzyl alcohol	206		113	78
1-Butanol	117		64	57
2-Butanol	99		75	65
2-Buten-1-ol	122			
Cinnamyl alcohol	257	33	121	90
2-Chloroethanol	130			51
1-Chloro-2-propanol	127		83	
Cholesterol		148		168
Cyclohexanol	161		112	82
Cyclopentanol	141		115	132
Diphenylmethanol		69	141	140
Ethanol	78		93	52
2-Ethyl-1-butanol	149		52	
1-Heptanol	177		47	68
2-Heptanol	160		49	
1-Hexanol	156		58	42
Methanol	65		108	47
4-Methoxybenzyl alcohol	260	25		92
1-Methoxy-2-propanol	119		85	
4-Methylbenzyl alcohol		60	118	79
2-Methyl-1-butanol	129		70	31
3-Methyl-1-butanol	132		61	57
3-Methyl-2-butanol	113		76	68
2-Methyl-3-buten-2-ol	98			
2-Methyl-1-pentanol	148		51	
3-Methyl-2-pentanol	134		43	
4-Methyl-2-pentanol	132		65	143
2-Methyl-1-propanol	108		87	86
1-Octanol	192		61	74
2-Octanol	179		32	114
1-Pentanol	138		46	46
2-Pentanol	119		61	
3-Pentanol	116		101	48
2-Phenoxyethanol	237			
1-Phenylethanol	203		95	94
2-Phenylethanol	219		108	80
1-Phenyl-1-propanol	219			
2-Phenyl-2-propanol	202	34		
1-Propanol	97		74	51
2-Propanol	82		123	88
2-Propen-1-ol	97		48	70
2-Propyn-1-ol	115			

Table 46.4 Aldehydes

COMPOUND	Bp, °C	Mp, °C	DERIVATIVE, Mp, °C	
			Semicarbazone	*2,4-Dinitrophenylhydrazone*
o-Anisaldehyde (*o*-methoxybenzaldehyde)	246	38	215	254
p-Anisaldehyde	247		210	254
Benzaldehyde	179		222	237
Butanal	75		106	123
2-Butenal	103		199	190
o-Chlorobenzaldehyde	208		225	207
p-Chlorbenzaldehyde	214	47	230	270 (*d*)
Cinnamaldehyde	252		215	255
Citral	228		164	116
Citronellal	206		82	77
2,5-Dimethoxybenzaldehyde		52	—	—
3,4-Dimethoxybenzaldehyde		44	177	263
p-Dimethylaminobenzaldehyde		74	222	325
o-Ethoxybenzaldehyde	248		219	
2-Ethylbutanal	116		99	134
Furfural	161		202	230
Heptanal	156		109	108
Hexanal	131		106	104
o-Hydroxybenzaldehyde (salicylaldehyde)	197		231	248
p-Hydroxybenzaldehyde		115	224	280 (*d*)
2-Methylbutanal	93		103	120
3-Methylbutanal	92		107	123
a-Methylcinnamaldehyde	270		208	
5-Methylfurfural	187		211	212
1-Naphthaldehyde	292	34	221	
p-Nitrobenzaldehyde		106	221	
o-Nitrobenzaldehyde		44	256	250
Phenylacetaldehyde	194		156	121
Piperonal (3,4-methylenedioxybenzaldehyde)	264	36	230	266
o-Tolualdehyde	200		212	195
p-Tolualdehyde	204		215	234

Table 46.5 Amides

COMPOUND	Mp, °C
Acetamide	82
Acetanilide	114
Acetoacetanilide	85
o-Acetoacetanisidide	84
p-Acetoacetanisidide	115
o-Acetoacetotoluide	104
o-Acetotoluide	112
m-Acetotoluide	66
p-Acetotoluide	153
Benzamide	130
p-Bromoacetanilide	167
p-Bromobenzamide	155
o-Bromobenzanilide	141
m-Bromobenzanilide	136
p-Chloroacetanilide	179
p-Chloroacetoacetanilide	134
m-Chlorobenzamide	134
p-Chlorobenzanilide	194
Cinnamanilide	153
o-Ethoxybenzamide	133
m-Ethoxybenzamide	139
o-Methoxybenzamide	129
N-Methylacetanilide	102
o-Toluamide	142
m-Toluamide	97
p-Toluamide	158

Table 46.6 Amines

COMPOUND	Bp, °C	Mp, °C	DERIVATIVE, Mp, °C		
			Acetamide	*Benzamide*	*Phenylthiourea*
Aniline	183		114	160	154
Benzylamine	184		60	105	156
N-Benzylaniline		37			
o-Bromoaniline	229		99	116	146
p-Bromoaniline		66	167	204	148
n-Butylamine	77			42	65
iso-Butylamine	69			57	82
sec-Butylamine	63			76	101
tert-Butylamine	46			134	120
o-Chloroaniline	207		87	99	156
m-Chloroaniline	230		72	120	124
p-Chloroaniline		70	179	192	152
Cyclohexylamine	134		104	149	148
Di-*n*-butylamine	160				86
2,4-Dichloroaniline		63	145	117	
2,5-Dichloroaniline		50	132	120	
Diethylamine	55			42	34
Di-*n*-propylamine	110				69
Di-*iso*-propylamine	86		[picrate, 140]		
p-Ethoxyaniline	250		137	173	136
Ethyl *p*-aminobenzoate		89	110	148	
N-Ethylaniline	205		54	60	
o-Ethylaniline	216		111	147	
n-Hexylamine	128			40	77
o-Methoxyaniline	225		87	84	
p-Methoxyaniline		58	128	155	154
2-Methoxy-5-methylaniline		50	110		
4-Methoxy-2-methylaniline		30	134		
N-Methylaniline	196		102	63	87
o-Nitroaniline		71	92	94	142
m-Nitroaniline		114	155	155	160
p-Nitroaniline		147	210	199	
α-Phenylethylamine	185		57	120	
β-Phenylethylamine	198		114	116	135
Piperidine	105			48	101
o-Toluidine	199		112	143	136
m-Toluidine	203		65	125	94
p-Toluidine		45	153	158	141

Table 46.7 Aromatic Halides and Hydrocarbons

			DERIVATIVE		
			Nitration Product		*Carboxylic Acid*
COMPOUND	Bp, °C	Mp, °C	*Positions*	*MP, °C*	*Mp, °C*
Anthracene		216			
Biphenyl		70	4,4′	233	
Bromobenzene	157		2,4	75	
4-Bromobiphenyl		89			
p-Bromochlorobenzene		67	2	72	
o-Bromotoluene	181				147
p-Bromotoluene	185	28	2	47	251
Chlorobenzene	132				
o-Chlorotoluene	159				140
m-Chlorotoluene	162		4,6	91	158
p-Chlorotoluene	162		2	38	242
p-Cymene (isopropyltoluene)	175		2,6	54	
o-Dibromobenzene	224		4,5	114	
p-Dibromobenzene		89	2	84	
2,5-Dibromotoluene					157
o-Dichlorobenzene	179		4,5	110	
p-Dichlorobenzene		53	2	54	
2,4-Dichlorotoluene	195				160
2,6-Dichlorotoluene	199				130
Diphenylmethane		26			
Ethylbenzene	135				122
Fluorene		115			
Mesitylene	164		2,4	86	
1-Methylnaphthalene	240				162
2-Methylnaphthalene		32	1	81	
α-Methylstyrene	169				122
2-Phenylethyl chloride	190				122
Toluene	111		2,4	70	122
o-Xylene	142		4,5	71	
m-Xylene	139		2,4	83	
p-Xylene	137		2,3,5	137	

Table 46.8 Esters

COMPOUND	Bp, °C	Mp, °C
Diethyl ethylmalonate	75 (5 mm)	
Diethyl glutarate	237	
Diethyl maleate	225	
Diethyl malonate	198	
Diethyl oxalate	185	
Diethyl phthalate	296	
Diethyl phenylmalonate	170 (14 mm)	16
Diethyl succinate	216	
Diethyl suberate	268	
Ethyl acetate	77	
Ethyl acetoacetate	181	
Ethyl *p*-anisate	270	
Ethyl benzoate	213	
Ethyl benzoylacetate	270	
Ethyl cinnamate	271	
Ethyl cyanoacetate	208	
Ethyl *p*-hydroxybenzoate		116
Ethyl 2-methylacetoacetate	187	
Ethyl *p*-nitrobenzoate		57
Ethyl phenylacetate	229	
Ethyl propionate	98	
Ethyl salicylate	234	
Ethyl *p*-toluate	241	
Isopropenyl acetate	96	
Isopropyl acetate	91	
Isopropyl formate	68	
Isopropyl benzoate	218	
Isopropyl salicylate	255	
Methyl acetate	57	
Methyl acetoacetate	169	
Methyl *p*-anisate		49
Methyl benzoate	198	
Methyl *n*-butyrate	102	
Methyl *iso*-butyrate	92	
Methyl *o*-chlorobenzoate	230	
Methyl *m*-chlorobenzoate	231	
Methyl cinnamate		35
Methyl heptanoate	173	
Methyl hexanoate	150	
Methyl *p*-hydroxybenzoate		130
Methyl mandelate		57
Methyl *p*-nitrobenzoate		95
Methyl pentanoate	130	
Methyl phenylacetate	218	
Methyl propionate	79	
Methyl *o*-toluate	213	
Methyl *p*-toluate		30
Phenyl acetate	197	
Phenyl benzoate		69
Phenyl salicylate		42

Table 46.9 Ethers (Aryl)

			DERIVATIVE, Mp, °C	
COMPOUND	Bp, °C	Mp, °C	*Nitro*	*Picrate*
Anisole	154		87 (di)	81
o-Bromoanisole	218		106	
p-Bromoanisole	223		88	
o-Chloroanisole	195		95	
p-Chloroanisole	200		98	
o-Dimethoxybenzene	206		92 (dibromo)	57
m-Dimethoxybenzene	214		140 (dibromo)	58
p-Dimethoxybenzene		55	142 (dibromo)	119
o-Methylanisole	171		63 (bromo)	119
m-Methylanisole	177		91 (tri)	114
p-Methylanisole	176			89
Phenetole (ethoxybenzene)	172		58	92

Table 46.10 Ketones

COMPOUND	Bp, °C	Mp, °C	DERIVATIVE, Mp, °C: *Semicarbazone*	DERIVATIVE, Mp, °C: *2,4-Dinitrophenylhydrazone*
Acetone	56		187	126
2-Acetonaphthone		54	234	262
Acetophenone	200		198	250
Benzophenone		48	167	239
p-Bromoacetophenone		51	208	235
Butanone	80		146	117
Butyrophenone	230		187	190
Chloroacetone	119		164	125
p-Chloroacetophenone	232		201	231
p-Chloropropiophenone		36	176	
Cyclohexanone	156		167	162
Cyclopentanone	131		203	146
3,3-Dimethyl-2-butanone	106		158	125
2,4-Dimethyl-3-pentanone	125		160	95
Fluorenone		83		283
2-Heptanone	151		127	89
3-Heptanone	148		103	
4-Heptanone	145		133	75
Hexane-2,5-dione	188		220 (di)	255 (di)
2-Hexanone	129		122	110
5-Hexen-3-one	129		102	108
4-Hexen-3-one	139		157	
p-Hydroxypropiophenone		148		229
Isobutyrophenone	222		181	163
p-Methoxyacetophenone		38	197	220
p-Methoxypropiophenone		28		
p-Methylacetophenone	226	28	205	258
3-Methyl-2-butanone	94		113	120
2-Methylcyclohexanone	163		195	137
4-Methylcyclohexanone	169		199	130
5-Methyl-3-heptanone	160		102	
6-Methyl-3-heptanone	160		132	
5-Methyl-2-hexanone	145		147	95
Methylcyclohexyl ketone	180		177	140
4-Methyl-2-pentanone	119		135	95
4-Methyl-3-penten-2-one	130		164	203
m-Nitroacetophenone		81	257	228
p-Nitroacetophenone		80		
2-Octanone	173		123	58
2,4-Pentandione	139		122 (mono)	209
2-Pentanone	102		112	144
3-Pentanone	102		139	156
Phenylacetone	216		198	156
4-Phenyl-2-butanone	235		142	
4-Phenyl-3-buten-2-one		41	187	
Propiophenone	218		174	191

Table 46.11 Nitriles

COMPOUND	Bp, °C	Mp, °C
Acetonitrile	81	
Acrylonitrile	78	
Adiponitrile	295	
Benzonitrile	191	
p-Bromobenzonitrile		112
Butyronitrile	118	
Chloroacetonitrile	127	
o-Chlorobenzonitrile	232	47
m-Chlorobenzonitrile		41
p-Chlorobenzonitrile		92
o-Chlorophenylacetonitrile	242	24
p-Chlorophenylacetonitrile	265	30
Glutaronitrile	286	
Isobutyronitrile	108	
Malononitrile	219	
p-Methoxybenzonitrile		62
1-Naphthaleneacetonitrile		35
1-Naphthonitrile	299	35
2-Naphthonitrile	306	66
o-Nitrobenzonitrile		110
m-Nitrobenzonitrile		118
p-Nitrobenzonitrile		147
p-Nitrophenylacetonitrile		116
Phenylacetonitrile	234	
o-Tolunitrile	205	
m-Tolunitrile	212	
p-Tolunitrile	217	27

Table 46.12 Nitro Compounds

COMPOUND	Bp, °C	Mp, °C
o-Bromonitrobenzene	261	43
m-Bromonitrobenzene	256	54
p-Bromonitrobenzene		126
4-Bromo-3-nitrotoluene		33
o-Chloronitrobenzene	246	32
m-Chloronitrobenzene	235	44
p-Chloronitrobenzene		83
2,5-Dibromonitrobenzene		85
2,4-Dichloronitrobenzene		52
2,4-Dimethylnitrobenzene	238	
2,5-Dimethylnitrobenzene	234	
2,6-Dimethylnitrobenzene	226	15
2,4-Dinitroanisole		89
1,3-Dinitrobenzene		90
1,4-Dinitrobenzene		172
2,4-Dinitrobromobenzene		72
2,4-Dinitrochlorobenzene		52
2,4-Dinitrotoluene		70
2,6-Dinitrotoluene		66
o-Nitroanisole	265	
p-Nitroanisole		54
Nitrobenzene	210	
4-Nitrobiphenyl		114
o-Nitrotoluene	224	
m-Nitrotoluene	231	16

Table 46.13 Phenols

			DERIVATIVE, Mp, °C		
COMPOUND	Mp, °C	Bp, °C	*Benzoate*	*3,5-Dinitro-Benzoate*	*Phenyl-urethane*
4-*t*-Butylphenol	100		81		
4-Chloro-3,5-dimethylphenol	115		[acetate, 48]		
o-Chlorophenol		176			121
p-Chlorophenol	43		88	186	148
o-Cresol (methylphenol)		190		138	142
m-Cresol		202	55	165	128
p-Cresol	36		70	189	146
2,4-Dichlorophenol	45		97	142	
3,5-Dichlorophenol	68		55		
2,4-Dimethylphenol	27	212	38	165	103
2,5-Dimethylphenol	75		61	137	161
2,6-Dimethylphenol	49			159	133
3,4-Dimethylphenol	62		59	182	120
3,5-Dimethylphenol	68			195	151
4-Ethylphenol	47		60	132	120
o-Hydroxyphenol (catechol)	104		84 (di)	152 (di)	169 (di)
m-Hydroxyphenol (resorcinol)	110		117 (di)	201 (di)	164 (di)
p-Hydroxyphenol (hydroquinone)	169		199 (di)	317 (di)	
2-Isopropyl-5-methylphenol (thymol)	51		33	103	107
2-Isopropylphenol		212	[aryloxyacetic acid, 133]		
4-Isopropylphenol	61		71		
2-Methoxyphenol	30	205	58	141	148
4-Methoxyphenol	56		87		137
4-Methyl-2-nitrophenol	34				
5-Methyl-2-nitrophenol	53				
1-Naphthol	94		56	217	177
2-Naphthol	122		107	210	155
o-Nitrophenol	45			155	
p-Nitrophenol	114		143	186	
Phenol	42	180	68	146	126

Appendix A

Useful Chemical Data

Selected Atomic Weights

Aluminum	26.98	Magnesium	24.31
Barium	137.34	Manganese	54.94
Boron	10.81	Mercury	200.59
Bromine	79.91	Nitrogen	14.01
Calcium	40.08	Oxygen	16.00
Carbon	12.01	Phosphorus	30.97
Chlorine	35.45	Potassium	39.10
Chromium	52.00	Selenium	78.96
Copper	63.54	Silicon	28.09
Fluorine	19.00	Silver	107.87
Hydrogen	1.008	Sodium	22.99
Iodine	126.90	Sulfur	32.06
Iron	55.85	Tin	118.69
Lithium	6.94	Zinc	65.37

Common Acids and Bases

	moles/liter	sp.gr.	g/100 mL
Hydrochloric acid, concentrated (37%)	12.0	1.19	44.0
, 10%	2.9	1.05	10.5
, 1*M*	1.0	1.02	3.6
Ammonium hydroxide, concentrated	15	0.90	25.6 (NH_3)
Sodium hydroxide, 10%	2.8	1.11	11.1
Sodium bicarbonate, saturated, 20°	1.1	1.06	9.5
Sodium carbonate, saturated, 20°	2.1	1.19	21
Sulfuric acid, concentrated (96%)	18	1.84	177
Nitric acid, concentrated (71%)	16	1.42	101
Hydrobromic acid, concentrated (48%)	8.8	1.49	71
Hydriodic acid, concentrated (57%)	7.6	1.7	97
Acetic acid, glacial	17.5	1.05	105

Boiling Points and Densities of Common Organic Liquids

	Boiling Point 760 mm (mp)	Density g/mL
Acetic acid	118° (16.6°)	1.05
Acetic anhydride	140°	1.08
Acetone	56°	0.79
Acetyl chloride	52°	1.10
Aniline	184°	1.02
Benzene	80° (5.5°)	0.88
Benzoyl chloride	197°	1.21
l-Butanol	118°	0.81
t-Butyl alcohol	82° (25.5°)	0.79
Carbon disulfide	45°	1.26
Carbon tetrachloride	77°	1.59
Chlorobenzene	132°	1.11
Chloroform	61°	1.49
Cyclohexane	81° (6.5°)	0.78
Cyclohexanol	161° (25°)	0.96
1,2-Dichloroethane	84°	1.26
Diethyl ether	35°	0.71
1,2-Dimethoxyethane (glyme)	85°	0.87
Dimethylformamide	153°	0.94
Ethanol	78°	0.79
Ethyl acetate	77°	0.90
Heptane	98°	0.68
Hexane	68°	0.66
Methanol	65°	0.79
Methyl acetate	57°	0.97
Methylene chloride	40°	1.34
Nitrobenzene	211° (5.7°)	1.20
Pentane	36°	0.63
l-Propanol	97°	0.78
2-Propanol	82°	0.79
Pyridine	116°	0.98
Tetrahydrofuran	65°	0.89
Toluene	111°	0.87

Index

J

K

L

M